Brazing Manual

BRAZING MANUAL

Third Edition, Revised

Prepared by
AWS Committee on Brazing and Soldering

Under the Direction of
AWS Technical Activities Committee

Approved by
AWS Board of Directors
February 28, 1975

AMERICAN WELDING SOCIETY, INC.
2501 N.W. 7th Street, Miami, Florida 33125

Library of Congress Number: 76-15811
International Standard Book Number: 0-87171-133-8

American Welding Society, 2501 N.W. 7th Street, Miami, FL 33125

First Edition published 1955. Second Edition 1963
Third Edition 1976

Printed in the United States of America

CONTENTS

PERSONNEL

R. L. Peaslee, Chairman	Wall Colmonoy Corporation
G. M. Slaughter, Chairman	Oak Ridge National Laboratory
M. M. Schwartz, Vice-Chairman	Rohr Corporation
D. J. Spillane, 1st Vice-Chairman	General Electric Company
R. E. Ballentine, 2nd Vice-Chairman	Westinghouse Electric Corporation
T. J. Olivera, Secretary	American Welding Society
W. G. Bader	Bell Telephone Laboratories
R. E. Beal	ITT Research Institute
C. R. Behringer	Western Gold & Platinum Company
J. R. Bonnar	Handy and Harman
J. P. Broderick	Eutectic and Castolin Institute
E. S. Charmer	Handy and Harman
G. D. Cremer	International Harvester Company
A. S. Cross, Jr.	Engelhard Minerals & Chemicals
D. C. Dilley	Consultant
F. C. Disque	Alpha Metals Incorporated
R. M. Evans	Batelle Memorial Institute
R. G. Gilliland	Pelton Steel
K. Gustafson	Westinghouse-Hanford
A. N. Kugler	Consultant
A. H. Lentz	Reynolds Metals Company
L. F. Lockwood	Taber Metals
J. B. Long	Tin Research Institute
R. O. McIntosh	National Electronics
J. A. Mehaffey	Stanley Flagg and Company
M. T. Merlo	Chemetron Corporation
F. Miller	Wall Colmonoy Corporation
E. J. Minarcik	NL Industries
W. J. Reichenecker	Westinghouse Electric Corporation
M. N. Ruoff	General Electric Company
J. F. Sith	Lead Industries Association
G. K. Sosnin	The Pipe Company
H. A. Sosnin	Consultant
H. W. Spaletta	Aerojet Nuclear Company
J. R. Terrill	Aluminum Company of America
D. Wireman	Aerobraze Corporation

Advisory Members

N.C. Cole	Combustion Engineering
G.S. Gschwind	United Aircraft of Canada, Ltd.
T. Hikido	Pyromet Industries
W.S. Lyman	Copper Development Association
M. Prager	Consultant
H.S. Sayre	U.S. Naval Ship Engineering Center

INTRODUCTION

Brazing is an ancient art, but knowledge of that art is continuously being supplemented by an ever-increasing amount of technical information about metals and their behavior, so that today brazing must be considered both an art and a science. This third edition of the *Brazing Manual* includes the fundamental concepts of brazing and incorporates the many advances made since the *Manual* was first published.

The American Welding Society defines brazing as "a group of welding processes which produces coalescence of materials by heating them to a suitable temperature and by using a filler metal having a liquidus above 840 °F (450 °C) and below the solidus of the base materials. The filler metal is distributed between the closely fitted surfaces of the joint by capillary attraction."

Brazing then must meet each of three criteria:

1. The parts must be joined without melting the base metals.
2. The filler metal must have a liquidus temperature above 840°F (450°C).
3. The filler metal must wet the base metal surfaces and be drawn into or held in the joint by capillary attraction.

To achieve a good joint using any of the various brazing processes described in this *Manual,* the parts must be properly cleaned and must be protected, either by fluxing or atmosphere during the heating process to prevent excessive oxidation. The parts must be designed to afford a capillary for the filler metal when properly aligned, and a heating process must be selected that will provide the proper brazing temperature and heat distribution.

No analysis of a subject that is continuously being improved can hope to be complete, nor can the subject be covered with a thoroughness that would satisfy the specialist. Thus, for unusual applications the information available must often be supplemented by trial and error. It is hoped, however, that the trials and errors will be fewer for having this *Manual* as a guide.

Comments, inquiries, and suggestions for future revisions of the *Manual* are welcome. Address them to Secretary, AWS Committee on Brazing and Soldering, American Welding Society, Inc., 2501 N.W. 7th Street, Miami, Florida 33125.

CHAPTER 1

BRAZING PROCESSES, EQUIPMENT, AND PROCEDURES

BRAZING PROCESSES AND EQUIPMENT

Brazing processes customarily are designated according to the sources or methods of heating. Those methods currently of industrial significance are:

(1) Torch brazing, (2) Furnace brazing, (3) Induction brazing, (4) Resistance brazing, (5) Dip brazing, and (6) Infrared brazing. Whatever the process used, the filler metal has a melting point above 840 °F (450 °C)* but below that of the base metal, and is distributed in the joint by means of capillary attraction. Chapter 3 discusses the commonly used brazing filler metals and processes for different base metals.

Torch Brazing

As the name implies, torch brazing is accomplished by heating with a gas torch or torches. Depending upon the temperature and the amount of heat required, the fuel gas (acetylene, propane, city gas, etc.) may be burned with air, compressed air, or oxygen.

Brazing filler metal may be preplaced in the forms of rings, washers, strips, slugs, powder, etc., or it may be face fed, that is, fed from hand-held filler metal, usually in the form of wire or rod. In any case, proper cleaning and fluxing are essential.

For manual torch brazing the torch may be equipped with a single tip, either single or multiflame, (see Fig. 1.1). Manual torch brazing is particularly useful on assemblies involving sections of unequal mass. Machine operations can be set up where the rate of production warrants using one or more torches equipped with single or multiflame tips (Fig. 1.2). The machine may be designed to move either the work or the torches or both. For premixed city gas-air flames the refractory-type burner is used (Fig. 1.3).

Furnace Brazing

Furnace brazing, as illustrated in Fig. 1.4, is used extensively where the parts to be brazed can be assembled with the filler metal preplaced near or in the joint. This process is particularly applicable for high production brazing. The preplaced brazing filler metal may be in the form of wire, foil, filings, slugs, powder, paste, tape, etc. Fluxing is employed except when an atmosphere is specifically introduced in the furnace to perform the same function. Most of the high production brazing is done in a reducing gas atmosphere, such as hydrogen, and either exothermic or endothermic combusted gas. Pure, dry, inert gases, such as argon or helium, are used to obtain special atmospheric properties.

A large volume of furnace brazing is being performed in a vacuum, which prevents oxidation and often eliminates the need for flux. Vacuum brazing has found wide application in the aerospace and nuclear fields, where reactive metals are joined or where entrapped fluxes would be intolerable. If the vacuum is maintained by continuous pumping it will remove volatile constituents liberated during brazing. There are several base metals and filler metals that can be harmed by brazing in a vacuum because their low boiling point or high vapor pressure constituents cause a part of the metal to be lost. The types of furnaces generally used are either batch or con-

*Note: Owing to the transition to the SI system, the AWS committee on Definitions, Symbols, and Metric Practice has redefined the dividing line between brazing and soldering as 840 °F (450 °C).

Fig. 1.1—Manual torch brazing.

tinuous. These furnaces are usually heated by electrical resistance elements, gas or oil, and should have automatic time and temperature controls. Cooling is sometimes accomplished by cooling chambers, which either are placed over the hot retort or are an integral part of the furnace design. Forced atmosphere injection is another method of cooling. Parts may be placed in the furnace singly, in batches, or in a continuous conveyor.

Vacuum is a relatively economical method of providing an accurately controlled brazing atmosphere. Vacuum provides the surface cleanliness needed for good wetting and flow of filler metals without the use of fluxes. Base metals containing chromium and silicon can be easily vacuum brazed whereas a very pure, low dew point atmosphere gas would otherwise be required.

Induction Brazing

The heat necessary for brazing with this process is obtained from an electric current induced in the parts to be brazed, hence the name induction brazing. For induction brazing, the parts are

Fig. 1.2—Machine brazing with multiple tips.

Fig. 1.3—Refractory type burner using city gas-air mixture.

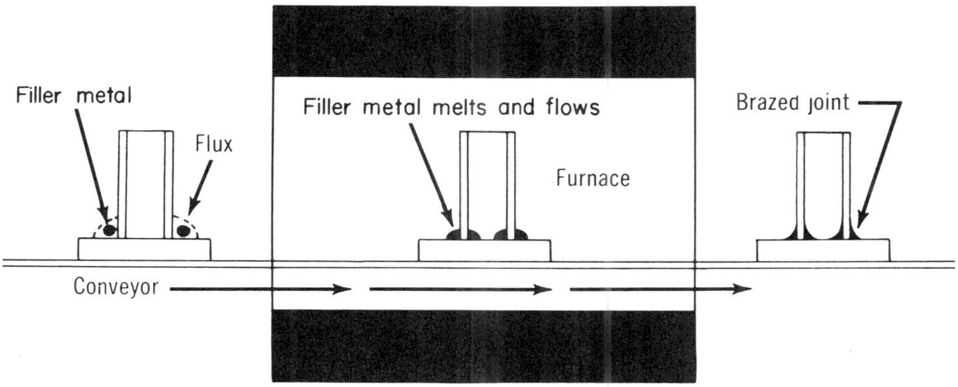

Fig. 1.4—Illustration of furnace brazing operation.

Fig. 1.5—Induction brazing setup.

placed in or near a coil carrying alternating current and do not form a part of the electrical circuit (see Fig. 1.5). The brazing filler metal is usually preplaced. Careful design of the joint and the coil setup are necessary to assure that the surfaces of all members of the joint reach the brazing temperature at the same time. Typical coil designs are shown in Fig. 1.6. Flux is employed except when an atmosphere is specifically introduced to perform the same function. The three common sources of high frequency electric current used for induction brazing are the motor-generator, resonant spark gap, and vacuum tube oscillator.

Resistance Brazing

The heat necessary for resistance brazing is obtained from the resistance to the flow of an electric current through the electrodes and the joint to be brazed. The parts comprising the joint form a part of the electric circuit. The brazing filler metal, in some convenient form, is preplaced or face fed. Fluxing is done with due attention to the conductivity of the fluxes. (Most fluxes are insulators when dry.) Flux is employed except when an atmosphere is specifically introduced to perform the same function. The parts to be brazed are held between two electrodes, and proper pressure and current are applied. The pressure should be maintained until the joint has solidified. In some cases both electrodes may be located on the same side of the joint with a suitable backing to maintain the required pressure.

The equipment consists of tongs or clamps with the electrodes attached at the end of each arm (see Fig. 1.7). The arms are current-carrying conductors attached by leads to a transformer. Direct current may be used but is comparatively expensive. The tongs should preferably be water-cooled to avoid overheating. Resistance welding machines are also used. The electrodes may be carbon, graphite, refractory metals, or copper alloys according to the required conductivity.

Dip Brazing

There are two methods of dip brazing: chemical bath dip brazing and molten metal bath dip brazing. (See Figs. 1.8, 1.9, 1.10.) In chemical bath dip brazing the brazing filler metal, in suitable form, is preplaced and the assembly is immersed in a bath of molten salt. The salt bath furnishes the heat necessary for brazing and usually pro-

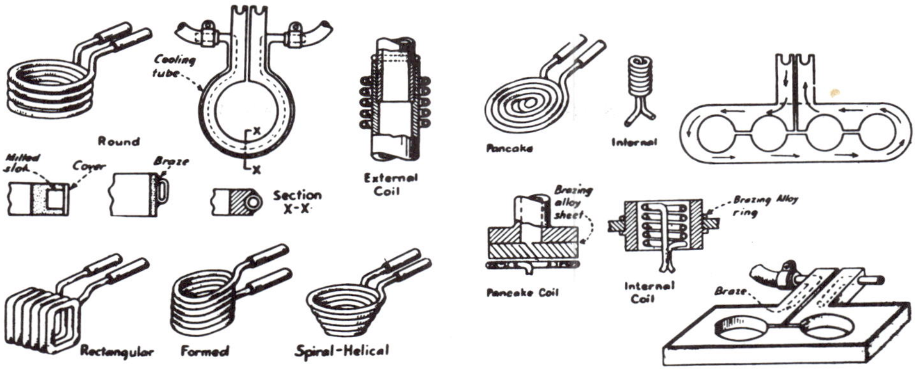

Fig. 1.6—Typical induction brazing coils and plates.

Fig. 1.7—Resistance brazing with tongs.

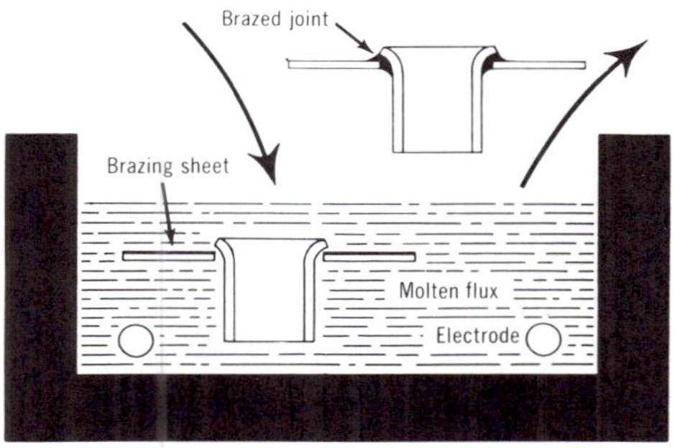

Fig. 1.8—Illustration of chemical bath dip brazing operation.

vides the necessary protection from oxidation; if not, a suitable flux should be used. The salt bath is contained in a metal or other suitable pot, also called the furnace (Fig. 1.9), which is heated (1) from the outside through the wall of the pot, (2) by means of electrical resistance units placed in the bath, or (3) by the I^2R loss in the bath itself.

In molten metal bath dip brazing the parts are immersed in a bath of molten brazing filler metal contained in a suitable pot (Fig. 1.10). The parts must be cleaned and fluxed if necessary. A cover of flux should be maintained over the molten bath. This method is largely confined to brazing small parts, such as wires or narrow strips of metal. The ends of the wires or parts must be held firmly together when they are removed from the bath until the brazing filler metal has fully solidified.

Infrared Brazing

Infrared has been used for a long time as a heat source, but with the development of the high intensity quartz lamp its use has become commercially important. Infrared heat is radiant heat obtained below the red rays in the spectrum. While with every "black" source there is some visible light, the principal heating is done by the invisible radiation. Heat sources (lamps) capable of delivering up to 5000 watts of radiant energy are commercially available. The lamps do not necessarily need to follow the contour of the part to be heated even though the heat input varies inversely as the square of the distance from the source unless reflectors are used to concentrate the heat.

Assemblies to be brazed are supported in a position that enables the energy to impinge on the part. In some applications only the assembly itself is enclosed. There are, however, applications

Fig. 1.9—Chemical bath dip brazing.

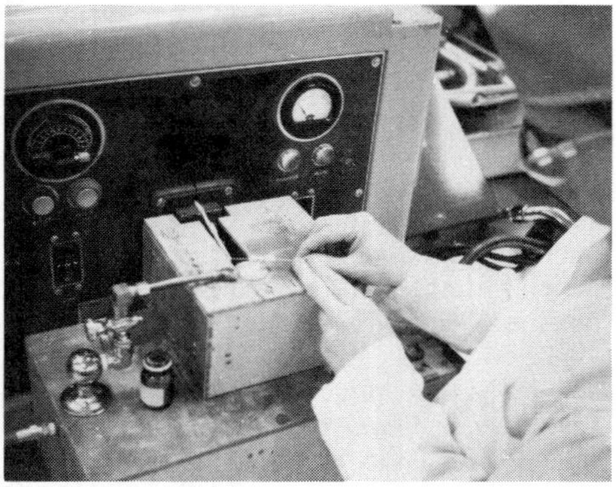

Fig. 1.10—Molten metal bath dip brazing.

whereby the assembly and the lamps are placed in a bell jar or retort that can be evacuated, or in which an inert gas atmosphere can be maintained. The assembly is then heated to a controlled temperature, as indicated by thermocouples. Figure 1.11 shows an experimental apparatus.

Blanket brazing is another of the processes used for brazing. A blanket is resistance heated, and most of the heat is transferred to the parts by two methods, conduction and radiation, the latter being responsible for the majority of the heat transfer.

Special Processes

In several industrial applications, such as the manufacture of electron tubes, the usual brazing processes described previously are used with certain particular refinements (see chapter 26). The use of a vacuum furnace is an illustration of this. The refinements used are determined according to the special requirements of each application and so cannot be adequately covered by the basic information given in this book. The special features of such cases are discussed elsewhere in this book.

Exothermic brazing is another special process used. The use of exothermic reactions for industrial brazing applications has been limited because little information has been available concerning exothermic brazing materials and the

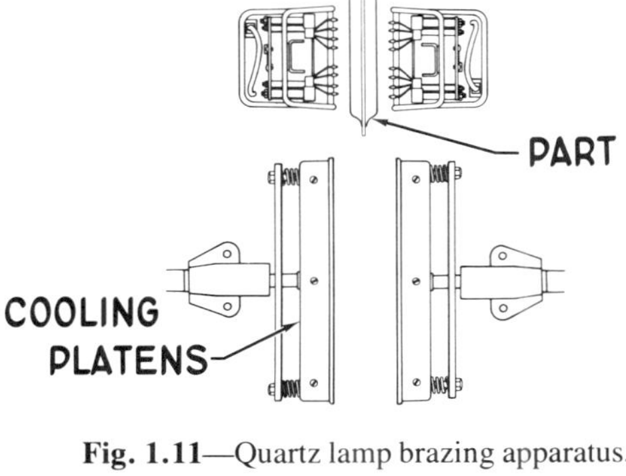

Fig. 1.11—Quartz lamp brazing apparatus.

Brazing Procedure No. ____________________

Joint ______________________________

(Insert sketch here showing shape and thickness of parts, joint clearance, placement of filler metal and assembly of joint including jigging or fixtures.)

____

Base metal(s): ____
(Specify Specification number or other identification.)

Precleaning: ____

____

____

(Specify cleaner to be used: wire brush, emery, or name or composition of cleaner. Also specify cleaning procedure including time, temperature, rinsing, etc. Specify plating operations, if used.)

Filler metal: ____
(Specify Classification Number or other identification and size and form.)

Flux or atmosphere: ____
(Specify AWS Type No. or other identification and location and means of application.)

Brazing Process: ____

____

____

____

(Specify process and other details such as brazing temperature; preheat temperature, if required; tip size and fuel gas for torch brazing; current and pressure for resistance brazing; frequency and coil type and number for induction brazing; etc.)

Postbraze cleaning: ____
(Specify cleaner to be used and rinse.)

Postbraze heat treatment: ____
(Other than coincident with brazing; specify rate of heating, temperature, cooling rate, etc.)

Inspection requirements: ____

____

____

(Specify test requirements, frequency of sampling for inspection, tests to be used, and details of tests such as pressure for pressure test, extent of permissible defects for radiography, size of fillet, etc.)

Remarks:*
(Insert here special precautions, techniques, and other pertinent information not otherwise covered.)

*This item appears on the reverse side of the form.

Fig. 1.12—Proposed form for recording brazing procedures.

Brazing Procedure No. A

Joint 102

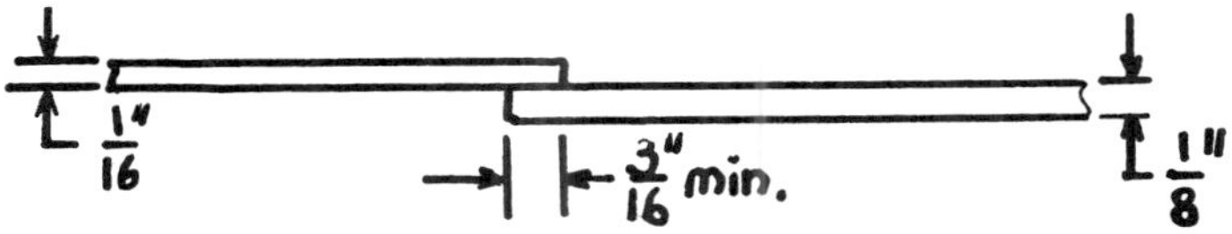

Base metal(s): Top: Deoxidized copper; Bottom: 1020 Steel
(Specify Specification number or other identification.)

Precleaning: Immersion in Greaseoff "A" for 19 min; dry; remove scale completely with 000 emery cloth.

(Specify cleaner to be used: wire brush, emery, or name or composition of cleaner. Also specify cleaning procedure including time, temperature, rinsing, etc. Specify plating operations, if used.)

Filler metal: BAg-1
(Specify Classification Number or other identification and size and form.)

Flux or atmosphere: AWS Type 3A Flux
(Specify AWS Type No. or other identification and location and means of application.)

Brazing Process: Torch braze using oxyacetylene flame without preheat. Use #4 tip, 4 psi oxygen, and 4 psi acetylene pressure. (Specify make of torch.) Concentrate heat on copper member. Complete penetration required.
(Specify process and other details such as brazing temperature; preheat temperature, if required; tip size and fuel gas for torch brazing; current and pressure for resistance brazing; frequency and coil type and number for induction brazing; etc.)

Postbraze cleaning: Permit joint to cool to 900 °F, then immerse in cold water. Brush joint to remove any flux residue. Complete flux removal essential.
(Specify cleaner to be used and rinse.)

Postbraze heat treatment: None
(Other than coincident with brazing; specify rate of heating, temperature, cooling rate, etc.)

Inspection requirements: Visual inspection of each joint to ascertain complete penetration with fillet on each side. Destructive tension test of 1 in each 100 pieces. Joint must fail in copper at not less than 28 000 psi.
(Specify test requirements, frequency of sampling for inspection, tests to be used, and details of tests such as pressure for pressure test, extent of permissible defects for radiography, size of fillet, etc.)

Remarks:* None
(Insert here special precautions, techniques, and other pertinent information not otherwise covered.)

*This item appears on the reverse side of the form.

Fig. 1.13—Use of form for recording brazing procedures. (For illustration only.)

economics of their use. Exothermic brazing refers to a process by which the heat required to melt and flow a commercial filler metal is generated by a solid state exothermic chemical reaction. An exothermic chemical reaction is defined as any reaction between two or more reactants in which heat is given off due to the free energy of the system. Nature has provided us with countless numbers of these reactions; however, only the solid state or nearly solid state metal-metal oxide reactions are suitable for use in exothermic brazing units.

Exothermic brazing utilizes simplified tooling and equipment. The process utilizes the reaction heat in bringing an adjoining or nearby metal interface to a temperature where an already preplaced brazing filler metal will melt and wet the metal interface surfaces. The brazing filler metal can be commercially available brazing filler metal having suitable flow temperatures. The only limitations may be due to the thickness of the metal that must be heated through and the effects of this heat, or any previous heat treatment, on the metal properties.

BRAZING PROCEDURES

The function of a recorded brazing procedure is to establish all the conditions and limits of the variables of the procedure to the end that the brazed products will be uniform and in compliance with the design requirements. A recorded brazing procedure is valuable in that it provides a positive means of conveying the intentions of the designer to the production department with assurance that the instructions will be clearly understood. A recorded brazing procedure can also be used for inspection purposes in determining whether the instructions are being followed in production. For example, an inspector can, by referring to the recorded brazing procedure, determine what base and filler metals are to be used and can verify that they are being used in production. A recorded brazing procedure can be filed upon completion of a job and referred to again later when a similar production job occurs. A suggested form for recording brazing procedures is shown in Fig. 1.12. The factors listed there may not all apply to every job, in which case they are merely omitted. Those items which do apply should be covered in such detail as is necessary to accomplish the desired results.

The following is a discussion of each item shown in Fig. 1.12. It is intended only as a guide to proper and complete recording of brazing procedures:

Sketch

The sketch should show all geometric details of the joint, including tolerances where required. The sketch may be of the complete assembly or of only the joint area.

Base Metal

Use ASTM, AISI, SAE, MIL, or other standard specification numbers. If standard specifications are not applicable, give chemical analysis, trade name, or other positive identification. For dissimilar combinations each base metal should be identified. Special conditions of the base metal (temper, finish, etc.) should also be specified.

Precleaning

Specify complete precleaning procedure. State time between cleaning and brazing, if important, and any special handling requirements to avoid contamination. Plating is to be specified only if used to facilitate brazing.

Filler Metal

Preferably use classification numbers in AWS A5.8, Specifications for Brazing Filler Metal. Give size and form. Also, specify how applied, i.e., face fed, preplaced, etc. Refer to sketch if necessary.

Flux or Atmosphere

In addition to positive identification as to type specify how applied and other special conditions, such as dryness of atmosphere. If both flux and atmosphere are used, both should be specified.

Brazing Process

First the brazing process should be specified, i.e., torch, furnace, induction, dip, etc. All the detailed information necessary to control the process should then be given, including type and size of tip for torch brazing, type of furnace for furnace brazing, etc. If heat is to be applied at a particular area or to be avoided at some other area, this factor should be carefully described.

Postbraze Cleaning

It is usually necessary to remove all traces of flux residue, particularly when corrosive fluxes are used, such as for aluminum and magnesium. In other cases it may be a matter of appearance or of importance for purposes of finishing.

Postbraze Heat Treatment

If a heat treatment is performed concurrent with the brazing operation, the necessary information should be included under the heading "Brazing Process." Other heat treatments should be specified here.

Inspection Requirements

Full details should be specified as to the test or tests to be used, the details of testing, and the test results. The frequency of sampling for testing should be specified, and the extent to which defective joints may be repaired or are rejected should also be defined. Figure 1.13 shows an example of a completed form for recording a brazing procedure. It is important to note that this procedure is used for illustration.

CHAPTER 2

PROPERTIES OF BASE METALS

COMPOSITIONS, MECHANICAL PROPERTIES, PHYSICAL CONSTANTS

The compositions, mechanical properties, and physical constants of many engineering materials that can be brazed are listed in the appendix. The information is necessarily condensed and is included to provide a source for quick reference. Where more exact information is required, reference to the supplier of the material, to competent engineering handbooks, or to applicable specifications should be made. The data in the appendix should not be used for critical design or specification purposes.

The metals and alloys are grouped together according to their main constituents. The names given are those most commonly used in industry. The composition shows the average values or the ranges of the main constituents. No attempt has been made to include the average value or permissible ranges of minor constituents. The mechanical properties are average values obtained for some forms in which the materials are available and are expressed in units commonly used for these materials.

METALLURGICAL CONSIDERATIONS

All the metals and alloys listed in the appendix, and many more, can be brazed when the proper procedure is employed. Some metals and alloys exhibit metallurgical phenomena that influence the behavior of brazed joints and in some cases necessitate special procedures. Some such phenomena are discussed in the following paragraphs and may be classified as base metal effects (carbide precipitation, oxide stability, hydrogen embrittlement, heat-affected zone, sulfur embrittlement), filler metal effects (vapor pressure), or base metal-filler metal interaction effects (alloying, phosphorus embrittlement, stress cracking).

Carbide Precipitation

Some stainless steels and other alloys which contain chromium and carbon are subject to carbide precipitation if heated to temperatures between 800 and 1500 °F (427 and 816 °C). Carbide precipitation means that the carbon combines preferentially with the chromium and is rejected as chromium carbide, usually at the grain boundaries. The alloy material immediately adjacent to the carbide particle, being depleted in chromium content, is no longer as corrosion resistant as the original metal. In certain corrosive environments, the mechanical properties may be impaired with little or no apparent surface attack.

Precipitated carbides in stainless steels may be redissolved by heat treating at 1850-2050 °F (1010-1121 °C) followed by rapid cooling.

Another stabilizing treatment that disperses the unprecipitated chromium uniformly throughout the structure is to heat to 1600 °F (871 °C) for 2 hours followed by furnace cooling to 1000 °F (538 °C) and subsequent air cooling.

If the brazing can be done very rapidly, no appreciable amount of carbides will be precipitated with the normal types of stainless steels. Where this cannot be done and it is necessary to braze stainless steels for corrosive service, one of the stabilized compositions, such as Types 347 or

321, or the extra low carbon grades, such as 304L, should be used.

Oxide Stability

Despite the usual precleaning operations, there are residual oxides present on metals and alloys. Most of these residual oxides are easily removed by the proper flux or reducing atmosphere. Oxides of chromium, aluminum, titanium, silicon, magnesium, manganese, and beryllium are more difficult to remove, and alloys containing these elements usually require special treatment.

Chromium oxide may be removed with some fluoride-bearing fluxes, but cannot be reduced by hydrogen unless the atmosphere is very dry (about −70 °F [−57 °C] dew point) and only at high temperatures (in the range of 1800-2000 °F [982-1093 °C]).

Aluminum, titanium, silicon, magnesium, manganese, and beryllium oxides can also be removed by special fluxes, but cannot be reduced in hydrogen atmospheres with usual brazing techniques (see chapter 4). It is important, therefore, that special fluxes or procedures be used when brazing materials that contain appreciable quantities of these elements.

In many high temperature brazing applications where precipitation hardening alloys containing aluminum or titanium are used, the oxides are extremely stable and difficult to reduce. Brazing can be conducted in hydrogen (−70 °F [−57 °C] dew point or lower), helium, argon, or in a vacuum. In order to make good quality joints consistently, the parts to be brazed may be electroplated prior to brazing, usually with nickel.

Hydrogen Embrittlement

Hydrogen is able to diffuse through many metals quite rapidly because of its small atomic size. The rate of diffusion increases as the temperature increases. When hydrogen diffuses into a metal that has not been completely deoxidized it may reduce the oxide of the metal if the temperature is high enough. Metallic sponge and water vapor are the end products of this reaction.

Because the molecular size of water vapor is too large to permit diffusion to the surface in the same manner as the hydrogen entered, a pressure is developed which has been calculated for tough pitch copper to be as high as 90 000 psi (620 MPa). These extremely high pressures literally tear the metal apart by starting many small fissures or blisters, mainly at the grain boundaries. Metal that has been hydrogen embrittled exhibits lowered tensile properties.

Electrolytic tough pitch copper, silver, and palladium, when they contain oxygen, are subject to hydrogen embrittlement if heated in the presence of hydrogen. If tough pitch copper is to be brazed without embrittlement, hydrogen must not be present in the heating atmosphere. A better practice is to use deoxidized copper or oxygen-free copper where brazing is to be performed. Oxygen-free copper, if improperly heated, may be oxidized and become subject to hydrogen embrittlement. It is impractical to salvage hydrogen-embrittled copper.

Steel is also subject to a form of hydrogen embrittlement, but of a different type. Hydrogen diffuses into steel in the same manner as into copper, but it tends to accumulate in small voids, such as those found around nonmetallic inclusions, and at grain boundaries. Water vapor is not formed, as in copper, but a high pressure is sometimes developed because the hydrogen diffuses in the atomic form and combines with other hydrogen atoms to form molecules which are less mobile, and remain trapped at the discontinuities. Steel which has been treated in this manner may show lower ductility when stressed. Fortunately, however, steel and other ferrous alloys may be salvaged by allowing the hydrogen to diffuse out by baking at slightly elevated temperatures (200-400 °F [93-204 °C]), or by permitting the steel to stand for long periods of time until the ductility is regained.

Some of the newer metals cannot be brazed in hydrogen because they become permanently embrittled. Titanium and zirconium and their alloys are typical examples. Columbium and tantalum are also embrittled when brazed in hydrogen, but fortunately ductility can be restored if proper thermal treatments are followed after brazing.

Most other metals and alloys whose oxides may be reduced by hydrogen contain an excess amount of deoxidizing elements and are not subject to hydrogen embrittlement.

Heat-Affected Zone

The heat employed for brazing affects the properties of the base metals being joined. Base metals whose mechanical properties were obtained by cold working (hard tempers) may soften or the grain size may increase if the brazing temperature is above their recrystallization temperatures. Where mechanical properties are obtained by thermal treatment, they may be altered by the brazing operation. Materials in the annealed condition will generally experience no appreciable change due to brazing.

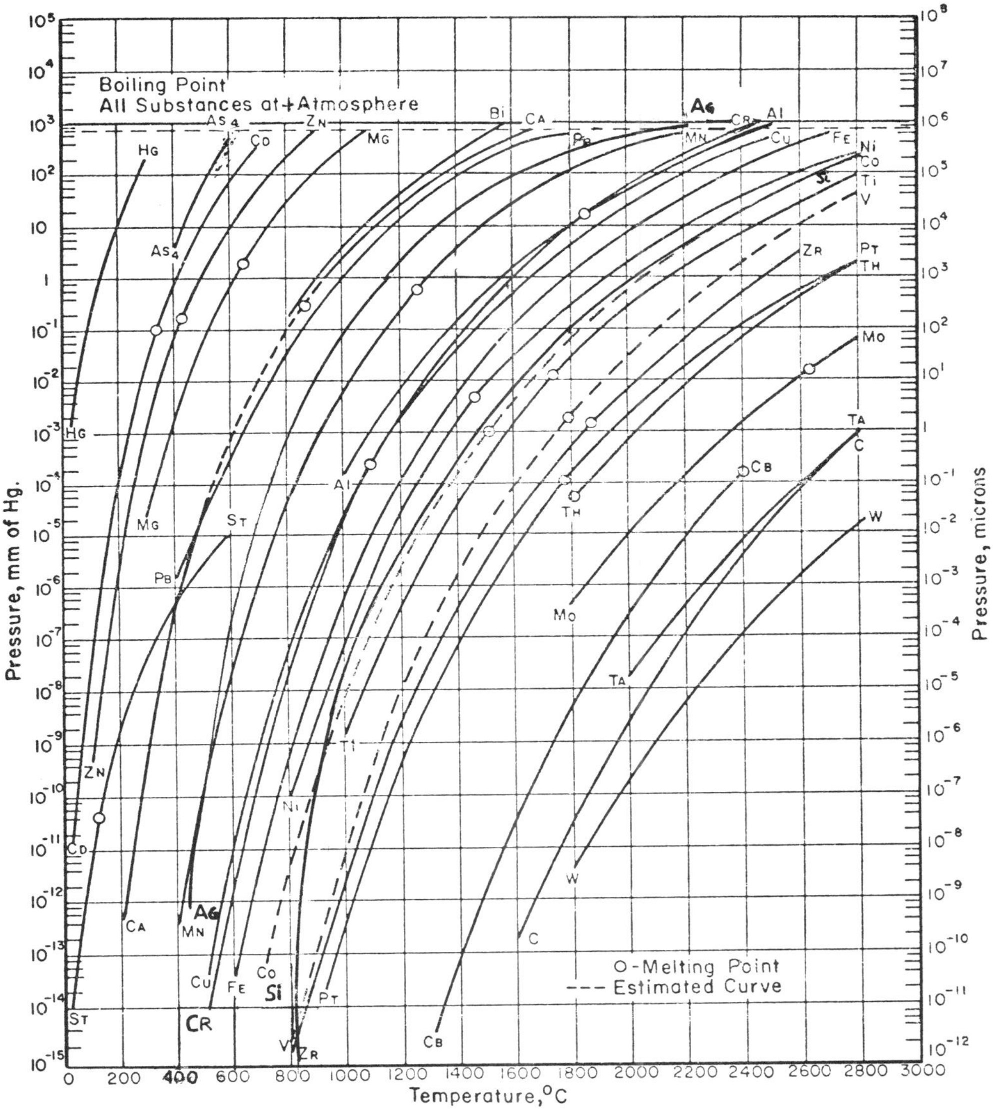

Fig. 2.1—Relationship between vapor pressure and temperature for common elements.

The width of the zone through which these changes may occur will vary with the process used. In some processes (torch and induction) only a localized zone is heated, while in others (furnace and dip) the whole part is heated to attain a braze. In general, the heat-affected zone from brazing is wider and less sharply defined than that obtained by other welding processes.

Sulfur Embrittlement

Nickel and certain alloys containing appreciable amounts of nickel, if heated in the presence of sulfur or compounds containing sulfur, may become embrittled. A low melting nickel sulfide is formed preferentially at the grain boundaries and, being brittle and weak, will crack if subsequently stressed. Material so damaged cannot be salvaged and is usually scrapped.

Nickel and nickel-copper alloys are most susceptible to this attack whereas those containing chromium are less susceptible. It is important for alloys in which nickel is the major component to be clean and free of sulfur-containing materials (such as oil, grease, paint, crayon marks, and drawing lubricants) prior to heating and that heating be done in a relatively sulfur-free atmosphere.

Vapor Pressure

Special attention must be given to the selection of brazing filler metals when the brazed joint is to function in a vacuum. This condition prevails in several industries, but applies particularly to electron tubes of larger size, such as power tubes, and to continuously pumped structures of which particle accelerators, vacuum furnaces, and coating units are examples. In all these devices, a high vacuum is either necessary for satisfactory operation or required prior to the introduction of a desired gas atmosphere. Any components in the system which give off undesirable gases or vapors at any temperature encountered during processing or operation will spoil the vacuum and will thereby shorten the life of the device.

Figure 2.1 shows the relationships between the vapor pressures and temperatures of some elements in common brazing filler metals and base metals. For vacuum tube applications, such elements as zinc and cadmium, which have high vapor pressures at service temperatures, cannot be permitted as constituents of the filler metals. Accordingly, special vacuum-tube-grade filler metals have become commercially available.

Alloying

The metallurgical effects just discussed are concerned with either the base metal or the filler metal, but not with interactions between the two. Base metal-filler metal interactions are most important in determining the behavior of brazed joints, and a discussion of the basic mechanisms involved is warranted. The extent of interaction varies greatly depending on the compositions of the base metal and the filler metal and on thermal cycles. There is always some interaction, except where mutual insolubility permits practically none.

The term "alloying" is a general term covering almost all aspects of interaction. We can be more specific, however, about the components of the term "alloying."

First, the molten filler metal can dissolve base metals. Second, constituents of the filler metal can diffuse into the base metal, either through the bulk of the grains or along the grain boundaries, or can penetrate the grain boundaries as a liquid. The results of such base metal dissolution or filler metal diffusion may be to raise or lower the liquidus or solidus temperatures of the filler metal layer, depending upon composition and thermal cycle.

In brazing nickel, Monel, or the cupronickels with copper, enough dissolution and diffusion occur so that the solidus of the copper filler metal is increased and flow is terminated. This also means that the remelt temperature of the filler metal layer is higher than its original solidus temperature.

If the base metal parts are relatively thin in cross section, the dissolution of the base metal and diffusion into it are especially damaging. Holes may appear, and geometrical defects, such as undercutting, may result. This action is sometimes referred to as erosion. It may occur when brazing aluminum and magnesium sheets, as well as in the brazing of high temperature alloys.

In the brazing of ferrous base high temperature alloys with filler metals containing boron, grain boundary penetration of the base metal by a low melting metal complex can cause joint degradation. The effect is particularly damaging if the base metal is thin, as in the case of brazed sandwich panels. Producers and experienced users of high temperature brazing filler metals should be consulted during the design stages of parts where the use of these filler metals is anticipated.

The brazing of gold to base metals in the production of gold-filled jewelry must be done rapidly so that the filler metal will not diffuse through the gold, causing discoloration and blemishes.

Whenever excessive base metal dissolution and diffusion are liable to occur, the brazing should be done in as short a time and at as low a temperature as possible. Sufficient filler metal should be present to fill the joint completely, but excess filler metal is both undesirable and uneconomical.

Another factor included under the general term "alloying" is the formation of intermetallic compounds as a result of interaction between constituents of the base and filler metals. These compounds are usually brittle. Whether or not the compounds form depends on base and filler metal compositions, time, and temperature. Just because intermetallic compounds do form, it does not necessarily follow that the joint is so embrittled as to lose engineering utility. This depends on the nature of the specific compound, its quantity, and its distribution. One example of deleterious compound formation, phosphorus embrittlement, is discussed in the next section.

Phosphorus Embrittlement

Phosphorus combines with many metals to form brittle compounds known as phosphides. For this reason, copper-phosphorus (BCuP) filler metals are not usually used with any iron or nickel base alloys; however, the nickel filler metals BNi-6 and BNi-7 are used in some cases for brazing heat resisting alloys.

Stress Cracking

Many high strength materials like stainless steels, nickel alloys, and copper-nickel alloys have a tendency to crack during brazing when in a highly stressed condition and in contact with molten brazing filler metal. Materials with high annealing temperatures, and particularly those that are age hardenable, are subject to this stress cracking phenomenon. The cracking occurs almost instantaneously during the brazing operation and is usually readily visible since the molten brazing filler metal follows the crack and completely fills it.

The process may be described as stress corrosion cracking where the molten filler metal is considered as the corrosive medium. Stressed steel in a caustic solution or stressed brass in an ammonia solution are widely known examples of stress corrosion cracking. Sufficient stress to cause stress cracking can be produced by cold work prior to brazing or by an externally applied stress from mechanical or thermal sources during the brazing operation.

Stresses set up by changes in the volume of parts made from some special alloys when they pass through a phase change on cooling may also cause cracking. This cracking is not usually catastrophic, but it may weaken the brazed joint.

When stress is encountered, its cause can usually be determined from a critical analysis of the brazing procedure. The usual remedy is to remove the source of stress. Stress cracking has been eliminated by:

1. Using annealed temper rather than hard temper material
2. Annealing cold worked parts prior to brazing
3. Removing the source of externally applied stress, such as parts that do not fit properly, jigs that exert stress on the parts, or overhanging unsupported weights (see chapter 8)
4. Redesigning parts or revising joint design (see chapter 6)
5. Heating at a slower rate; heavy parts can be heated so rapidly that stresses are set up by steep thermal gradients
6. Heating the fluxed and assembled parts in a torch brazing application to a temperature high enough to effect stress relief, cooling to the brazing temperature, and then hand feeding the filler metal
7. Selecting a brazing filler metal that is less likely to induce this type of damage (see chapter 3)

Figure 2.2 shows stress corrosion cracking of Inconel alloy 718 brazed with silver (BAg) brazing filler metal. Note that the cracks emerge from the braze filler metal over into the base metal. The cracks are readily visible since they are outlined in the color of the silver filler metal.

The age hardenable high nickel alloys are very susceptible to stress corrosion cracking. These alloys should be brazed in the annealed or solution-treated condition with a relatively high melting brazing filler metal (preferably above 1600 °F [871 °C]) that has sufficient strength to withstand handling during the age hardening treatment.

POSTBRAZING THERMAL TREATMENT

It is frequently desirable to give brazed assemblies a post brazing thermal treatment to improve mechanical properties. In ferrous alloys, this treatment entails quenching from an elevated temperature followed by tempering at some lower temperature. In other alloys, such as beryllium copper, Duranickel, "K" Monel, Inconel "X," and Stainless 17-7PH, the treatment consists of heating for a period of time to some intermediate temperature followed by a controlled rate of cooling.

When a thermal treatment is performed subsequent to brazing, it is important that the brazing filler metal selected have sufficient strength at the thermal treatment temperatures to withstand the necessary handling. Postbrazing thermal treatments may be a source of residual stress generation in brazed joints. Microfissuring may occur, for example, and result in lowered joint strength.

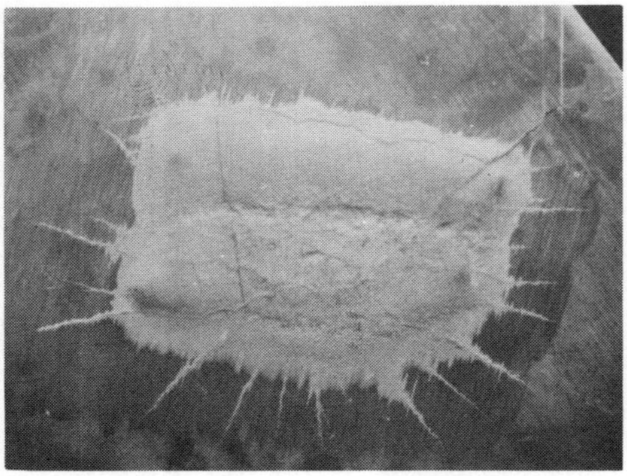

Fig. 2.2—Stress corrosion cracking of Inconel.

DISSIMILAR METAL COMBINATIONS

Many dissimilar metal combinations may be brazed. In fact, brazing can often be used where metallurgical incompatibility precludes the use of other welding processes.

One of the most important criteria when brazing dissimilar metals is a consideration of the difference in thermal expansion between them. If a metal having a high thermal expansion surrounds a low expansion metal, clearances that are satisfactory for promotion of capillary flow at room temperature will be too great at brazing temperature. Conversely, if the low expansion metal surrounds the high expansion metal, no clearance may exist at brazing temperature. For example, when brazing a molybdenum plug in a copper block, the parts must be press fit at room temperature. But, if a copper plug is to be brazed in a molybdenum block, a properly centered loose fit at room temperature is required.

Figure 2.3 is a nomograph useful in learning the actual change in clearance in the "ring and plug" type of joints between dissimilar metals. In cases where more accuracy is important or one of the variables is off the nomograph scale, the equation may be used. In more complex joint configurations, it is usually best to prepare preproduction samples to establish ideal clearances. The appendix contains information on the coefficients of thermal expansion for some common metals and alloys. For design purposes, more detailed information should, however, be obtained from the manufacturers of the metals and alloys, or from comprehensive tables.

A technique often used when brazing materials with different coefficients of expansion is "sandwich brazing." A common application of this technique is in the manufacture of carbide-tipped metal cutting tools. A relatively ductile metal is coated on each side with the brazing filler metal, and the composite is used in the joint. This places a third material in the joint which will creep during cooling and reduce the stresses caused by differential contraction. In some variations of the technique wire mesh is used in place of the foil.

There are other criteria that must also be considered for successful brazing of dissimilar metals. The filler metal must be compatible with both the base metals. Wide differences in base metal melting points must be considered when choosing the filler metal. Where corrosion or oxidation resistance is needed, the filler metal should have properties at least equal to the poorest of the two metals being brazed. In addition, under conditions of the application, galvanic couples that may promote crevice corrosion should be avoided. Brazing filler metals that form low melting phases with the base metals are not recommended unless, as in some special cases, such a technique is used to form the final filler metal in situ.

The metallurgical reactions occurring during the brazing or subsequent thermal treatments between the brazing filler metal and the base metal are important. One example is the brazing of aluminum to copper. The copper reacts with the aluminum to form a low melting brittle compound. Such problems can be overcome by coating one of the base metals with a compatible metal. In the case of aluminum to copper, the copper can be coated with silver or a high silver alloy and then the joints made with a standard aluminum brazing filler metal.

CORROSION RESISTANCE

Corrosion resistance is a subject so complex and requiring so much space that it cannot be covered in this manual. Most corrosion data contained in handbooks may not be directly applicable when referring to brazed assemblies due to the dissimilar metals and joint configurations involved.

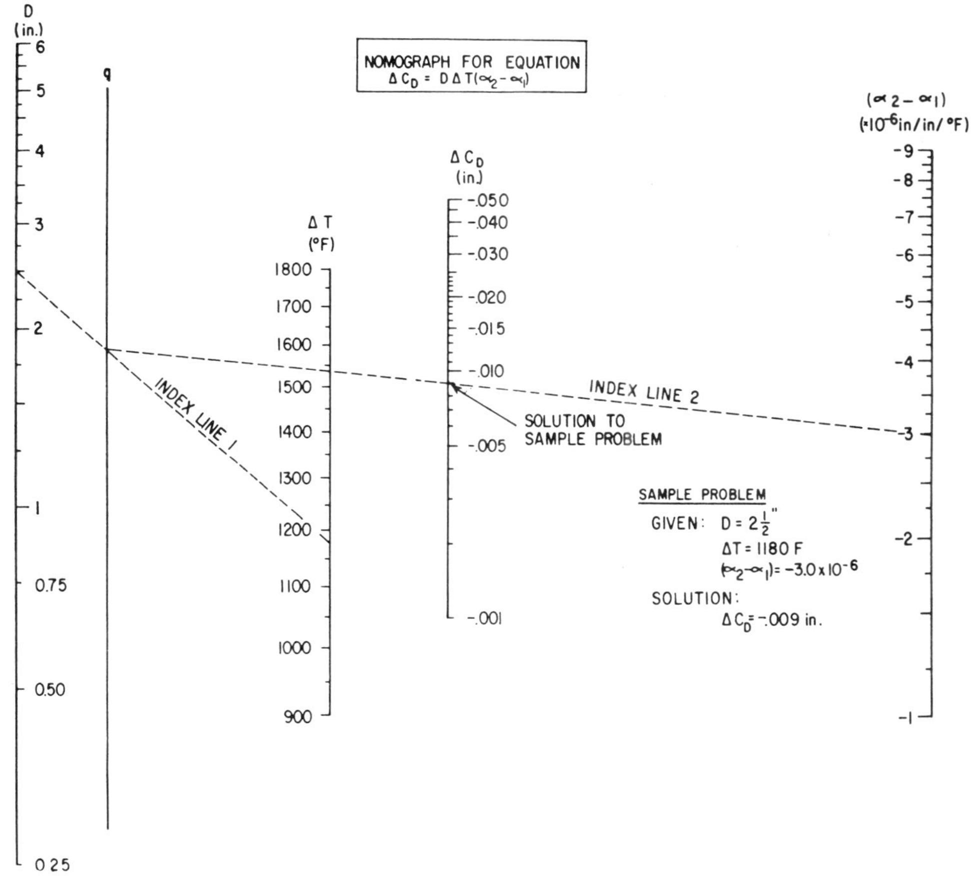

NOTES:

1. This nomograph gives change in diameter caused by heating. Clearance to promote brazing filler metal flow must be provided at brazing temperature.
2. D $=$ nominal diameter of joint, in.
   ΔC_D $=$ change in clearance, in.
   ΔT $=$ brazing temperature minus room temperature, °F
   α_1 $=$ mean coefficient of thermal expansion, male member, in./in./°F
   α_2 $=$ mean coefficient of thermal expansion, female member, in./in./°F
3. This nomograph assumes a case where α_1 exceeds α_2; thus, scale value for $(\alpha_2 - \alpha_1)$ is negative. Resultant values for ΔC_D are therefore also negative, signifying that the joint gap reduces upon heating. Where $(\alpha_2 - \alpha_1)$ is positive, values of ΔC_D are read as positive, signifying enlargement of the joint gap upon heating.

Fig. 2.3—Nomograph for finding the change in diametrical clearance in joints of dissimilar metals for a variety of brazing situations.

CHAPTER 3

BRAZING FILLER METALS

The American Welding Society defines a brazing filler metal as a metal to be added when making a braze. Brazing filler metals are metals or alloys that have melting temperatures above 840 °F (450 °C) but below those of the metals being joined. The clearly defined term "brazing filler metal" has replaced some of the terms formerly used, i.e., hard solder, silver solder, gold solder, and brazing alloy.

For satisfactory use as a brazing filler metal, a metal or alloy should have the following characteristics:

1. Ability to wet and make a strong, sound bond on the base metal on which it is used
2. Suitable melting and flow properties to permit distribution by capillary attraction in properly prepared joints
3. A composition of sufficient homogeneity and stability to minimize separation by liquation under the brazing conditions to be encountered; excessively volatile constituents may be objectionable
4. Ability to produce a braze joint that will meet service requirements, such as strength and corrosion resistance
5. Depending on the requirements, ability to produce or avoid base metal-filler metal interactions.

MELTING OF FILLER METALS

A better understanding of the nature of brazing filler metals and how to select one for a specific application can be obtained by observing the melting characteristics of metals and alloys. Melting of the pure metals is easy to describe: upon heating pure metals transform from the solid state to the liquid state at one temperature.

Most brazing filler metals are alloys rather than pure metals. As alloys may not transform from solid to liquid forms, or vice versa, at one temperature, melting usually progresses from solid to liquid over a temperature range. While at temperatures within this "melting range," both solid and liquid phases are present, and the filler metal is usually a semisolid (also called the "plastic" form). The melting characteristics of a filler metal may be an important consideration when selecting the filler metal for a specific application.

An alloy series can best be studied by drawing a chart to show melting characteristics in relation to chemical composition. Technically, a chart of this type is called a constitutional diagram. Such a diagram is made by preparing a series of alloys that cover the entire range of compositions to be investigated. After careful chemical analysis, one can make cooling curves for each alloy from which can be determined the temperature at which the alloy begins to freeze and the temperature at which it becomes completely solid. These points are plotted and connected by lines to form a series of curves. The customary practice is to represent the composition on the horizontal axis and temperature on the vertical axis. To better illustrate this principle, let us consider the silver-copper constitutional diagram shown in Fig. 3.1.

In order to understand the diagram more fully, the terms used may be defined as follows:

The **solidus** is the highest temperature at which a metal or alloy is completely solid (curve ADCEB, Fig. 3.1, is a plot of the solidus for compositions of the Ag-Cu alloy system).

The **liquidus** is the lowest temperature at which a metal or alloy is completely liquid (curve ACB, Fig. 3.1, is a plot of the liquidus for compositions of the Ag-Cu alloy system).

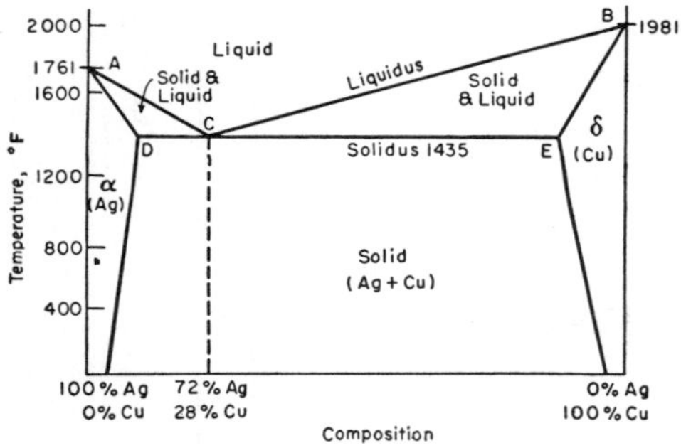

Fig. 3.1—Silver-copper constitutional diagram.

Eutectic compositions are those specific alloy compositions that melt at one temperature, not over a range. In this respect, eutectic alloys behave like pure metals. Technically, in the diagram shown in Fig. 3.1, or in any system having a eutectic, the eutectic is that composition where two descending liquidus curves meet; thus, the composition at that point has a lower liquidus than its neighboring compositions (point C in Fig. 3.1).

As shown in Fig. 3.1, 100% Ag (point A) has a melting point of 1761 °F (961 °C); 100% Cu (point B) has a melting point of 1981 °F (1083 °C). The solidus for any composition between 91.2 Ag-8.8Cu and 8 Ag-92Cu (line DE) is 1435 °F (779 °C). From the diagram it can be seen that the eutectic composition is 72Ag-28Cu (point C). This composition melts and becomes completely liquid at 1435 °F (779 °C). Any composition other than the eutectic will not become completely liquid until a higher temperature is reached. For instance, 50Ag-50Cu will have a solidus temperature of 1435 °F (779 °C) and a liquidus temperature of 1580 °F (860 °C). In other words, this composition of Ag-Cu will begin to melt at 1435 °F (779 °C) and will become completely liquid at 1580 °F (860 °C).

The difference between the solidus (ADCEB) and liquidus (ACB) or the area bounded by curves ACBECDA is called the melting range. In this range some constituents of the alloy are in the liquid phase and others are in the solid phase.

Melting point and flow point are terms that have been in common use, but they have not always been applied with the same meaning, and confusion has developed. For this reason the terms solidus and liquidus, which can be more clearly defined, will be used.

Brazing filler metals with narrow melting ranges and eutectic compositions (both of which behave during melting like a pure metal) are widely used for brazing in applications where the joint clearance is adequately controlled. They are suitable for processes with either a slow or rapid heating rate. When brazing with compositions having a wide melting range, some precautions may be necessary to avoid separation of the solid and liquid portions during a slow heating process.

Transformation from the solid to the liquid phase does not occur uniformly. A relatively large percentage of some compositions will become fluid at a temperature slightly above the solidus and may become essentially fluid before the true liquidus is attained. Other compositions may remain essentially solid until heated well within the melting range. Recommended brazing temperatures recognize this difference in melting of each filler metal. Heating rates, methods of applying the brazing filler metal, and other features of the brazing process may have to be adapted to conform with the melting behavior of the filler metal.

TEMPERATURE REQUIRED FOR BRAZING

Generally, a brazing filler metal is completely molten before it flows into a joint and is distributed by capillary attraction. Therefore, the liquidus normally may be considered the lowest temperature which should be used for brazing, and all sections of the joint must be heated to this temperature or higher. A few alloys are used as brazing filler metals, however, which become sufficiently fluid below the actual liquidus. With these certain compositions, successful brazing can be accomplished even though the liquidus is not attained.

In practice, one cannot be sure of heating internal and external sections of an entire joint only to the minimum temperature. The location and rate of heat input, the mass of the parts, and the thermal conductivity of the base metal are factors that influence the distribution of heat within the work. With localized rapid heating methods, temperature distribution may not be uniform and the temperature at the heated surfaces may be greater than in inner areas. With slower heating methods, where work is heated evenly and slowly, temperature distribution in the assembly is more uniform. (See tables in this chapter for the recommended brazing temperature range for each brazing filler metal.)

It will be noted in the tables that the brazing temperature ranges begin at the liquidus, or for a few compositions below the liquidus. A good practice is to use the lowest temperature within the recommended brazing range consistent with the time needed to complete the braze. This may depend upon (1) the materials being joined, (2) the filler metals, (3) the brazing process used, and (4) the joint design used.

BRAZING FILLER METAL-BASE METAL INTERACTION

When a brazing filler metal wets the base metal on which it is applied, interactions between the filler metal and the base metal occur (see chapter 2). The interactions between the filler metal and the base metal may affect the physical and mechanical properties of both the joint and the base metal.

Factors influencing the extent of these interactions are (1) temperature, (2) time, (3) joint geometry, and (4) base metal and filler metal compositions. For minimizing the interactions, the brazing cycle should be kept as short as practical, and the temperature kept low and within the recommended brazing temperature range for the filler metal being used.

SELECTION OF A FILLER METAL FOR A BRAZING APPLICATION

Characteristics desirable in a brazing filler metal are listed in the introduction to this chapter. Additional factors to be considered when selecting a brazing filler metal for a specific application include (1) the base metals being joined, (2) the method of heating to be used, (3) the method of applying the filler metal, (4) the brazing temperature required, and (5) service requirements for the brazed assembly.

BRAZING FILLER METALS FOR USE ON VARIOUS BASE METALS

The filler metal must interact or otherwise make a suitable bond with the base metal on which it is to be used. Refer to Table 3.1 for an outline of suitable filler metal-base metal combinations. For more detailed metal information, refer to the chapters on the base metals being used. For comments on the specific compositions of each classification of brazing filler metal, see the discussion of each composition beginning on page 42.

HEATING RATE AND BRAZING FILLER METAL FLOW

The melting range of a filler metal is discussed on p. 21. The data for each classification tabulated on pp. 44-45 give the melting range for each.

Pure filler metals and those filler metals with narrow melting ranges are most suitable for use in applications where the joint clearances are small. The heating rate with these filler metals is not critical.

In contrast, some of the compositions with wide melting ranges may be adversely affected by slow heating. See the section "Melting of Filler Metals" previously discussed. When a filler metal of this type is preplaced at the joint and the work is heated within the melting range for more than a brief period, there may be time to allow the fluid portion to flow into the joint from the preplaced filler metal, leaving behind the portion with a higher melting temperature as a skeletal residue or skull. This process has been called "liquation." The fluid portion of the filler metal may be lost from the preplaced form by flow into the joint capillary or by reaction within the base metal.

Liquation, separation of the solid and liquid portion, can be minimized by utilizing heating methods that will heat the filler metal rapidly through the melting range. Liquation is more often encountered when preplaced filler metal is used. With manual face feeding, the filler metal can be withheld until the work is heated to the recommended brazing temperature and the filler metal will flow promptly into the joint. The tendency of a filler metal with a wide melting range to liquate is dependent upon the proportions of the alloy that melt and the time when this melting occurs. Compositions that remain virtually solid until high in the melting range may not liquate excessively, whereas the reverse may be true for compositions of which part become very fluid early in the melting process.

Filler metals with wider melting ranges, which do not liquate excessively, have an additional use when applied by face feeding, where the fit of parts is not ideal and especially where uniform clearances cannot consistently be maintained in production. In face-feeding operations, when parts have suitable clearances, the filler metal can be applied above its liquidus and the joint filled by capillary attraction. However, when an assembly with excessive clearance is encountered, the filler metal may be applied with care at a temperature within the melting range so that

Table 3.1—Base metal-filler metal combinations

	Al & Al alloys	Mg & Mg alloys	Cu & Cu alloys	Carbon & low alloy steels	Cast iron	Stainless steels	Ni & Ni alloys	Ti & Ti alloys	Be, Zr, Vo, & alloys (reactive metals)	W, Mo, Ta, Cb & alloys (refractory metals)	Tool steels
Al & Al alloys	BA1Si										
Mg & Mg alloys	X	BMg									
Cu & Cu alloys	X	X	BAg, BAu, BCuP, RBCuZn								
Carbon & low alloy steels	BA1Si	X	BAg, BAu, RBCuZn	BAg, BAu, BCu, RBCuZn, BNi							
Cast iron	X	X	BAg, BAu, RBCuZn	BAg, RBCuZn	BAg, RBCuZn, BNi						
Stainless steel	BA1Si	X	BAg, BAu	BAg, BAu, BCu, BNi	BAg, BAu, BCu, BNi	BAg, BAu, BCu, BNi					
Ni & Ni alloys	X	X	BAg, BAu, RBCuZn	BAg, BAu, BCu, RBCuZn, BNi	BAg, BCu, RBCuZn	BAg, BAu, BCu, BNi	BAg, BAu, BCu, BNi				
Ti & Ti alloys	BA1Si	X	BAg	BAg	BAg	BAg	BAg	Y			
Be, Zr, Va & alloys (reactive metals)	X BA1Si(Be)	X	BAg	BAg, BNi*	BAg, BNi*	BAg, BNi*	BAg, BNi*	Y	Y		
W, Mo, Ta, Cb & alloys (refractory metals)	X	X	BAg	BAg, BCu, BNi*	BAg, BCu, BNi*	BAg, BCu, BNi*	BAg, BCu, BNi*	Y	Y	Y	
Tool steels	X	X	BAg, BAu, RBCuZn, BNi	BAg, BAu, BCu, RBCuZn, BNi	BAg, BAu, RBCuZn, BNi	BAg, BAu, BCu, BNi	BAg, BAu, BCu RBCuZn, BNi	X	X	X	BAg, BAu, BCu, RBCuZn, BNi

NOTE: Refer to text for information on the specific compositions within each classification.
X — Not recommended; however, special techniques may be practicable for certain dissimilar metal combinations.
Y — Generalizations on these combinations cannot be made. Refer to chapters 20 and 21 for usable filler metals.
* — Special brazing filler metals are available and are used successfully for specific metal combinations. Refer to chapters 20 and 21.

larger gaps can be filled. In the same manner, fillets may be built up which would be impossible without the wide melting range. However, the use of a brazing filler metal in such a manner should be considered an expedient and avoided if possible.

FILLER METAL APPLICATION

Brazing filler metals may be manually face fed after the work is heated. Table 3.2 lists the standard forms available for the various filler metals. Rod and wire forms are usually used for manual face-feeding. Many special shapes and forms designed for the particular application are utilized as preplaced or preformed filler metals. Rings, washers, formed wires, shims, and powder are the most common forms. Such preplacing of the filler metal assures that there is a uniform amount of the filler metal in the correct position on each assembly. Preplaced filler metal may be useful with manual brazing, but is usually an important part of any mechanized brazing procedure. Where joints areas are large, brazing filler metal may be located between the faying surfaces. Brazing rings are sometimes inserted into grooves machined into the work. Totally enclosed rings in such grooves may be necessary for long sleeve joints or for salt bath dip brazing where it is desirable to avoid melting the ring before the work is heated to the brazing temperature.

Brazing Temperature and the Selection of the Filler Metal

First consideration in selecting filler metal for a specific brazing application is that the melting temperature of the metals being joined must not be exceeded in the process. Generally, a filler metal with a low brazing temperature is preferred. Several advantages of lower brazing temperatures are:

1. To minimize heating effects on base metals such as annealing, grain growth, and distortion.
2. To minimize base metal and filler metal interaction during brazing, particularly to avoid erosion of the base metal.
3. To increase the life of fixtures or other tooling.
4. To economize on the heating energy required in the process.

On the other hand, there may also be important considerations for selecting a higher temperature filler metal:

1. A higher melting filler metal is usually more economical to purchase in the original form than low melting filler metals.
2. Brazing temperatures are sometimes selected to coincide with other heat treatment with the base metal such as stress relief and annealing (especially to avoid stress corrosion cracking).
3. Higher temperatures promote base metal filler metal interactions, especially to increase the remelting temperatures of the braze joint.
4. Higher brazing temperatures may be used to permit later processing at elevated temperatures to develop physical properties of the base metal or for sequential brazing with a lower temperature braze filler metal.
5. Atmospheric or vacuum brazing without a flux will more effectively remove oxides when higher temperatures are employed.

Elevated Service Temperatures

The data on each class of filler metal describe some service limitations on various compositions. The most frequent requirements are corrosion or oxidation resistance and strength at elevated temperatures. Chapters on various base metals discuss commonly encountered corrosion problems, but tests under actual service conditions are recommended to confirm the performance of the base metal-filler metal combination being used. Generalized elevated temperature recommendations are shown in Table 3.3. Conditions which may modify these recommendations are:

1. Service temperature
2. Life expectancy of the assembly
3. Thermal cycling
4. Stress conditions
5. Corrosive conditions
6. Radiation stability

Subzero Service Temperatures

Although it is not specifically covered in Table 3.3, available data indicate that the tensile, shear, and impact properties of brazed joints are not adversely affected by low service temperatures. Brazed applications have found acceptable use at cryogenic temperatures. When in doubt, samples should be tested to determine whether or not the necessary requirements of time, temperature, and strength can be met by the joint at the intended service temperature.

AWS A5.8

Specification for Brazing Filler Metal

Issued, 1952; Revised, 1956, 1962, 1969, 1976

Scope: This specification prescribes requirements for filler metals which are added when making a braze. The compositions are selected to include those having different brazing properties as well as those having important commercial applications.

Note 1: The values stated in U.S. Customary Units are to be regarded as the standard.

Note 2: Vertical lines in the margins of this specification indicate where changes (additions, modifications, corrections, deletions) from the 1969 edition were made. This was done as a convenience only, and the American Welding Society assumes no liability whatsoever for any inaccuracy in these notations.

1. Classification and Acceptance

1.1 Classification.[1] The brazing filler metals are classified on the basis of their chemical compositions as manufactured.

1.2 Acceptance. At the option and expense of the purchaser any or all of the test requirements of this specification may be used as a basis for acceptance of the products covered by this specification.

1.3 Chemical Composition Requirements

1.3.1 The brazing filler metals shall conform to the chemical composition shown in Table 1 for the classification in which they are to be included.

1. For an explanation of the method of classification of the filler metals in this specification, see A1.6 in the Appendix.

Table 1 — Chemical composition requirements, percent (see Notes 1 and 2)

Note 1. Analysis shall be made for the elements for which specific values are shown in this table. If, however, the presence of other elements is indicated in the course of routine analysis, further analysis shall be made to determine that the total of these other elements is not present in excess of the limits specified for "other elements total" in the last column in the table.

Note 2. Single values shown are maximum percentages, except where otherwise specified.

Silver

AWS classification	Ag	Cu	Zn	Cd	Ni	Sn	Li	P	Other elements total
BAg-1	44.0 - 46.0	14.0 - 16.0	14.0 - 18.0	23.0 - 25.0	---	---	---	---	0.15
BAg-1a	49.0 - 51.0	14.5 - 16.5	14.5 - 18.5	17.0 - 19.0	---	---	---	---	0.15
BAg-2	34.0 - 36.0	25.0 - 27.0	19.0 - 23.0	17.0 - 19.0	---	---	---	---	0.15
BAg-2a	29.0 - 31.0	26.0 - 28.0	21.0 - 25.0	19.0 - 21.0	---	---	---	---	0.15
BAg-3	49.0 - 51.0	14.5 - 16.5	13.5 - 17.5	15.0 - 17.0	2.5 - 3.5	---	---	---	0.15
BAg-4	39.0 - 41.0	29.0 - 31.0	26.0 - 30.0	---	1.5 - 2.5	---	---	---	0.15
BAg-5	44.0 - 46.0	29.0 - 31.0	23.0 - 27.0	---	---	---	---	---	0.15
BAg-6	49.0 - 51.0	33.0 - 35.0	14.0 - 18.0	---	---	---	---	---	0.15
BAg-7	55.0 - 57.0	21.0 - 23.0	15.0 - 19.0	---	---	4.5 - 5.5	---	---	0.15
BAg-8	71.0 - 73.0	remainder	---	---	---	---	---	---	0.15
BAg-8a	71.0 - 73.0	remainder	---	---	---	---	0.25 - 0.50	---	0.15
BAg-13	53.0 - 55.0	remainder	4.0 - 6.0	---	0.5 - 1.5	---	---	---	0.15
BAg-13a	55.0 - 57.0	remainder	---	---	1.5 - 2.5	---	---	---	0.15
BAg-18	59.0 - 61.0	remainder	---	---	---	9.5 - 10.5	---	0.025	0.15
BAg-19	92.0 - 93.0	remainder	---	---	---	---	0.15 - 0.30	---	0.15
BAg-20	29.0 - 31.0	37.0 - 39.0	30.0 - 34.0	---	---	---	---	---	0.15
BAg-21	62.0 - 64.0	27.5 - 29.5	---	---	2.0 - 3.0	5.0 - 7.0	---	---	0.15

Table 1 (continued) — Chemical composition requirements, percent (see Notes 1 and 2)

Copper - Phosphorus [a]

AWS classification	P	Ag	Cu	Other elements total
BCuP-1	4.8 - 5.2	---	remainder	0.15
BCuP-2	7.0 - 7.5	---	remainder	0.15
BCuP-3	5.8 - 6.2	4.8 - 5.2	remainder	0.15
BCuP-4	7.0 - 7.5	5.8 - 6.2	remainder	0.15
BCuP-5	4.8 - 5.2	14.5 - 15.5	remainder	0.15
BCuP-6	6.8 - 7.2	1.8 - 2.2	remainder	0.15
BCuP-7	6.5 - 7.0	4.8 - 5.2	remainder	0.15

[a] Chemical analysis for phosphorus contents of these ranges should be made by any suitable method agreed upon by the supplier and purchaser. In case of dispute, the procedure in the latest edition of the ASTM E156, Standard Photometric Method for Determination of Phosphorus in High-Phosphorus Brazing Alloys shall be the referee method.

Nickel [d]

AWS classification	Ni	Cr	B	Si	Fe	C	P	S	Al	Ti	Mn	Cu	Zr	Other elements total
BNi-1	remainder	13.0 - 15.0	2.75 - 3.50	4.0 - 5.0	4.0 - 5.0	0.6 - 0.9	0.02	0.02	0.05	0.05	---	---	0.05	0.50
BNi-1a	remainder	13.0 - 15.0	2.75 - 3.50	4.0 - 5.0	4.0 - 5.0	0.06	0.02	0.02	0.05	0.05	---	---	0.05	0.50
BNi-2	remainder	6.0 - 8.0	2.75 - 3.50	4.0 - 5.0	2.5 - 3.5	0.06	0.02	0.02	0.05	0.05	---	---	0.05	0.50
BNi-3	remainder	---	2.75 - 3.50	4.0 - 5.0	0.5	0.06	0.02	0.02	0.05	0.05	---	---	0.05	0.50
BNi-4	remainder	---	1.5 - 2.2	3.0 - 4.0	1.5	0.06	0.02	0.02	0.05	0.05	---	---	0.05	0.50
BNi-5	remainder	18.5 - 19.5	0.03	9.75 - 10.50	---	0.10	0.02	0.02	0.05	0.05	---	---	0.05	0.50
BNi-6	remainder	---	---	---	---	0.10	10.0 - 12.0	0.02	0.05	0.05	---	---	0.05	0.50
BNi-7	remainder	13.0 - 15.0	0.01	0.10	0.2	0.08	9.7 - 10.5	0.02	0.05	0.05	0.04	---	0.05	0.50
BNi-8	remainder	---	---	6.0 - 8.0	---	0.10	0.02	0.02	0.05	0.05	21.5 - 24.5	4.0 - 5.0	0.05	0.50

[d] Cobalt 0.1 percent max (unless otherwise specified) if determined.

Table 1 (continued)—Chemical composition requirements, percent (see Notes 1 and 2 on p. 28)

Copper and Copper-Zinc

AWS classi-fication	Cu	Zn	Sn	Fe	Mn	Ni	P	Pb	Al	Si	Other elements total
BCu-1	99.90 min.	---	---	---	---	---	0.075	0.02	0.01	---	0.10
BCu-1a	99.0 min.	---	---	---	---	---	---	---	---	---	0.30[a]
BCu-2[b]	86.5 min	---	---	---	---	---	---	---	---	---	0.50[c]
RBCuZn-A[e]	57.0 - 61.0	remainder	0.25 - 1.00	*	*	---	---	0.05*	0.01*	*	0.50[d]
RBCuZn-C[e]	56.0 - 60.0	remainder	0.8 - 1.1	0.25 - 1.20	0.01 - 0.50	---	---	0.05*	0.01*	0.04 - 0.15	0.50[d]
RBCuZn-D[e]	46.0 - 50.0	remainder	---	---	---	9.0 - 11.0	0.25	0.05*	0.01*	0.04 - 0.25	0.50[d]

[a] Total other elements requirement pertains only to the metallic elements for this filler metal. Balance is essentially oxygen.
[b] These chemical requirements pertain only to the copper oxide and do not include requirements for the organic vehicle in which the copper oxide is suspended.
[c] Total other elements requirement pertains only to metallic elements for this filler metal.

The following limitations are placed on the nonmetallic elements:

Constituent	percent (max)
Chlorides	0.4
Sulfates	0.1
Oxygen	remainder
Nitric acid insoluble	0.3
Acetone soluble matter	0.5

[d] Total other elements, including the elements marked with an asterisk (*), shall not exceed the value specified.
[e] This AWS classification is intended to be identical with the same classification that appears in AWS A5.7, Specification for Copper and Copper-Alloy Welding Rods.

Magnesium

AWS classi-fication	Al	Mn	Zn	Si	Cu	Ni	Fe	Be	Mg	Other elements total
BMg-1	8.3 - 9.7	0.15 min.	1.7 - 2.3	0.05	0.05	0.005	0.005	0.0002 - 0.0008	remainder	0.30
BMg-2a	11.0 - 13.0	---	4.5 - 5.5	---	---	---	---	0.0002 - 0.0008	remainder	0.30

Table 1 (continued)—Chemical composition requirements, percent (see Notes 1 and 2 on p. 28)

Precious Metals

AWS classification	Au	Cu	Pd	Ni	Other elements total
BAu-1	37.0 - 38.0	remainder	---	---	0.15
BAu-2	79.5 - 80.5	remainder	---	---	0.15
BAu-3	34.5 - 35.5	remainder	---	2.5 - 3.5	0.15
BAu-4	81.5 - 82.5	---	---	remainder	0.15
BAu-5	29.5 - 30.5	---	33.5 - 34.5	35.5 - 36.5	0.15

Aluminum - Silicon

AWS classification	Si	Cu	Fe	Zn	Mg	Mn	Cr	Ti	Al	Other elements each	Other elements total
BAlSi-2	6.8 - 8.2	0.25	0.8	0.20	---	0.10	---	---	remainder	0.05	0.15
BAlSi-3	9.3 - 10.7	3.3 - 4.7	0.8	0.20	0.15	0.15	0.15	---	remainder	0.05	0.15
BAlSi-4	11.0 - 13.0	0.30	0.8	0.20	0.10	0.15	---	---	remainder	0.05	0.15
BAlSi-5	9.0 - 11.0	0.30	0.8	0.10	0.05	0.05	---	0.20	remainder	0.05	0.15
BAlSi-6	6.8 - 8.2	0.25	0.8	0.20	2.0 - 3.0	0.10	---	---	remainder	0.05	0.15
BAlSi-7	9.0 - 11.0	0.25	0.8	0.20	1.0 - 2.0	0.10	---	---	remainder	0.05	0.15
BAlSi-8	11.0 - 13.0	0.25	0.8	0.20	1.0 - 2.0	0.10	---	---	remainder	0.05	0.15

Cobalt

AWS classification	Co	Cr	Ni	Si	W	Fe	B	C	P	S	Al	Ti	Zr	Other elements total
BCo-1	remainder	18.0 - 20.0	16.0 - 18.0	7.5 - 8.5	3.5 - 4.5	1.0	0.7 - 0.9	0.35 - 0.45	0.02	0.02	0.05	0.05	0.05	0.50

1.3.2 The method of chemical analysis shall conform to the appropriate requirements of the latest edition of the Annual Book of ASTM Standards, Part 12, Methods for Chemical Analysis of Metals.[2]

1.4 Mechanical Properties.[3] This specification does not cover the mechanical properties of either the filler metal or the brazed joint.

2. Manufacture

2.1 Method of Manufacture. The brazing filler metals may be made by any method that will yield a product conforming to the requirements of this specification.

2.2 Standard Forms, Sizes, and Tolerances

2.2.1 Standard forms and sizes are as shown in Table 2.

2.2.2 Standard tolerances are as shown in Table 3.

2.2.3 Standard sieve analyses for powdered brazing filler metals are shown in Table 4. Sieve analyses shall be made in accordance with ASTM B214, Standard Method for Sieve Analysis of Granular Metal Powders. Special sieve analyses other than the standard ones shall be as agreed upon by the supplier and the purchaser.

2.3 Packaging. The brazing filler metals shall be suitably wrapped, boxed or otherwise packaged to protect them against injury during shipment or under normal dry storage conditions.

2.4 Marking

2.4.1 All packages containing rods, containers containing powder or formed shapes, and coils of wire or strip shall be legibly labeled or tagged with the following information:

2.4.1.1 Specification and classification numbers including the issue date.

2.4.1.2 Supplier's name and trade designation.

2.4.1.3 Size, length (if rods), and net weight or number of pieces.

2.4.1.4 Lot, control, or heat number.

2.4.2 Wire and strip coils shall be identified on the outside of the package with the classification number. If the filler metal is supplied in special containers (spools, mandrels, etc.) these containers shall be marked with the classification number.

---

2. ASTM Standards can be obtained from the American Society for Testing and Materials, 1916 Race Street, Philadelphia, PA 19103.

3. See A1.3 in the Appendix.

Table 2—Standard forms and sizes (see Notes a through d on p. 36)

AWS classification	Standard form	Nominal size in.	Actual size in.	Actual size mm
BAlSi-2[a]	sheet		0.010	0.25
			0.015	0.38
			0.020	0.51
BAlSi-3	wire in coils rod, 36 in. (910 mm) lengths	1/16	0.063	1.6
		3/32	0.094	2.4
		1/8	0.125	3.2
		5/32	0.157	4.0
		3/16	0.188	4.8
	sheet		0.010	0.25
			0.015	0.38
			0.020	0.51
BAlSi-4	wire in coils rod, 36 in. (910 mm) lengths	1/16	0.063	1.6
		3/32	0.094	2.4
		1/8	0.125	3.2
		5/32	0.157	4.0
		3/16	0.188	4.8
	sheet		0.010	0.25
			0.015	0.38
			0.020	0.51
	powder		100 mesh (sieve analysis)	
BAlSi-5[a]	sheet		0.010	0.25
			0.015	0.38
			0.020	0.51
BAlSi-6[a]	sheet		There are no standard sizes. Sizes will be furnished according to customer requirements subject to mill limitations.	
BAlSi-7[a]	sheet	There are no standard sizes. Sizes will be furnished according to customer requirements subject to mill limitations.		
BAlSi-8	sheet, wire	There are no standard sizes. Sizes will be furnished according to customer requirements subject to mill limitations.		
BCuP-1	[d] strip, 1 in. (25 mm), and 1.5 in. (38 mm) widths		0.010	0.25
			0.015	0.38
			0.030	0.8

Table 2—Standard forms and sizes (see Notes a through d on p. 36)

AWS classification	Standard form	Nominal size in.	Actual size in.	Actual size mm
BCuP-2 BCuP-3 BCuP-4 BCuP-6 BCuP-7	wire in coils and on spools	round, square or rectangular 3/64 - 1/8, incl.	0.047 - 0.125, incl.	1.2 - 3.2, incl.
	[d] rod	round, square or rectangular 1/16 - 1/4, incl. by 18, 20 or 36	0.063 - 0.250, incl. by 18, 20 or 36	1.6 - 6.4, incl. by 460, 500 of 910
	powder		100 mesh 140 C mesh 140 F mesh 325 mesh	
BCuP-5	strip, widths, 0.25 in. (6.4 mm) to 6 in. (150 mm) by multiples of 0.25 in.		0.003 0.005 0.010 0.025	0.08 0.13 0.25 0.6
	[d] wire in coils and on spools	round, square or rectangular 3/64 - 1/8 incl.	0.047 - 0.125 incl.	1.2 - 3.2 incl.
	[d] rod	1/16 - 1/4 incl. by 18, 20 or 36	0.063 - 0.250 incl. by 18, 30 or 36	1.6 - 6.4 incl. by 460, 500 or 910
	[d] powder		100 mesh 140 C mesh 140 F mesh 325 mesh	
BAg-1 BAg-1a BAg-2 BAg-2a BAg-3 BAg-4 BAg-5 BAg-6 BAg-7 BAg-8 BAg-8a BAg-13 BAg-13a BAg-18 BAg-19 BAg-20 BAg-21	[d] strip, widths, 0.25 in. (6.4 mm) to 6 in. (150 mm) by multiples of 0.25 in.		0.002 0.003 0.005 0.010 0.020	0.05 0.08 0.13 0.25 0.51
	[d] wire in coils and rods	1/32 3/64 1/16 3/32 1/8	0.031 0.047 0.063 0.094 0.125	0.8 1.2 1.6 2.4 3.2

Table 2—Standard forms and sizes (see Notes a through d on p. 36)

AWS classification	Standard form	Nominal size in.	Actual size in.	Actual size mm
BAg-1 BAg-1a BAg-2 BAg-2a BAg-3 BAg-8 BAg-13 BAg-21	[d] powder		100 mesh 140 C mesh 140 F mesh 325 mesh	
BAu-1 BAu-2 BAu-3 BAu-4 BAu-5	sheet wire in coils powder		There are no standard sizes. Sizes will be furnished according to customer requirements subject to mill limitations.	
BCu-1 RBCuZn-A RBCuZn-C RBCuZn-D	strip		There are no standard sizes. Sizes will be furnished according to customer requirements subject to mill limitations.	
	rod, 18 or 36 in. (460 or 910 mm) lengths	1/32	0.031	0.8
		1/16	0.063	1.6
		3/32	0.094	2.4
		1/8	0.125	3.2
		5/32	0.157	4.0
		3/16	0.188	4.8
		1/4	0.250	6.4
		5/16	0.312	8.0
		3/8	0.375	9.5
	wire in coils		Round, same as "strip" above.	
BCu-1a	powder		140 C mesh 140 F mesh	
BCu-2	---	---	Available as a suspension in a volatile medium, commonly called a paste.	
BMg-1	wire in coils rod, 36 in. (910 mm) lengths	1/16	0.063	1.6
		3/32	0.094	2.4
		1/8	0.125	3.2
		5/32	0.157	4.0
		3/16	0.188	4.8

Table 2 (continued) — Standard forms and sizes

AWS classification	Standard form	Nominal size in.	Actual size in.	Actual size mm
BMg-2a	wire in coils; rod, 36 in. (910 mm) lengths	1/16	0.063	1.6
		1/8	0.125	3.2
	strip		0.010	0.25
BNi: All classifications	rod[c], length, 10 in. (250 mm)	1/8	0.125	3.2
		1/16	0.063	1.6
	strip[b]; transfer tape[b]		There are no standard sizes. Sizes will be furnished according to customer requirements subject to mill limitations.	
	powder		140 C mesh; 140 F mesh; 325 mesh	
BCo-1 [d]	rod[c], length, 10 in. (250 mm)	1/16	0.063	1.6
	strip[b]		There are no standard sizes. Sizes will be furnished according to customer requirements subject to mill limitations.	
	powder		140 C mesh; 140 F mesh; 325 mesh	

[a] Brazing sheet cladding alloy: cladding varies from five to ten percent of the thickness of the base metal. Several aluminum alloy base metals are available and cladding is supplied on one or both sides.
[b] These filler metal forms are plastically bonded powder rather than rolled or drawn. Permissible size variations as listed in Table 3 do not apply.
[c] These rods are cast. Permissible size variations listed in Table 3 do not apply.
[d] Within this table, general note 2 regarding revision from the 1969 edition does not apply. Departures are indicated by this footnote.

2.4.3 All packages of brazing filler metal except the BAg-1, BAg-1a, BAg-2, BAg-2a and BAg-3 classifications or all individually packaged units enclosed within a larger package shall carry the following caution label, as a minimum, prominently displayed and in legible type on the package. Continuous lengths of brazing filler metals except the BAg-1, BAg-1a, BAg-2, BAg-2a, and BAg-3 classifications which are supplied in special containers (spools, mandrels, etc.) shall have the following caution label, as a minimum, permanently affixed to the body of the container in a prominent position and in legible type:[4]

> **CAUTION**
> **Welding including brazing may produce fumes and gases hazardous to health. Avoid breathing these fumes and gases. Use adequate ventilation. See American National Standard Z49.1, Safety in Welding and Cutting, published by the American Welding Society.**

2.4.4 All packages of brazing filler metal of the BAg-1, BAg-1a, BAg-2, BAg-2a, and BAg-3 classifications or all individually packaged units enclosed within a larger package shall carry the following warning label, as a minimum, prominently displayed and in legible type on the package. Continuous lengths of brazing filler metal of the BAg-1, BAg-1a, BAg-2, BAg-2a, and BAg-3 classifications which are supplied in special containers (spools, mandrels, etc.) shall have the following warning label, as a minimum, permanently affixed to the body of the container in a prominent position and in legible type:

> **WARNING**
> **CONTAINS CADMIUM**
> **POISONOUS FUMES MAY BE FORMED ON HEATING**
> **Do not breathe fumes. Use only with adequate ventilation such as fume collectors, exhaust ventilators, or air-supplied respirators. See American National Standard Z49.1, Safety in Welding and Cutting, published by the American Welding Society. If chest pain, cough or fever develops after use call physician immediately. Keep children away when using.**

4. For a discussion of labelling practices see Appendix B2.

Table 3 - Permissible variations in dimensions for wire, rod, strip, and sheet

Nominal size	Actual size		Permissible variations, plus and minus			
			Round		Square	
in.	in.	mm	in.	mm	in.	mm
Drawn wire:						
Diameter (round) or width (square):	0.010 - 0.020 incl.	0.25 - 0.51	0.0003	0.008	---	---
	over 0.020 - 0.030 incl.	0.51 - 0.8	0.0005	0.013	---	---
	over 0.030 - 0.040 incl.	0.8 - 1.0	0.0007	0.018	---	---
	over 0.040 - 0.050 incl.	1.0 - 1.3	0.0008	0.020	---	---
	over 0.050 - 0.060 incl.	1.3 - 1.5	0.0010	0.025	---	---
	over 0.060 - 0.080 incl.	1.5 - 2.0	0.0015	0.038	---	---
	over 0.080 - 0.250 incl.	2.0 - 6.4	0.0020	0.051	0.004	0.10
Drawn rod:						
Diameter (round):						
5/32 and under	0.157 and under	4.0	0.003	0.08	---	---
3/16 and over	0.188 and over	4.8	0.004	0.10	---	---
Rolled or extruded rod and wire:						
Diameter (round) or width (square):						
3/64 - 1/16 incl.	0.047 - 0.063	1.2 - 1.6	0.005	0.13	---	---
over 1/16 - 1/8 incl.	over 0.063 - 0.125	1.6 - 3.2	0.006	0.15	---	---
over 1/8 - 3/16 incl.	over 0.125 - 0.188	3.2 - 4.8	0.007	0.18	0.009	0.23
over 3/16 - 1/4 incl.	over 0.188 - 0.250	4.8 - 6.4	0.008	0.20	0.010	0.25
			8 in. (200 mm) and under in width		8 in. (200 mm) in width	
Strip or sheet:						
Thickness:	0.006 and under	0.15	0.0006	0.015	---	---
	over 0.006 - 0.013 incl.	0.15 - 0.33	0.0010	0.025	standard strip	
	over 0.013 - 0.021 incl.	0.33 - 0.53	0.0015	0.038	sheet tolerances	
	over 0.021 - 0.026 incl.	0.53 - 0.66	0.0020	0.05	apply	

Table 4 — Standard sieve[a] analyses

100 mesh	Thru No. 60 sieve - 100 percent min Thru No. 100 sieve - 95 percent min
140 C mesh	On No. 100 sieve - trace On No. 140 sieve - 10 percent max Thru No. 325 sieve - 20 percent max
140 F mesh	On No. 100 sieve - trace On No. 140 sieve - 10 percent max Thru No. 325 sieve - 55 percent max
325 mesh	On No. 200 sieve - trace On No. 325 sieve - 10 percent max Thru No. 325 sieve - 90 percent min

[a] These are standard ASTM sieve sizes selected from Table 1 of ASTM E11, Standard Specification for Wire-Cloth Sieves for Testing Purposes. Sieve tests are conducted in accordance with the latest edition of ASTM B214, Standard Method of Test For Sieve Analysis of Granular Metal Powders.

2.5 Rounding Off Procedures. For purposes of determining conformance with this specification, an observed or calculated value shall be rounded off to the nearest 1000 psi (7 MPa) tensile and yield strength, and "to the nearest unit" in the last right-hand place of figures used in expressing the limiting value for other values in accordance with the rounding off method of ASTM E29, Recommended Practice for Indicating Which Places of Figures are to be Considered Significant in Specified Limiting Values.

2.6 Warranty. The supplier shall warrant that the filler metals, in all sizes and classifications, conform to this specification, and any other warranty is specifically disclaimed. The supplier's responsibility shall be limited to replacement of all filler metals that do not conform to requirements of this specification.

3. Special Filler Metal Grades

Note: This section gives special tests and requirements for the vacuum grade brazing filler metals for vacuum devices. All pertinent requirements in Sections 1 and 2 are applicable.

3.1 Classification. Vacuum grade filler metals for vacuum devices in a non-oxidizing atmosphere are classified on the basis of chemical composition and spattering characteristics when melted. (Grade A — spatter free, and Grade B — not spatter free).

3.2 Chemical Composition Requirements

3.2.1 The vacuum grade brazing filler metals shall conform to the chemical composition shown in Table 5 for the classification in which they are to be included.

3.2.2 The chemical analysis shall be as stipulated in 1.3.2.

3.3 Melting Test and Requirements

3.3.1 The melting test shall be performed on an "as received" sample. Approximately one gram of the sample shall be cut with clean, dry tools and placed into a clean, dense polycrystalline (99.9 percent) alumina crucible or a clean, fused silica crucible or boat which has been precleaned by air firing at a minimum temperature of 2012 °F (1100 °C) and stored in a dust-free location.

3.3.2 The crucible or boat with sample shall be placed in a dense polycrystalline alumina or fused silica combustion tube muffle or the equivalent. It shall be purged with dry hydrogen (-40 °F [-40 °C] dewpoint or below) heated to 36 °F (20 °C) above the liquidus, held for ten minutes and then cooled to 150 °F (65 °C) before stopping the hydrogen flow and removing the sample for inspection.

Note: If the sample does not melt under these conditions, the temperature measurement is incorrect or the chemical composition is wrong.

3.3.3 The sample shall be examined at a 5X magnification. A light smokiness with no discrete black specks is the maximum residue that is permissible.

3.4 Spatter Test and Requirements

3.4.1 Spattering shall be tested by bridging the crucible or boat with a nickel channel, the legs of which are designed to allow a small clearance, 0.06 in. (1.6 mm) maximum, above the crucible. The bridge shall be no more than 0.38 in. (9.5 mm) above the filler metal.

3.4.2 Upon the completion of the test, the bottom side of the nickel channel shall be examined at a 5X magnification for evidence of any spatter. No evidence of spatter is permissible.

Table 5 – Chemical composition requirements for special filler metal grades, percent[a]

	Vacuum grade filler metals								
AWS classification	Ag	Cu	Ni	Pd	Zn	Cd	Pb	P	C
BVAg-0	99.95	0.05	---	---	0.001	0.001	0.002	0.002	0.005
BVAg-6b	49.0 - 51.0	49.0 - 51.0	---	---	0.001	0.001	0.002	0.002	0.005
BVAg-8	71.0 - 73.0	27.0 - 29.0	---	---	0.001	0.001	0.002	0.002	0.005
BVAg-8b	70.0 - 72.0	27.0 - 29.0	0.5 - 1.0	---	0.001	0.001	0.002	0.002	0.005
BVAg-8c	67.0 - 69.0	26.0 - 28.0	---	4.5 - 5.5	0.001	0.001	0.002	0.002	0.005
BVCu-1x	0.05	99.95	---	---	0.002	0.002	0.002	0.002	0.005

Notes:

1. All other metallic impurities having a vapor pressure higher than 10^{-7} torr at 932°F (500°C) are limited to 0.002 percent max each. Those having a vapor pressure lower than 10^{-7} torr at 932°F (500°C) are limited to a total of 0.05 percent.
2. All finished material shall be reasonably smooth and bright and free from dirt, oil, grease or other foreign materials.
3. Filler metals specified Grade "A" are spatter free. Grade "B" filler metals are not spatter free. Complete designation of specified material must include the grade specification letter (i.e. BVAg-6b, Grade "A").

[a] Material shall be made from virgin metal or known composition-clean scrap.

Appendix A: Guide to AWS Classification of Brazing Filler Metals

A1 Introduction

A1.1 This guide is appended to the specification as a source of information; it is not mandatory and does not form a part of the specification. It has been prepared as an aid to users of brazing filler metals, to help them determine which classification of brazing filler metal is best for a particular job. The AWS *Brazing Manual* should be consulted for more detailed information.

A1.2 Brazing filler metals are metals that are added when making a braze. They have melting points below those of the metals being brazed and properties suitable for making joints by capillary attraction between closely fitted surfaces.

A1.3 The classification method for brazing filler metals is based on chemical composition rather than on mechanical property requirements. The mechanical properties of a brazed joint depend, among other things, on the base metal and filler metal used. Therefore, a classification method based on mechanical properties would be misleading since it would only apply if the brazing filler metal were used on a given base metal. If a user of brazing filler metal desires to determine the mechanical properties of a given base metal and filler metal combination, tests should be conducted using the latest edition of AWS C3.2, Standard Method for Evaluating the Strength of Brazed Joints.

A1.4 Brazing filler metals are standardized into eight fairly well-defined groups of classifications as follows: aluminum-silicon, copper-phosphorus, silver, precious metals, copper and copper-zinc, magnesium, nickel, and cobalt filler metals. Many of these classifications are used for joining applications for vacuum tubes and vacuum apparatus. For these critical applications it is desirable to hold the high vapor pressure elements to a minimum, as they usually become "emitters" during operation of the device. Filler metals for vacuum devices have been incorporated as an additional "vacuum grade" classification within this specification.

A1.5 This specification is intended to provide both the supplier and the user of brazing filler metal with a means of production control and a basis of acceptance through mutually acceptable standard requirements. This specification classifies only those filler metals in common use at the time of issue. Other brazing filler metals have been omitted since their volume of use is considered insufficient to warrant inclusion. As usage increases some of these may be included in later revisions of this specification.

A1.6 The nine basic groups (including vacuum grade) of classifications of brazing filler metal are identified, as shown in Tables 1 and 5 of this specification, by the principal element or elements in their chemical composition. In a typical example, such as BCuP-2, the B is for brazing filler metal (like the "E" is for electrodes and the "R" for welding rods in other AWS specifications). The RB in RBCuZn-A, RBCuZn-C and RBCuZn-D indicates that the filler metal is suitable as a welding rod and as a brazing filler metal. CuP is for copper-phosphorus, the two principal elements in this particular brazing filler metal. (Similarly, in other brazing filler metals, Si is for silicon, Ag for silver, etc., using standard chemical symbols.) The numeral 2 is for one particular chemical analysis within a group (1 is another analysis within the same group, 3 another, etc.). The vacuum grade nomenclature follows the examples above with two exceptions. The first exception is the addition of the letter "V," yielding the generic letters BV for vacuum grade. The second exception is the use of the grade suffix, "Grade A" to indicate spatter free or "Grade B" to indicate non-spatter free grades. Thus a vacuum grade spatter free filler metal could be indicated by BVAg-6b, Grade A. Table 5 illustrates vacuum grade filler metals. To avoid confusion, solidus and liquidus are specified instead of melting and flow points. The terms solidus and liquidus are well established in metallurgical use with definitions as follows:

A1.6.1 solidus. The highest temperature at which the metal is completely solid; in other words, the temperature at which melting starts.

A1.6.2 liquidus. The lowest temperature at which the metal is completely liquid; in other words, the temperature at which freezing starts.

A1.7 The solidus and liquidus for a particular metal are definite and fixed.

A1.8 Tables A1 and A2 in this appendix list the solidus, liquidus, and also the recommended brazing temperature range for the various brazing filler metals.

A1.9 When brazing with some brazing filler metals (particularly those with a wide temperature range between solidus and liquidus), the several constituents of the filler metals tend to separate during the melting process. The lower melting constituent will flow, leaving behind a "skull" of the high melting constituent. This occurrence, called liquation, is undesirable since the unmelted skull does not contribute significantly to the actual brazed joint; it does not flow into the joint. However, where fit-up is poor, a filler metal with a wide temperature range will usually fill the joint more easily.

A1.10 When brazing with any of the various processes, prebraze cleaning of the base metal is essential. The surface should be free from grease, oxides, scale, and dirt of any kind. Cleaning can be done chemically or mechanically. Grease or oil can be removed by a reliable degreasing solution such as trisodium phosphate. Oxides should be removed from some materials by a chemical bath or mechanical means after degreasing. Steel grit blasting, shot blasting, or grinding with a fine wheel can be used to remove surface scale.

A1.11 Ventilation During Welding Including Brazing

A1.11.1 Three major factors in welding and brazing which govern the

Table A1 — Solidus, liquidus, and brazing temperature ranges[a]

AWS classification	Solidus °F	Solidus °C	Liquidus °F	Liquidus °C	Brazing temperature range °F	Brazing temperature range °C
Silver						
BAg-1	1125	607	1145	618	1145 - 1400	618 - 760
BAg-1a	1160	627	1175	635	1175 - 1400	635 - 760
BAg-2	1125	607	1295	702	1295 - 1550	702 - 843
BAg-2a	1125	607	1310	710	1310 - 1550	710 - 843
BAg-3	1170	632	1270	688	1270 - 1500	688 - 816
BAg-4	1240	671	1435	779	1435 - 1650	779 - 899
BAg-5	1250	677	1370	743	1370 - 1550	743 - 843
BAg-6	1270	688	1425	774	1425 - 1600	774 - 871
BAg-7	1145	618	1205	652	1205 - 1400	652 - 760
BAg-8	1435	779	1435	779	1435 - 1650	779 - 899
BAg-8a	1410	766	1410	766	1410 - 1600	766 - 871
BAg-13	1325	718	1575	857	1575 - 1775	857 - 968
BAg-13a	1420	771	1640	893	1600 - 1800	871 - 982
BAg-18	1115	602	1325	718	1325 - 1550	718 - 843
BAg-19	1435	779	1635	891	1610 - 1800	877 - 982
BAg-20	1250	677	1410	766	1410 - 1600	766 - 871
BAg-21	1275	691	1475	802	1475 - 1650	802 - 899
Aluminum-Silicon						
BAlSi-2	1070	577	1135	613	1110 - 1150	599 - 621
BAlSi-3	970	521	1085	585	1060 - 1120	571 - 604
BAlSi-4	1070	577	1080	582	1080 - 1120	582 - 604
BAlSi-5	1070	577	1095	591	1090 - 1120	588 - 604
BAlSi-6[b] . .	1038	559	1125	607	1110 - 1150	599 - 621
BAlSi-7[b] . .	1038	559	1105	591	1090 - 1120	588 - 604
BAlSi-8[b] . .	1038	559	1075	579	1080 - 1120	582 - 604
Precious Metals						
BAu-1	1815	991	1860	1016	1860 - 2000	1016 - 1093
BAu-2	1635	891	1635	891	1635 - 1850	891 - 1010
BAu-3	1785	974	1885	1029	1885 - 1995	1029 - 1091
BAu-4	1740	949	1740	949	1740 - 1840	949 - 1004
BAu-5	2075	1135	2130	1166	2130 - 2250	1166 - 1232

[a] Solidus and liquidus shown are for the nominal composition in each classification.

[b] The solidus and liquidus temperatures shown for this filler metal were obtained in air. The solidus and liquidus temperatures of this filler metal in vacuum are different.

Table A1 (continued) – Solidus, liquidus, and brazing temperature ranges

AWS classification	Solidus °F	Solidus °C	Liquidus °F	Liquidus °C	Brazing temperature range °F	Brazing temperature range °C
Copper and Copper-Zinc						
BCu-1	1980	1082	1980	1082	2000 - 2100	1093 - 1149
BCu-1a	1980	1082	1980	1082	2000 - 2100	1093 - 1149
BCu-2	1980	1082	1980	1082	2000 - 2100	1093 - 1149
RBCuZn-A . .	1630	888	1650	899	1670 - 1750	910 - 954
RBCuZn-C . .	1590	866	1630	888	1670 - 1750	910 - 954
RBCuZn-D . .	1690	921	1715	935	1720 - 1800	938 - 982
Copper-Phosphorus						
BCuP-1	1310	710	1695	924	1450 - 1700	788 - 927
BCuP-2	1310	710	1460	793	1350 - 1550	732 - 843
BCuP-3	1190	643	1495	813	1325 - 1500	718 - 816
BCuP-4	1190	643	1325	718	1275 - 1450	691 - 788
BCuP-5	1190	643	1475	802	1300 - 1500	704 - 816
BCuP-6	1190	643	1450	788	1350 - 1500	732 - 816
BCuP-7	1190	643	1420	771	1300 - 1500	704 - 816
Magnesium						
BMg-1	830	443	1110	599	1120 - 1160	604 - 627
BMg-2a	770	410	1050	566	1080 - 1130	582 - 610
Nickel						
BNi-1	1790	977	1900	1038	1950 - 2200	1066 - 1204
BNi-1a	1790	977	1970	1077	1970 - 2200	1077 - 1204
BNi-2	1780	971	1830	999	1850 - 2150	1010 - 1177
BNi-3	1800	982	1900	1038	1850 - 2150	1010 - 1177
BNi-4	1800	982	1950	1066	1850 - 2150	1010 - 1177
BNi-5	1975	1079	2075	1135	2100 - 2200	1149 - 1204
BNi-6	1610	877	1610	877	1700 - 2000	927 - 1093
BNi-7	1630	888	1630	888	1700 - 2000	927 - 1093
BNi-8	1800	982	1850	1010	1850 - 2000	1010 - 1093
Cobalt						
BCo-1	2050	1121	2100	1149	2100 - 2250	1149 - 1232

amount of airborne concentrations of substances to which welders and brazers may be exposed are:

A1.11.1.1 Dimensions of the space in which welding or brazing is to be done (with special regard to height of ceiling)

A1.11.1.2 Number of welders or brazers in that space

A1.11.1.3 Possible evolution of hazardous fumes, gases or dusts according to the base metal and filler metals being used.

A1.11.2 American National Standard Z49.1, Safety in Welding and Cutting, discusses the ventilation that is required during welding and brazing and should be referred to for details. Attention is particularly drawn to Section 8, Health Protection and Ventilation.

A2 Operating Characteristics and Usability

A2.1 BAlSi (Aluminum-Silicon) Classifications. Brazing filler metals of the BAlSi classifications are used for joining the following grades of aluminum and aluminum alloys: 1060, EC, 1100, 3003, 3004, 3005, 5005, 5050, 6053, 6061, 6062, 6063, 6951, 7005, and cast alloys A612 and C612. BAlSi-2, -3, -4 and -5 filler metals are suitable for furnace and dip brazing processes. BAlSi-3, -4 and -5 filler metals are also suitable for manual and automatic torch brazing processes. BAlSi-2, -3, -4 and -5 filler metals are also used to a limited extent with other brazing methods. They are generally used with lap or T joints rather than butt joints. BAlSi-6, -7 and -8 filler metals are suitable for vacuum brazing aluminum. During the vacuum brazing process, magnesium is depleted from the BAlSi-6, -7 and -8 filler metals, causing their solidus temperature to increase and approach the solidus temperature of the BAlSi-2, -5 and -4 filler metals respectively.

A2.1.1 Joint clearances of 0.006 to 0.010 in. (0.15 to 0.25 mm) are common for members which overlap less than ¼ in. (6.4 mm). Joint clearances up to 0.025 in. (0.64 mm) are used for members which overlap more than ¼ in. (6.4 mm). Fluxing is essential for all processes. After brazing, the brazed parts must be cleaned thoroughly. Immersion in boiling water generally will remove the residue; if this is not adequate, the parts are usually immersed in a concentrated commercial nitric acid or other suitable acid solution and then rinsed thoroughly.

A2.1.2 BAlSi-2 brazing filler metal is available as sheet and as a standard cladding on one or both sides of brazing sheet having a core of either 3003 or 6951 aluminum alloy. It is used for furnace and dip brazing only.

A2.1.3 BAlSi-3 is a general-purpose brazing filler metal. It is used with all brazing processes, with some casting alloys, and where controlled flow is desired.

A2.1.4 BAlSi-4 is a general-purpose brazing filler metal. It is used with all brazing processes for complete flow and for good corrosion resistance.

A2.1.5 BAlSi-5 brazing filler metal is available as sheet and as a standard cladding on one or both sides of brazing sheet having a core of 6951 aluminum alloy. BAlSi-5 is used for furnace and dip brazing at lower temperatures than BAlSi-2. The core alloy employed in brazing sheet with this filler metal cladding can be heat treated and aged.

A2.1.6 BAlSi-6 is a vacuum brazing filler metal available as a standard cladding on one or both sides of brazing sheet having a core of 3003 alloy.

A2.1.7 BAlSi-7 is a vacuum brazing filler metal available as a standard cladding on one or both sides of brazing sheet having a core of 3003 or 6951 aluminum alloy. The 6951 alloy cored brazing sheet can be heat treated and aged after brazing.

A2.1.8 BAlSi-8 is a vacuum brazing filler metal available as wire and as a standard cladding on two-side clad brazing sheet having a core of 3003 aluminum alloy. BAlSi-8 is used for vacuum brazing at lower temperatures than BAlSi-6 or -7 filler metals.

A2.2 BCuP (Copper-Phosphorus) Classifications. Brazing filler metals of the BCuP classifications are used primarily for joining copper and copper alloys although they have some limited use on silver, tungsten, and molybdenum. These filler metals should not be used on ferrous or nickel-base alloys and copper-nickel alloys containing a nickel content in excess of ten percent. They are suitable for all brazing processes. Lap joints are recommended, though butt joints may be used if requirements are less stringent.

A2.2.1 These filler metals have self-fluxing properties when used on copper; however, a flux is recommended when used on all other metals, including the copper alloys. Corrosion resistance is satisfactory except when the joint is in contact with sulfurous atmospheres at elevated temperatures.

A2.2.2 These filler metals also have a tendency to liquate if heated slowly. It is important, therefore, especially if the filler metal is preplaced, to heat as rapidly as possible. The color after brazing is light gray, but immersion in ten percent sulfuric acid will restore the copper color.

A2.2.3 It will be noted in Table A1 that the brazing temperature ranges begin below the liquidus. It is a basic principle in brazing that the lowest temperature within the recommended range be used, consistent with the joint clearance and the time needed to complete the braze. This in turn will depend on (1) the materials joined, (2) the brazing process employed, and (3) the joint design used.

A2.2.4 BCuP-1 brazing filler metal is available in strip form 0.010 in. (0.25 mm) minimum and is designed for preplacing in a joint. It is particularly suited for resistance brazing applications. This filler metal is somewhat more ductile than other BCuP filler metals containing more phosphorus. It is also less fluid at the brazing temperature.

A2.2.5 BCuP-2 and -4 brazing filler metals are extremely fluid at brazing temperatures and will penetrate joints with very little clearance. Best results are obtained with clearances of 0.001 to 0.003 in. (0.03 to 0.08 mm).

A2.2.6 BCuP-3 and -5 brazing filler metals may be used where very close fits cannot be held. Joint clearances of 0.001 to 0.005 in. (0.03 to 0.13 mm) are recommended.

A2.2.7 BCuP-6 brazing filler metal combines some of the properties of BCuP-2 and BCuP-3. It has the ability to fill gaps at the lower end of its brazing range and is very fluid at the higher end. Joint clearances of 0.001 to 0.005 in. (0.03 to 0.13 mm) are recommended.

A2.2.8 BCuP-7 brazing filler metal is more fluid than BCuP-3 or BCuP-5 and has a lower liquidus temperature. It is used extensively in the form of preplaced rings in heat exchanger and tubing joints. Joint clearances of 0.001 to 0.005 in. (0.03 to 0.13 mm) are recommended.

A2.3 BAg (Silver) Classifications. Brazing filler metals of the BAg classifications are used for joining most ferrous and nonferrous metals except aluminum and magnesium. These filler metals have good brazing properties and are suitable for preplacement in the joint or for manual feeding into the joint. All methods of heating may be used. Although lap joints are generally used, butt joints may be used if requirements are less stringent. Joint clearances of 0.002 and 0.005 in. (0.05 to 0.13 mm) are recommended for proper capillary attraction. Flux is generally required; however, on most metals, if brazing is done in a suitable atmosphere with BAg-8, BAg-8a, BAg-18 or BAg-19, flux may not be necessary.

A2.3.1 BAg-1 brazing filler metal has the lowest brazing temperature range of the BAg filler metals. It also flows most freely into capillary joints. Its narrow melting range is suitable for rapid or slow methods of heating. BAg-1 is recommended for first consideration on applications requiring a low-temperature, free-flowing BAg filler metal. BAg-1 is more economical (five percent less silver) than BAg-1a.

A2.3.2 BAg-1a brazing filler metal has properties similar to BAg-1. Either composition may be used where low-temperature, free-flowing filler metals are desired.

A2.3.3 BAg-2 brazing filler metal is, like BAg-1, free-flowing and suited for general-purpose work. Its broader melting range is helpful where clearances are not uniform, but, unless heating is rapid, care must be taken that the lower melting constituents do not separate out by liquation.

A2.3.4 BAg-2a brazing filler metal is similar to BAg-2 but is more economical than BAg-2 since it contains five percent less silver.

A2.3.5 BAg-3 brazing filler metal is a modification of BAg-1a. It has good corrosion resistance in marine and caustic media and when used on stainless steel will inhibit interface corrosion. Because its nickel content improves its wetting ability on tungsten carbide tool tips, its largest use is to braze carbide tool assemblies. Melting range and low fluidity make BAg-3 suitable for building fillets or bridging large gaps.

A2.3.6 BAg-4 brazing filler metal is, like BAg-3, used extensively for carbide tip brazing, but it is freer flowing than BAg-3.

A2.3.7 BAg-5 and -6 brazing filler metals are used especially for brazing in the electrical industry. They are also used in the dairy and food industries where the use of cadmium-containing filler metals might be prohibited. Although BAg-6 has a broad melting range and is not as free flowing as BAg-1 and -2, it is a good filler metal for bridging gaps or forming fillets.

A2.3.8 BAg-7 brazing filler metal is low-melting with good flow and wetting properties. It is used (1) where lower zinc and cadmium content is helpful in low heating cycles, (2) for food equipment where cadmium must be avoided, (3) to minimize stress corrosion of nickel or nickel-base alloys at low brazing temperatures, and (4) where white color will improve appearance.

A2.3.9 BAg-8 brazing filler metal is suitable for use in controlled atmosphere furnace brazing (including vacuum brazing) without the use of a paste flux. It is usually used on copper or copper alloys. When molten, BAg-8 is very fluid and may flow out over the work surfaces during some furnace brazing applications. It can also be used on stainless steel, nickel-base alloys and on carbon steel, although its wetting action on these metals is slow.

A2.3.10 BAg-8a brazing filler metal is used for pure, dry atmosphere brazing and is of particular advantage when brazing precipitation-hardening and other stainless steels in the 1400 and 1600 °F (760 to 870 °C) range. The lithium content serves to promote wetting and to increase the flow of the filler metal on difficult-to-braze metals and alloys. Lithium is particularly helpful on base metals containing minor amounts of titanium or aluminum.

A2.3.11 BAg-13 brazing filler metal is used for service temperatures up to 700 °F (370 °C). Its low zinc content makes it suitable for furnace brazing.

A2.3.12 BAg-13a brazing filler metal is similar to BAg-13 except that it contains no zinc, which is advantageous where zinc volatilization is objectionable in furnace operations.

A2.3.13 BAg-18 brazing filler metal is similar to BAg-8 in its applications. Its tin content helps promote wetting on stainless steel, nickel-base alloys, and carbon steel. It has a lower liquidus than BAg-8 and is used in step brazing applications where fluxless brazing is important.

A2.3.14 BAg-19 brazing filler metal is used for the same applications as BAg-8a. BAg-19 is usually used where the precipitation-hardening heat treatment is combined with the brazing operation; BAg-8a is used where the higher brazing temperatures are not required.

A2.3.15 BAg-20 brazing filler metal possesses good wetting and flow characteristics and has a brazing temperature range higher than the popular AgCuZnCd compositions. Due to its good brazing properties, freedom from cadmium, and a more economical silver content, new uses for this filler metal are being developed.

A2.3.16 BAg-21 brazing filler metal is used in brazing AISI 300 and 400 series stainless steels as well as the precipitation hardening alloys. BAg-21 is particularly suited to atmosphere furnace brazing because of the absence of zinc and cadmium and does not require a flux for proper brazing unless the temperatures are low. It requires a rather high brazing temperature, and it flows in a sluggish manner. The nickel content makes it immune to crevice corrosion, particularly on the 400 series stainless steels, by imparting a nickel-rich layer to the fillet edge. It has been used for brazing stainless steel vanes of aircraft gas turbine engines.

A2.4 BAu (Precious Metals) Classifications. Brazing filler metals of the BAu classifications are used for the brazing of iron, nickel, and cobalt-base metals where resistance to oxidation or corrosion is required. Because of their low rate of interaction with the base metal, they are commonly used on thin base metals. These filler metals are usually used with induction, furnace, or resistance heating in a protective atmosphere which includes vacuum and in these cases no mineral flux is used. For other applications, a borax-boric acid flux is used.

A2.4.1 BAu-1, -2 and -3 brazing filler metals, when used for different joints in the same assembly, permit variation in brazing temperature so that step brazing can be used.

A2.4.2 BAu-4 brazing filler metal is used to braze a wide range of high temperature iron- and nickel-base alloys.

A2.4.3 BAu-5 brazing filler metal is primarily used for joining heat and corrosion resistant base metals where corrosion resistant joints with good strength at high temperatures are required. This filler metal is well suited for furnace brazing under protective atmospheres (including vacuum).

A2.5 BCu (Copper) Classifications. Brazing filler metals of the BCu classifications are used for joining various ferrous and nonferrous metals. They can also be used with various brazing processes. Lap and butt joints are commonly used.

A2.5.1 BCu-1 brazing filler metal is used for joining ferrous metals, nickel-base and copper-nickel alloys. It is very free flowing and is often used in furnace brazing, with a combusted city gas, hydrogen or dissociated ammonia atmosphere and generally no flux. On metals that have constituents with difficult-to-reduce oxides (chromium, manganese, silicon, titanium, vanadium, aluminum, and zinc), a flux may be required. However, pure dry hydrogen, argon or dissociated ammonia and vacuum atmosphere will adequately handle chromium, manganese and silicon containing base metals. Vacuum atmospheres or electronic plating or both are used for base metals containing minor amounts of titanium and aluminum.

A2.5.2 BCu-1a brazing filler metal is a powder form of BCu-1 and its application and use are similar to those of BCu-1.

A2.5.3 BCu-2 brazing filler metal is supplied as a copper-oxide suspension in an organic vehicle. Its applications are similar to BCu-1 and -1a.

A2.6 RBCuZn (Copper-Zinc) Classifications. Brazing filler metals of the RBCuZn classifications are used for the same materials and under the same circumstances as the BCu classifications. However, with the RBCuZn filler metals, overheating must be particularly guarded against since voids may be formed in the joint by entrapped zinc vapors. The corrosion resistance of these filler metals is generally inadequate for joining copper, silicon bronze, copper-nickel or stainless steel.

A2.6.1 RBCuZn-A brazing filler metal is used on steels, copper, copper alloys, nickel, nickel-base alloys, and stainless steel where corrosion resistance is not of importance. It is used with torch, furnace, and induction processes. Fluxing is generally required and a borax-boric acid flux is commonly used. Joint clearances from 0.002 to 0.005 in. (0.05 to 0.13 mm) are suitable.

A2.6.2 RBCuZn-C brazing filler metal is used on steels, copper, copper alloys, nickel, nickel alloys, and stainless steel. It is used with the torch, furnace, and induction processes. Fluxing is required, and a borax-boric acid flux is commonly used. Joint clearances from 0.002 to 0.005 in. (0.05 to 0.13 mm) are suitable.

A2.6.3 RBCuZn-D brazing filler metal (called white brass) is used with steel, nickel, and nickel-base alloys. It can be used with all brazing processes.

A2.7 BMg (Magnesium) Classifications. Brazing filler metal BMg-1 is used for joining AZ10A, K1A, and M1A magnesium-base metals while BMg-2a is used for joining these alloys and the AZ31B and ZE10A compositions. BMg-1 and -2a brazing filler metals are suitable for the torch, dip, or furnace brazing processes. The BMg-2a classification, because of its lower melting range, is usually preferred in most brazing applications. Heating must be closely controlled with all filler metals to prevent melting of the base metal. A flux is used with all processes. Joint clearances of 0.004 to 0.010 in. (0.10 to 0.25 mm) are best for most applications. Corrosion resistance is good if the flux is completely removed after brazing.

A2.8 BNi (Nickel) Classifications. Brazing filler metals of the BNi classifications are generally used for their corrosion and heat resistant properties. Base metal alloys most commonly brazed with the BNi filler metals are AISI 300 series stainless steels, AISI 400 series stainless steels, and nickel-, and cobalt-base alloys. Other base metals such as carbon steel, low alloy steels, and copper are also brazed with the BNi filler metals when specific properties are desired.

A2.8.1 The BNi filler metals retain their heat resistance at service temperatures up to 1800 °F (980 °C), depending on the specific filler metal. They are also satisfactorily used for room temperature applications and where the service temperatures are equal to the temperatures of liquid oxygen, helium, or nitrogen.

A2.8.2 Best quality can be obtained by brazing in an atmosphere which is reducing to both the base metal and the brazing filler metal. A vacuum (below 5×10^{-3} torr [0.67 Pa]), a pure, dry hydrogen (-60 °F [-51 °C] dew point or below) atmosphere or a pure, dry argon (-80 °F [-63 °C] dew point or below) atmosphere are examples which are commonly used.

A2.8.3 With the application of a suitable flux, any brazing process (torch, furnace, induction, etc.) can be used to braze an assembly with BNi filler metals. The most used filler metals for torch brazing are the BNi-2 for its lower melting and good flow characteristics and BNi-4 for its lower hardness and better ductility.

A2.8.4 The BNi filler metals are particularly suited to vacuum systems and vacuum tube applications because of their very low vapor pressure. Chromium is the limiting element in those alloys in which it is employed in vacuum applications. For the BNi classifications, no silver, copper, zinc, or other high vapor pressure elements are involved. It should be noted that when phosphorus is combined with some other elements, these compounds have very low vapor pressures and can be readily used in a vacuum brazing atmosphere of 1×10^{-3} torr (0.13 Pa) at 1950 °F (1066 °C) without removal of the phosphorus.

A2.8.5 BNi-1 brazing filler metal is widely used for high strength and heat resistant joints. It is used for joining turbine blades, jet engine parts, highly stressed sheet metal structures, and other hightly stressed components.

A2.8.6 BNi-1a brazing filler metal is a low carbon grade of BNi-1 with identical chemical composition except that the carbon content is 0.06 percent maximum. This filler metal is also used in high temperature jet engine parts as well as in room temperature and cryogenic applications.

A2.8.7 BNi-2 brazing filler metal has properties and uses which are similar to BNi-1, except that brazing can be conducted at lower temperatures with better flow and diffusion.

A2.8.8 BNi-3 brazing filler metal is a good heat resistant metal for lower temperature brazing of highly stressed parts. It is used for applications similar to BNi-1 and will flow well in marginal atmospheres, or where joints of close tolerances or large areas are encountered.

A2.8.9 BNi-4 brazing filler metal is similar to BNi-3. It is used for forming larger, more ductile fillets. It can be used to form joints where fairly large gaps are present.

A2.8.10 BNi-5 brazing filler metal is used for high-strength and oxidation-resistant joints for elevated temperature service. Its applications are similar to BNi-1 except that it can be used in certain nuclear applications where boron cannot be tolerated.

A2.8.11 BNi-6 brazing filler metal is extremely free flowing and exhibits a minimum amount of erosion with most nickel- and iron-base metals. It is good for use in marginal atmospheres and for brazing low chromium steels in exothermic atmospheres.

A2.8.12 BNi-7 brazing filler metal is used for the brazing of honeycomb structures, thin-walled tube assemblies and other structures which are used at high temperatures. Erosion can be controlled because of low solubility with iron- and nickel-base alloys. It produces strong, leakproof joints with heat resistant base metals at relatively low brazing temperatures. It is recommended for nuclear applications where boron cannot be used. The best results are obtained when it is used in the furnace brazing process. Ductility is improved by increasing time at brazing temperature.

A2.8.13 BNi-8 brazing filler metal was originally developed for its low interaction and good brazing properties on nickel. It has also been found useful in the aircraft field in honeycomb brazements and on other stainless and corrosion resistant materials. Since this filler metal contains a reasonably large percentage of manganese, special brazing procedures should be observed. As manganese oxidizes more readily than chromium, the hydrogen, argon and helium brazing atmospheres must be pure and very dry and a vacuum atmosphere must have low pressure and a low leak rate to insure a very low partial pressure of oxygen. It should be noted that the chemical composition and the melting characteristics of this filler metal will change when the manganese is oxidized in a gas atmosphere or when it is vaporized or oxidized in a vacuum atmosphere. However, the effect of manganese is not a problem in an atmosphere of proper quality.

A2.9 BCo (Cobalt) Classification. Brazing filler metals of the BCo classification are generally used for their high temperature properties and their compatibility with cobalt-base metals. BCo-1 brazing filler metal is used for its special high temperature properties. For optimum results, brazing should be performed in a high quality atmosphere. Special high temperature fluxes are available. The brazing technique requires a degree of skill. By using diffusion brazing procedures, the filler metal is currently being used in jet engine applications with service temperatures up to 1900 °F (1040 °C) and excursions to 2100 °F (1150 °C) and above.

A2.10 Vacuum Grade Filler Metals. These brazing filler metals are specially controlled to fabricate high quality electron tubes on electron devices where the service life and operating characteristics are of prime importance. The letter "V" following "B" in the classification designation identifies this grade. The solidus, liquidus and the brazing temperature range for each classification are shown in Table A2.

Table A2 — Vacuum grade filler metals

Solidus, liquidus and brazing temperature range						
AWS	Solidus		Liquidus		Brazing range	
classification	°F	°C	°F	°C	°F	°C
BVAg-0	1761	(961)	1761	(961)	1761 - 1900	(961 - 1038)
BVAg-6b	1435	(779)	1602	(872)	1600 - 1800	(871 - 982)
BVAg-8	1435	(779)	1435	(779)	1435 - 1650	(779 - 899)
BVAg-8b	1435	(779)	1463	(795)	1470 - 1650	(799 - 899)
BVAg-8c	1485	(807)	1490	(810)	1490 - 1700	(810 - 927)
BVCu-1x	1981	(1083)	1981	(1083)	2000 - 2100	(1093 - 1149)

A2.10.1 Vacuum grade brazing filler metals should be used in a high purity atmosphere in order to maintain the purity of the filler metal and to assure proper brazing and final brazement quality.

A2.10.2 It may be very important in some applications that the brazing filler metal not spatter onto areas near the joint area. For this reason, the non-spattering grade is generally preferred.

A2.10.3 BVAg-0 brazing filler metal is excellent for use as the first braze in step-brazing operations because of its clearly defined single melt and flow point, well above most other filler metals. It is an excellent filler metal for use where very close joint tolerances must be maintained.

A2.10.4 BVAg-6b brazing filler metal has better wetting characteristics on ferrous materials than BVAg-8 because of the increase in copper content. BVAg-6b shows a much broader melt range than BVAg-8 and is sometimes used in step brazing with BVAg-8.

A2.10.5 BVAg-8 brazing filler metal is the eutectic composition in the silver-copper binary system. Since it melts and flows at one temperature, it is very fluid and useful in tight, close tolerance joints. It performs best on copper and copper alloys.

A2.10.6 BVAg-8b brazing filler metal is similar to BVAg-8 except that it contains a small amount of nickel. The addition of nickel renders this alloy a little more sluggish in flow characteristics than BVAg-8, but it has improved wettability on ferrous alloys. On either silver base or copper base alloys it exhibits a decrease in fluidity and an increase in remelt temperature, after brazing, due to the solution of either silver or copper in the eutectic.

A2.10.7 BVAg-8c brazing filler metal is similar to BVAg-8 but has an addition of palladium, which greatly increases its wettability on ferrous and nickel base alloys.

A2.10.8 BVCu-1x brazing filler metal, like BVAg-0, is excellent for use in step brazing, and in applications where very tight joint tolerances must be held. It has excellent wetting and flow characteristics on ferrous metals as well as on nickel, or on cobalt-base alloys.

Appendix B: Marking and Labelling

B1 Marking

The problem of marking brazing filler metals is an important one for which no satisfactory solution has been developed. While it is recognized that the marking of filler metals is desirable, and provided for in welding filler metals, it is more difficult to apply to brazing filler metals. The matter of filler metal cleanliness, which is important in welding filler metals, is critically important in brazing. Many of the methods proposed for marking brazing filler metals have proved to be a source of contamination which violated this requirement for cleanliness. The marking systems which have been proposed involve one or more of the following techniques: color coding with paint; applying marks or letters by controlled oxidation of the metal surface; notching the filler metal in wire form; and printing or stamping designation on sheet forms.

B1.1 Color coding is accomplished by the application of one or more colors in the form of paints or dyes on the ends of the cut lengths of filler metal. Such paints or dyes are a source of contamination as far as brazing is concerned and must be completely removed before the filler metal is used for brazing. Therefore, color coding of brazing filler metals is not a practical solution.

B1.2 The application of markings or letters by the system of controlled oxidation or other chemical reaction leaves the surfaces of the brazing filler metal unsuited for brazing — in short, dirty. Therefore, such markings must be removed so as not to contaminate the braze. Marking of brazing filler metals by controlled oxidation or other chemical processes has not proved to be a practical solution.

B1.3 Notching of wires and rods is another system proposed. In this connection, braze welding filler metals of 1/8 in. (3.2 mm) and larger in size are identified by means of letters and symbols stamped on the rods. This stamping cannot be successfully applied to smaller diameters which are more widely used in brazing. Notching of brazing filler metals in small wire sizes tends to cause the wires to break through the notches, which leads to waste of short pieces. Such notched wire is totally unsuited for the manufacture of preforms. Further, the notches tend to accumulate dirt and other contamination. Identification by notching has also not proved practical.

B1.4 A modification of the notching system is covered under Federal Specification QQ-B-654 dated May 8, 1970, which provides for identifying the preinserted rings of brazing filler metal as used on certain brass pipe fittings. These rings — usually of rectangular cross sections — are required to be identified by a system of dots pressed into the ring. Such marking is practical since it may be performed at the time the rings are formed. Since there is no need to bend these rings further, after insertion in the pipe fitting, there is little chance of fracture through these indentations.

B1.5 The final method of marking is limited to those filler metals which are produced in sheet form. This marking is accomplished by means of printing or stamping the identification on the sheets or rolls of filler metal. Again, the inks or other substances used for such marking may contaminate the surface of the filler metal. Since filler metals in sheet or foil form are frequently used in brazing in protective atmospheres (including vacuum), such contamination cannot be tolerated. Some of the inks used may leave a residue even after the surface is apparently clean. The only safe procedure is to cut out and discard the marked section. Identification printed or stamped on sheet or foil filler metal has not proved practical and in some cases has been detrimental.

B1.6 Brazing filler metals, as received from suppliers, will be identified by means of tags bearing the trade name and classification. In some cases, they may also be in boxes or other packages which provide identification. In those instances where further identification of brazing filler metals is required, it can be accomplished better in the store room or shop as a matter of good shop "housekeeping." In this manner, the identification can be tailored to fit the shop requirements.

B1.7 The matter of marking brazing filler metals has been the subject of intensive study. To date, no method has been found that does not provide a source of contamination for the brazing operation or lead to waste.

B2 Caution and Warning Labels

B2.1 The mandatory requirements for the use of caution and warning labels (2.4.3 and 2.4.4) were started with the 1969 edition of this specification. Two forms of labels are involved, depending upon the type of the brazing filler metal and the nature of the potential fume hazard.

B2.2 The generalized *caution* label shown in 2.4.3 is in keeping with a similar requirement appearing in other AWS filler metal specifications. It recognizes and cautions the user of filler metals that fumes and gases may be generated when welding or brazing. Since control of such fumes and gases is not a proper function for a filler metal specification, reference is made to American National Standard Z49.1, Safety in Welding and Cutting, where ventilation requirements for welding and brazing are discussed.

B2.3 Certain brazing filler metals, BAg-1, BAg-1a, BAg-2, BAg-2a and BAg-3, contain cadmium. If such brazing filler metals are improperly used and subjected to overheating, cadmium oxide fumes can be generated. Cadmium oxide fumes are hazardous and inhalation of these fumes can be fatal. Therefore, in recent years it has been a common practice to place a warning label on packages of silver brazing filler metals containing cadmium.

B2.4 Since these cadmium-bearing silver brazing filler metals are used in maintenance work on farms, and may on occasion get into homes, the U.S. Department of Health, Education and Welfare, Food and Drug Administration, on December 1, 1966, placed all "cadmium-containing solders and brazing rods" under the provisions of the Federal Hazardous Substances Act. It is, therefore, a requirement of this law that the brazing filler metals containing cadmium (BAg-1, BAg-1a, BAg-2, BAg-2a and BAg-3) carry the warning label set forth in 2.4.4. It should be noted that by the terms of the above law, the exact wording of the label in 2.4.4 must be used as this is the approved form. Again, the user is referred to American National Standard Z49.1, Safety in Welding and Cutting, for a discussion on ventilation.

B2.5 The cadmium-bearing silver brazing filler metals delivered to the purchaser in the original packaging will be labelled by the producer in accordance with the provisions of 2.4.4 of this specification. However, it is not uncommon for local suppliers to remove the cadmium-bearing brazing filler metal from its original package and furnish it to the consumer in new packages smaller than the standard package sizes offered by the original producer. Under the terms of the Federal Hazardous Substances Act it is necessary for the supplier to furnish a copy of the *warning* label with each quantity of cadmium-containing brazing filler metal sold, regardless of the size of the purchase.

Appendix C: Metric Equivalents

In the past few years, AWS filler metal specifications have become more widely used in locations where there is greater familiarity with the numerical units of the metric system rather than the U.S. customary units. Accordingly metric (SI) units have been placed in this specification in parentheses after the U.S. units.

CHAPTER 4

FLUXES AND ATMOSPHERES

The purpose of a brazing flux is to promote the formation of a brazed joint. In this sense a gas, or even a vacuum, such as may be used to surround the work and provide an active or protective atmosphere, is a flux. For this reason, both fluxes and atmospheres are discussed in this chapter, but to comply with the usual terminology and because the methods of application and characteristics are quite different, fluxes and atmospheres are discussed and tabulated separately.

When metals that are not noble are exposed to air, chemical reactions occur. The rate of these reactions is generally accelerated as the temperature is raised. The prevalent reaction is oxidation, though nitrides and even carbides are formed in some instances. The rate of oxide formation varies with each metal composition as does the nature of the oxide, i.e., tenacity, structure, resistance to removal, and/or further oxidation, etc. Oxide formation on some metals in air is for practical purposes instantaneous, even at or below room temperatures. In almost all conditions, such oxides or other compounds, if present, prevent the formation of properly brazed joints.

FLUXES

Most fluxes are not designed or intended for the primary removal of oxides, coatings, oil, grease, dirt, or other foreign materials from the parts to be brazed. All parts for brazing must be subjected to appropriate cleaning operations as dictated by the particular metals (see chapter 7).

In its operation in the making of a braze the flux is often needed to combine with, remove, or dissolve those unwanted residual compounds and/or products of the brazing operation which would otherwise impair or totally prevent formation of the braze.

Oxides are the principal source of surface contamination, and the dissolution and removal of the oxides during brazing is the most common function of a flux. So as not to impair metal flow, the flux must also be readily displaced by the molten brazing filler metal, and the viscosity of a flux at brazing temperature is therefore an important property.

Under some circumstances flux may be called upon to suppress the volatilization of high vapor pressure constituents in a filler metal. The variation with temperature of these physical characteristics is useful, particularly during torch brazing, as a temperature indicator.

Some filler metals, such as the lithium-bearing or phosphorous-copper filler metals, are self-fluxing on certain alloys. For instance, the Cu_3P constituent in the phosphorous-copper filler metals acts as a flux when used on copper or low alloys of copper. In most cases, use of the self-fluxing filler metals alone will result in joints of a quality equivalent to those to which additional flux has been added. These filler metals are self-fluxing only in the molten state and will themselves oxidize during the heating cycle. When large sections are to be brazed or where prolonged heating times are contemplated, the use of additional flux is advisable.

Constituents of Fluxes

The following discussion of brazing fluxes is provided for the purpose of enabling the user to

Table 4.1—Fluxes for brazing

AWS Brazing Flux Type No.	Recommended base metals	Recommended filler metals	Recommended useful temp. range	Ingredients	Forms supplied
1	All brazeable aluminum alloys (See appendix)	BA1Si	700-1190F 371- 643C	Chlorides Fluorides	Powder
2	All brazeable magnesium alloys (See appendix)	BMg	900-1200F 482- 649C	Chlorides Fluorides	Powder
3A	All except those listed under 1, 2, and 4*	BCuP, BAg	1050-1600F 566- 871C	Boric acid Borates Fluorides Fluoborates Wetting agent	Powder Paste Liquid
3B	All except those listed under 1, 2, and 4	BCu, BCuP, BAg, BAu, RBCuZn, BNi	1350-2100F 732-1149C	Boric acid Borates Fluorides Fluoborates Wetting agent	Powder Paste Liquid
4	Aluminum bronze, aluminum brass and iron or nickel base alloys containing minor amounts of Al and(or Ti†	BAg (all) BCuP (Copper base alloys only)	1050-1600F 566- 871C	Chlorides Fluorides Borates Wetting agent	Powder Paste
5	All except those listed under 1, 2, and 4	Same as 3B (excluding BAg-1 through -7)	1400-2200F 760-1204C	Borax Boric acid Borates Wetting agent	Powder Paste Liquid

NOTE: This table provides a guide for classification of most of the proprietary fluxes available commercially. The information given here is generally not adequate, when used alone, for a specific application. Chapter 4 should be read thoroughly. The AWS is currently preparing a brazing flux specification.

*Some Type 3A fluxes are specifically recommended for base metals listed under Type 4.

†In some cases Type 1 flux may be used on base metals listed under Type 4.

improve his understanding of their functions and thus permit a more effective use of brazing flux. It is not recommended that the user attempt to formulate his own flux, as this is a very complex operation requiring much experience and special equipment.

A large number of chemical compounds are useful in formulating fluxes. Proprietary fluxes are mixtures of specific compounds formulated in certain ways that have been found to give satisfactory results for a specific purpose. After the chemicals are mixed (and particularly when they are heated) reactions take place between the various ingredients so that the resulting solution of compounds at brazing temperature is quite different chemically and physically from the unreacted constituents. For instance, if a fluoborate is an ingredient in a flux, fluorides are formed after the ingredients react. During the brazing operation the chemistry is especially transient since, as mentioned previously, the reaction rates of the flux, base metals, and filler metals and any gaseous materials present increase with temperature. The ingredients of the flux must thus be carefully tailored to suit all the factors in the braze, including time. The attack of the flux on the metals must be limited but of sufficient intensity and duration at the brazing temperature to enable the joint to be satisfactorily formed. Active halides, such as chlorides and fluorides, are, for instance, necessary in a flux for aluminum or other highly electropositive metal-bearing alloys (see Table 4.1, Types 1, 2, and 4), and some particularly active fluxes must be mixed just before use to avoid reactions between ingredients during storage.

The following list includes the most common ingredients of fluxes. Some comments about their functions follow.

Borates (sodium, potassium, lithium, etc.)
Fused borax
Fluoborates (potassium, sodium, etc.)

Fluorides (sodium, potassium, lithium, etc.)
Chlorides (sodium, potassium, lithium)
Acids (boric, calcined boric)
Alkalies (potassium hydroxide, sodium hydroxide)
Wetting agents
Water (either as water of hydration or as an addition for paste fluxes)

Borates — Borates are useful in formulating the higher melting fluxes. They have good oxide dissolving power and provide protection against oxidation for long periods of time. Most of the borates melt and are effective at temperatures around 1400 °F (760 °C) or higher. In the molten condition they have a relatively high viscosity and therefore must be mixed with other salts.

Fused Borax — Fused borax is another high melting material that is active at high temperatures. It is little used in the lower melting brazing process.

Fluoborates — Fluoborates react similarly to other borates in many respects; however, while they do not provide protection from oxidation to the same extent as other borates, they flow better in the molten state and have greater oxide-dissolving properties. Fluoborates are used with other borates or with alkaline compounds, such as carbonates, to make a flux. A related class of compounds is fluosilica-borates, which have a somewhat higher melting point than the fluoborates and provide good coverage and surface adherence. The high melting point limits their use. (For safety recommendations for fluorine-bearing compounds see chapter 29.)

Fluorides — The fluorides react readily with most metallic oxides at elevated temperatures and are therefore used extensively in fluxes as cleaning agents. They are particularly important in applications where refractory oxides, such as those of chromium and aluminum, are encountered. The flourides are often added to increase the fluidity of molten borates, thereby facilitating their displacement and improving the capillary flow of the molten brazing filler metal. (For safety recommendations for fluorine-bearing compounds see chapter 29.)

Chlorides — The chlorides function in a manner similar to that of the fluorides but have a lower effective temperature range. Chlorides must be used judiciously since, at elevated temperatures, they tend to oxidize the work. Chlorides also melt at lower temperatures and are used to depress the melting temperatures of flouride-base fluxes.

Boric Acid — Boric acid is a principal constituent used in brazing fluxes. It is used in the hydrated form, or sometimes in the calcined form, which reacts similarly to boric acid but has a somewhat higher melting point. Boric acid has the property of facilitating the removal of the glasslike flux residue after brazing. The melting point of boric acid is best described as being below that of the borates but higher than that of the fluorides.

Alkalies — Potassium and sodium hydroxide are used sparingly, if at all, because of their deliquescent properties. Even small amounts involved with other flux agents can create difficulty in humid weather and may seriously limit the storage life of the flux. The function of alkalies is to elevate the useful working temperature of the flux. They are beneficial additions to fluxes that are used on molybdenum-containing tool steels.

Wetting Agents — Wetting agents are employed in paste and liquid fluxes to facilitate the flow or spreading of the flux onto the workpiece prior to brazing. Obviously, such wetting agents as are used must not adversely affect the normal functions of the brazing flux.

Water — Water is present in brazing fluxes either as water of hydration in the chemicals used in formulating the flux or as a separate addition for the purpose of making a paste or liquid. If the water of hydration is undesirable, it may be removed by calcining. Water used to form a paste must be evaluated for suitability, and hard waters should be avoided. If the available water is unsuitable, then a wise precaution is to substitute deionized or distilled water.

Bases for Choice of Flux

The cost of the flux is outside the scope of this discussion, but it must be kept in mind that the choice of flux and technique of its use greatly influence the time required to make a braze and the quality of the braze. There is no single universal flux that is best for all brazing applications. Since there are many variable conditions, i.e., base metal, filler metal, method of brazing, time required to braze, complexity of joint, etc., there are necessarily many useful formulations of flux. Each flux is compounded somewhat differently, and each has its optimum performance region.

These formulations can, however, be classified into six groups according to their characteristics. Table 4.1 lists these groups with an AWS Brazing Flux Type Number assigned to identify each group. Since the range of functions covered within each group is wide, generally no single flux formulation will best satisfy all the requirements listed for that Type Number, but, with few exceptions, each flux formulation can be assigned a definite single Type Number and will perform satisfactorily over most of the range stated. Table 4.1 is based on a comprehensive survey of fluxes as reported by their manufacturers. It is a guide to aid in selecting a flux for specific applications, and its use will be enhanced when manufacturers state the AWS Type Number in describing their fluxes. Table 4.1 is not a substitute for a thorough search for an optimum flux for a specific high production joint. Recommendations of the supplier should be sought and trial is recommended when experience is lacking. For successful use, a flux must be chemically compatible with all the base metals and filler metals involved in the braze and must be active throughout the brazing temperature range. If one knows the base metals involved and either the temperature at which it is desired to braze or the filler metal which is required, one or more flux types will be found in Table 4.1 for that combination. Where more than one type of flux is recommended, other considerations, such as safety (see chapter 29) and cost, will enter into the decision.

Within a particular AWS Type Number of flux, there are several criteria for choosing a specific flux for maximum efficiency:

1. For dip brazing, water (including water of hydration) must be avoided.

2. For resistance brazing, the flux mixture must allow the passage of current. This generally requires a dilute, wet flux.

3. The effective temperature range of the flux must include the brazing temperature for the specific brazing filler metal involved. This is important on both the high and low sides of the brazing temperature range. If the time during brazing is long (many minutes), a less active and longer-lived flux is desirable; a quick heating cycle (such as in induction brazing) needs a more active flux and long life is not required. If a nonoxidizing atmosphere is used, the life of the flux is prolonged.

4. Ease of flux residue removal should be considered.

5. Corrosive action on the base metal or filler metal should be minimized.

Specification for Flux

Some manufacturers and the federal government have established the practice of publishing detailed specifications for some of the flux products for particular applications. An example is Federal Specification O-F-499 (Ref. 1). Requirements of the flux, packaging, procedures for sampling, inspection, and testing are specified. Such a specification can be devised for any flux application for purposes of quality control and competitive acquisition of flux compounds.

Methods of Applying Fluxes

Fluxes for brazing are generally available in the form of powder, paste, or liquid. The form selected depends upon the individual work requirements, the brazing process, and the brazing procedure used.

Frequently, powdered flux is mixed with water or alcohol to make a paste. The powder may also be used in the dry form and sprinkled on the joint, although adherence in this form is not as good. Dry powdered flux may also be applied to the heated end of a filler metal rod simply by dipping the rod into the flux container. Dry powdered flux is also used to make the bath for chemical bath dip brazing.

Paste is probably the most commonly used form for applying brazing flux. It can be applied with good adherence to a joint and filler metal before brazing. A diluted paste flux is used where the flux is to be sprayed on a joint. Certain flux ingredients will completely dissolve in water to produce a liquid solution called "liquid flux" (see Types 3A and 3B).

The particle size of dry flux or paste flux is important since better fluxing action will result when all constituent particles of a flux are small and thoroughly mixed. Stirring, ball milling, or grinding of a flux mixture shortly before use is helpful if the flux has become lumpy. Preheating of the flux may facilitate application.

Liquid flux of Type 5 is used almost exclusively in torch brazing. The fuel gas used is passed through the liquid flux container thus entraining the flux in the fuel gas. With the application of the flame to the work, the flux is applied where needed. Usually it is used with a small amount of other preplaced flux and serves to flux the joint surroundings.

Mixtures of powdered filler metal in flux have been successfully used for concurrent preplacement of both flux and filler metal.

Flux Removal

Flux residue, if exposed to moisture under service conditions, should be removed to avoid corrosion from remaining active chemicals. The residue obtained from a flux, particularly when considerable oxide removal has occurred, is a form of glass. The less the flux is required to clean the metal, the less will be the formation of glass and the easier the task of removing the flux residue.

Flux removal from properly cleaned, brazed parts can usually be accomplished by washing in hot water. Preferably, this rinse should be done immediately after the brazing operation. Thorough drying following this rinse is recommended.

For residues containing a moderate amount of glass, removal may be accomplished by immersing the brazed joint in cold water before it has fully cooled from the brazing temperature, thereby creating a thermal shock that cracks off the residue. This method should not be used where the thermal cycle involved will impair the strength of the brazed joint.

For residues which are difficult to remove, a chemical dip may be used. Proprietary compounds are available for this purpose. See the specific chapters on base metals for details of pickling solutions and chemical cleaning.

Mechanical methods are used for the removal of flux residues but are usually limited to those cases where any of the methods just discussed will not be effective because of high glass content or the large size of the parts. Among the mechanical means used are fiber brush, wire brush, blast cleaning, and steam jet. Caution should be exercised to make sure that the method used is compatible with the base metal. Such soft metals as aluminum are susceptible to embedment of flux residue particles. Additional cleaning with fluids may be desirable subsequent to mechanical cleaning.

The use of an adequate amount of the proper flux will make flux removal easier because spent flux has a high glass content.

CONTROLLED ATMOSPHERES

Controlled atmospheres are used to prevent the formation of oxides during brazing and in most cases to reduce the oxides present so that the brazing filler metal can wet and flow on clean base metal. Controlled atmosphere brazing is widely used for the production of high quality joints. Large tonnages of assemblies of a wide variety of base metals are mass produced by this process.

Like fluxes, controlled atmospheres are not intended to perform the primary cleaning for the removal of oxides, coatings, grease, oil, dirt, or other foreign materials from the parts to be brazed. All parts for brazing must be subjected to appropriate prebraze cleaning operations as dictated by the particular metals (see chapter 7). Controlled atmospheres are most commonly employed in furnace brazing (see chapter 1); however, they may also be used with induction or resistance brazing. In applications where only a controlled atmosphere is used, postbraze cleaning is generally not necessary. When flux is required, the use of a controlled atmosphere may still be desirable to extend the useful life of a flux and to minimize postbraze cleaning.

The use of controlled atmospheres avoids formation of oxides and scale over the whole part and permits finish machining to be done before brazing in many applications. In some applications, such as the manufacture of electronic tubes, eliminating flux is tremendously important. Some types of equipment, e.g., metallic muffle furnaces and vacuum systems, may be damaged or contaminated by the use of flux.

Although brazing in a controlled atmosphere is a relatively simple process in many applications, use of the process in more complex situations is increasing. Supplying sufficiently pure reducing atmosphere to a brazing furnace which is in good condition does not guarantee that a proper environment for a braze is attainable. Two examples follow:

1. A purified hydrogen atmosphere of low dew point used to braze stainless steel at a temperature of 2000 °F (1093 °C), will still contain enough oxygen to react so that the stainless steel oxidizes as it is heated through the temperature range below that at which reduction of the oxides takes place. This may produce a thin film of spongy metal that aids in capillary flow, but if heating is slow a large amount of oxide may form. Sufficient time at an elevated temperature must subsequently be allowed for reduction of this oxide.

2. Oxygen-bearing copper- or silver-base filler metals can contaminate the hydrogen atmosphere surrounding nearby stainless steel surfaces to the extent that a heavy, green oxide will form. Further, the copper or silver parts would probably be embrittled and porous after the brazing attempt as a result of a reaction between the oxygen contained in the filler metal and the hydrogen atmosphere. The dew point of the hydrogen is increased by reaction of the hydrogen with the small amounts of oxygen in the copper or silver, even though no oxide might appear on their bright, metallic surfaces.

Table 4.2—Atmospheres for brazing

AWS brazing atmosphere type number	Source	Maximum dew point incoming gas	Approximate composition, %				Application		Remarks
			H_2	N_2	CO	CO_2	Filler metals	Base metals	
1	Combusted fuel gas (low hydrogen)	Room temp.	5-1	87	5-1	11-12	BAg,* BCuP, RBCuZn*	Copper, brass*	
2	Combusted fuel gas (decarburizing)	Room temp.	14-15	70-71	9-10	5-6	BCu, BAg,* RBCuZn,* BCuP	Copper,† brass,* low-carbon steel, nickel, monel, medium carbon steel§	Decarburizes
3	Combusted fuel gas, dried	−40F (−40C)	15-16	73-75	10-11		Same as 2	Same as 2 plus medium and high-carbon steels,monel, nickel alloys	
4	Combusted fuel gas, dried (carburizing)	−40F (−40C)	38-40	41-45	17-19		Same as 2	Same as 2 plus medium and high-carbon steels	Carburizes
5	Dissociated ammonia	−65F (−54C)	75	25			BAg,* BCuP, RBCuZn,* BCu, BNi	Same as for 1, 2, 3, 4 plus alloys containing chromium‡	
6	Cylinder hydrogen	Room temp.	97-100				Same as 2	Same as 2	Decarburizes
7	Deoxygenated and dried hydrogen	−75F (−59C)	100				Same as 5	Same as 5 plus cobalt, chromium, tungsten alloys and carbides‡	
8	Heated volatile materials	Inorganic vapors (i.e., zinc, cadmium, lithium, volatile fluorides)					BAg	Brasses	Special purpose. May be used in conjunction with 1 thru 7 to avoid use of flux
9	Purified inert gas	Inert gas (e.g. helium, argon, etc.)					Same as 5	Same as 5 plus titanium, zirconium, hafnium	Special purpose. Parts must be *very* clean and atmosphere must be pure

Table 4.2, cont.—Atmospheres for brazing

AWS brazing atmosphere type number	Source	Pressure	Approximate composition, %				Application		Remarks
			H_2	N_2	CO	CO_2	Filler metals	Base metals	
10	Vacuum	Vacuum above 2 torr					BCuP, BAg	Cu	
10A	Vacuum	0.5 to 2 torr					BCu, BAg	Low carbon steel, Cu	
10B	Vacuum	0.001 to 0.5 torr					BCu, BAg	Carbon and low alloy steels, Cu	
10C	Vacuum	1×10^{-3} torr and lower					BNi, BAu, BA1Si, Ti alloys	Ht. and corr. resisting steels, A1, Ti, Zr, refractory metals	

NOTE: AWS Types 6, 7, and 9 include reduced pressures down to 2 torr.
*Flux required in addition to atmosphere when alloys containing volatile components are used.
†Copper should be fully deoxidized or oxygen-free. See chapters 2 and 14.
‡Flux must be used in addition to the atmosphere if appreciable quantities of aluminum, titanium, silicon, or beryllium are present.
§Heating time should be kept a minimum to avoid objectionable decarburization.

Mixtures of some atmospheres with air are explosively combustible. Before heating a furnace chamber containing any such atmosphere one must purge all air. For example, mixtures of hydrogen with air ranging from 4 to 75 percent by volume hydrogen are explosive. Waste atmosphere must be either burned or rapidly diluted.

The precautions dictated in chapter 29 should be followed when metals or fluxes containing toxic elements are used. Some atmospheres contain carbon monoxide (CO), which is toxic. Burning of waste gases or adequate local ventilation is required.

Composition of Atmospheres for Brazing

Table 4.2 lists ten recommended atmosphere types, and approximate compositions are specified where applicable. The types are listed approximately in the order of increasing difficulty of attainment and maintenance.

The first four brazing atmospheres listed are generated by passing metered mixtures of hydrocarbon fuel gas and air into a retort for reaction. With natural gas, the ratio of air to gas may range between 9.5 to 1 and 2 to 1 for the production of useful atmospheres. For ratios between 5 to 1 and 9.5 to 1, the reaction is exothermic, and the heat liberated is sufficient to continue it. Mixtures richer in natural gas than about 5 (air) to 1 (gas) require the addition of heat and the presence of a catalyst in the retort for combustion to occur. This type of reaction is termed endothermic. Mixtures leaner than 9.5 to 1 result in a gas product that contains too much oxygen to be suitable for a controlled brazing atmosphere. Following the combustion of such air-gas mixtures, water may be removed either by cooling alone or by the additional use of an absorption-type dryer to reduce the dew point to an acceptable value. Carbon dioxide may be scrubbed out if required.

The AWS No. 5 brazing atmosphere is obtained by the dissociation of dry ammonia (NH_3) into hydrogen and nitrogen in a catalyst-filled heated retort. Drying of the dissociated ammonia may be necessary.

The recommended applications for atmospheres are shown in Table 4.2. Recommendations of suppliers of gases and gas-producing equipment should be consulted for specific applications. The atmosphere for brazing must be compatible with both the base and brazing filler metal for satisfactory results, and the equipment used for handling the atmosphere must be such that the atmosphere is applied at brazing temperature without objectionable contamination.

The following discussion deals with the action of some atmospheres and the atmosphere components.

Hydrogen (H_2) — Hydrogen is an active agent for the reduction of most metal oxides at elevated temperatures. Hydrogen can damage some base metals by hydrogen embrittlement, as described in chapter 2 and in the various chapters on specific base metals.

Carbon Monoxide (CO) — Carbon monoxide is an active agent for the reduction of some metal oxides, e.g., those of iron, nickel, cobalt, and copper, at elevated temperatures. Carbon monoxide can serve as a desirable source of carbon, as in some applications on carbon steels, or as an undesirable source of carbon and oxygen in other applications. Carbon monoxide can be generated from oil on the parts at brazing temperatures. Carbon monoxide is toxic, and adequate ventilation must be provided unless waste gas is trapped and burned.

Carbon Dioxide (CO_2) — Carbon dioxide serves as a diluent of the brazing atmosphere in recommended applications provided the proper ratio of carbon monoxide to carbon dioxide is maintained. Carbon dioxide may also provide useful carbon in applications on carbon steels. Its presence may be undesirable as a source of oxygen, carbon, and carbon monoxide when decomposed. In CO-CO_2 atmospheres the carbon dioxide content of a furnace atmosphere can be undesirably increased by air leakage.

Nitrogen (N_2) — The presence of nitrogen in a controlled atmosphere is tolerable as a diluent in recommended applications. The use of nitrogen must be avoided when the formation of nitrides occurs and is objectionable.

Water Vapor (H_2O) — The amount of water vapor in the atmosphere is specified by the dew point (temperature at which moisture in the gas will condense). The relationship between dew point temperature and moisture content of gases by volume is as follows:

Dew point temperature F	Dew point temperature C	Moisture content % by volume	Moisture content ppm
0	−18	0.150	1500
−30	−34	0.0329	329
−60	−51	0.0055	55
−80	−62	0.0014	14
−100	−73	0.0002	2

Water vapor is undesirable except in applications in which decarburization during brazing is

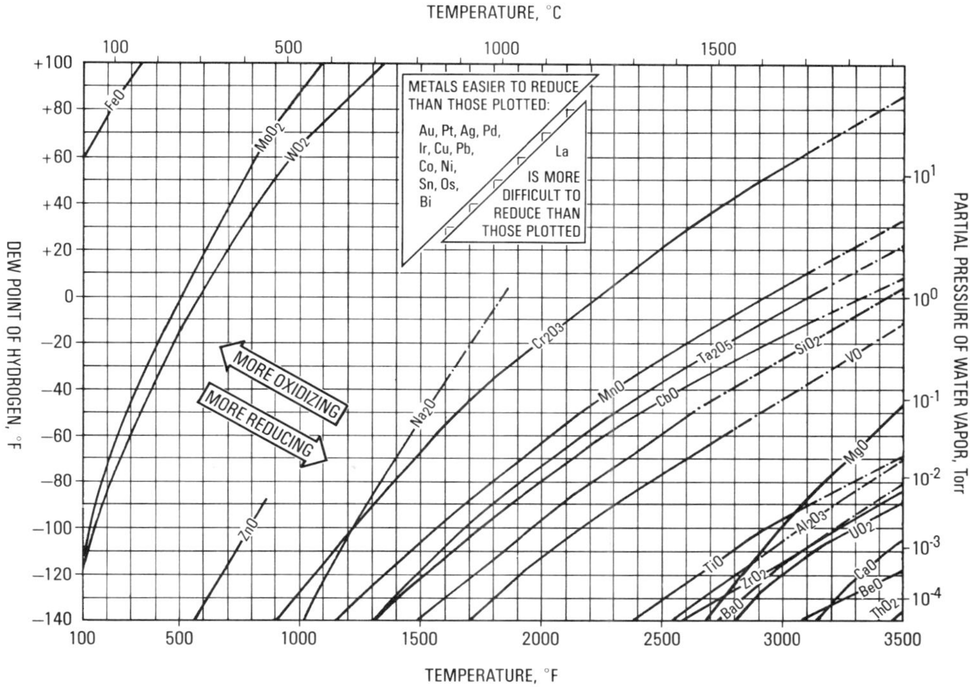

Fig. 4.1—Metal-metal oxide equilibria in hydrogen atmospheres.

useful. Water vapor may be unintentionally added to the atmosphere from air leakage, air carried into the furnace with the work, free oxygen in the atmosphere, reduction of metal oxides, leakage from water jackets, contaminated gas lines, diffusion of oxygen through inadequate flame curtain, and other less obvious sources. Water vapor is objectionable where its presence can inhibit wetting by promoting oxidation.

Oxygen (O_2) — In addition to sources already mentioned oxygen may come from gases occluded on surfaces in the heating chamber. The presence of free oxygen in the brazing atmosphere is always undesirable.

Methane (CH_4) — Methane may come from the atmosphere gas as generated or organic materials left on the parts by inadequate cleaning. Methane may also be deliberately added to the gas after it is generated. Methane can serve as a source of carbon and hydrogen.

Sulfur (S) — Sulfur or sulfur compounds may be an unintentional contaminant in the atmosphere, and can react with the metal involved in a braze to inhibit wetting. Sulfur compounds can be introduced from fuel gas used in the generation of the atmosphere, air burned with fuel gas, lubricants left on the surface of the parts as a result of inadequate cleaning, brickwork, and other thermal insulation used in the structure of the furnace.

Inorganic Vapors — In equipment designed for their use, vapors such as those of zinc, cadmium, lithium, and fluorine compounds can serve to reduce metal oxides and scavenge the atmosphere of oxygen. They are useful to replace constituents of alloys evolved during brazing. These vapors are toxic (see chapter 29).

Inert Gases — Inert gases such as helium and argon form no compounds with metals. In equipment designed for their use, they inhibit evaporation of volatile components during brazing and permit the use of weaker retorts than are required for a vacuum. Inert gases used at ambient pressure reduce the evaporation rate of volatile elements compared to vacuum.

Vacuum — An increasing amount of brazing is done in a vacuum. Some common units of vacuum measurements are as follows:

1 mm Hg = 1 torr (135 Pa)

1 micron Hg (μ) = 1×10^{-3} mm Hg (0.15 Pa)

= 1×10^{-3} torr (0.15 Pa)

By removal of gases to a suitably low pressure,

including gases which are evolved during heating to brazing temperature, very clean surfaces are obtainable. A vacuum is particularly useful in the aerospace, electronic tube, and nuclear fields where metals that react chemically with reducing atmospheres are used or where entrapped fluxes or gases are intolerable. The maximum tolerable pressure for successful brazing depends on a number of factors which are primarily determined by the composition of the base metals, the brazing filler metal, and the gas that remains in the evacuated chamber.

Vacuum brazing is economical for fluxless brazing of many similar and dissimilar base metal combinations. Vacuums are especially suited for brazing very large, continuous areas where (1) solid or liquid fluxes cannot be removed adequately from the interfaces during brazing, and (2) gaseous atmospheres are not completely efficient because of their inability to purge occluded gases evolved at close-fitting brazing interfaces. Vacuum is also suitable for brazing similar and dissimilar base metals, such as titanium, zirconium, columbium, molybdenum, and tantalum. The characteristics of these metals are such that even very small quantities of atmospheric gases may result in embrittlement and sometimes disintegration at brazing temperatures. These metals and their alloys may also be brazed in inert-gas atmospheres if the gases are of sufficiently high purity to avoid contamination and the resultant loss in properties of the metals. It is interesting to note, however, that a vacuum system evacuated to 10^{-5} torr (0.0015 Pa) contains only 0.00000132% residual gases based on a starting pressure of 1 atmosphere (760 torr).

Vacuum brazing has the following advantages and disadvantages compared with other high purity brazing atmospheres.

1. Vacuum removes essentially all gases from the brazing area, thereby eliminating the necessity of purifying the supplied atmosphere. Commercial vacuum brazing is generally done at pressure varying from 0.5 to 500 millitorr (0.065 to 65 Pa) and above, depending upon the materials brazed, the filler metals being used, the area of the brazing interfaces, and the degree to which gases are expelled by the base metals during the brazing cycle.

2. Certain oxides of the base metal will dissociate in vacuum at brazing temperatures. Vacuum is used widely to braze stainless steel, superalloys, aluminum alloys, and, with special techniques, refractory materials.

3. Difficulties sometimes experienced with contamination of brazing interfaces due to base metal expulsion of gases are negligible in vacuum brazing. Occluded gases are removed from the interfaces immediately upon evolution from the metals.

4. The low pressure existing around the base and filler metals at elevated temperatures removes volatile impurities and gases from the metals. Frequently the properties of the base metals themselves are improved. This characteristic is also a disadvantage where the filler, base metals, or elements of them volatilize at brazing temperatures because of the low surrounding pressures. This tendency may, however, be corrected by proper vacuum brazing techniques.

Metal-Metal Oxide Equilibria in Hydrogen

Figure 4.1 is a graphical presentation of metal-metal oxide equilibria. It is based on the thermodynamics of the metal-oxide formation.

Furnace brazing, bright annealing, sintering, and all other metallurgical processes where metallic oxides are reduced in hydrogen atmospheres require a fundamental knowledge of the factors involved in the oxide reducing process. The ability of pure hydrogen to reduce oxides on the surfaces of metals and alloys is determined by three basic factors: (1) temperature, (2) oxygen content of the hydrogen (measured as dew point), and (3) pressure of the gas.

Most applications are at a pressure of one atmosphere; therefore, only two variable factors remain: temperature and dew point. The diagram presented in Fig. 4.1 is a plot of the dew point at which the oxide and the metal are in equilibrium at various temperatures. The twenty curves shown in this diagram define the equilibrium conditions for the twenty pure metal-metal oxide systems. The position of thirteen additional elements whose curves fall outside the chart is also indicated. The oxide chosen for the calculation of this diagram represents the most difficult to reduce oxide of each metal.

The metal-metal oxide equilibrium curves (Fig. 4.1) slope upward and to the right for each metal. The region above and to the left of each curve represents conditions that are oxidizing for that metal. All points below and to the right of each curve cover the conditions required for reducing the oxides. The diagram therefore illustrates that the higher the processing temperature, the higher the dew point (or oxygen content) that can be used for any particular metal. Or, in other words, a given purity of hydrogen becomes progressively more reducing at progressively higher temperatures.

Use of the Diagram

Use of this diagram for practical purposes requires, first, that the correct curve be selected. For processing an alloy, the element having the most stable oxide (farthest to the right) is the governing curve. Example: for processing chromium stainless steels, the Cr_2O_3 curve applies since chromium oxides are more stable than those of iron or nickel. Generally, it has been found that, when the most difficult-to-reduce constituent of an alloy is present in more than about 1 atomic percent, a continuous film of its oxide is formed and its curve therefore is applicable. Alloys having a concentration progressively lower than 1 atomic percent of the most stable oxide former appear to lie progressively closer to the curve of the next most stable oxide former.

In practice it is necessary to use hydrogen that has a somewhat lower dew point than that indicated by the curve for any given metal. This is in part because the surface becomes oxidized during heating until the temperature is reached that corresponds to the equilibrium temperature for that dew point. The reduction of these oxides formed during heating requires that sufficient time be allowed at conditions sufficiently below, or to the right, of the equilibrium curve. It is also necessary in practice to provide a continuous flow of hydrogen into the work zone during the processing to maintain the necessary atmosphere purity at the metal surface.

Practice has shown that inert gases, if of similar oxygen content and flowed continually through the work zone, give results similar to hydrogen atmosphere. A continuously pumped vacuum roughly equivalent to the partial pressure of water vapor as presented on the right-hand ordinate of Fig. 4.1 also gives similar results in practice. Since the equilibrium diagram is presented only for H_2/H_2O atmospheres, all the oxygen (O_2) in the hydrogen atmospheres must be converted to H_2O before the dew point is determined.

Special caution is indicated when dealing with metals having a high affinity for H_2. Pure metals or alloys composed principally of titanium, zirconium, hafnium, tantalum, columbium (niobium), and certain others will immediately react with H_2 at sufficiently high temperatures, forming hydrides, or dissolving hydrogen in the base metal.

Such curves as shown in Fig. 4.1, and others which have been published (Ref. 2), aid in comprehending the actions of hydrogen and water vapor in respectively reducing and oxidizing metal oxides and metals, but they do not portray the complete story involved in the use of controlled atmospheres. They do not indicate the rate at which reduction will occur nor the physical form of the oxide. The curves show whether the removal of these oxides for the assumed chemical reaction is possible under the conditions of gas composition and temperature indicated. In each case, the oxide is assumed to be reduced to the metal in one operation, which may not be the case.

It is also possible that complex metal oxides may not behave in a manner similar to simple oxides. It will be noted that oxides of aluminum, magnesium, beryllium, calcium, and titanium are only reducible at very high temperatures and with very low dew points. Usually the equilibrium temperature is far above a useful temperature for brazing. If a base metal contains small amounts of these elements (below 0.5%), a satisfactory braze can usually be obtained. If greater amounts of these elements are present in the base metal, a flux must be used or a surface plating used to mask the unreducible oxides.

The predictions of Fig. 4.1 are in rough agreement with actual observations. A notable deviation is that copper oxidizes between 150 and 350 °F (66-177 °C) even in a hydrogen atmosphere having a dew point below 0 °F (−18 °C). Remember that experiments in brazing with a controlled atmosphere are subject to considerable error unless extreme precautions are taken to avoid contamination of the atmosphere. Some of the contaminations that are apt to occur are pointed out in this chapter under Composition of Atmospheres for Brazing. Other experimental difficulties include that of the analysis of the atmosphere gas at brazing temperature in the immediate vicinity of the brazing operation.

A rigorous theoretical treatment would include reaction rates, not only the reaction of the reducing gas with the metal oxide but all other possible reactions, such as metal with oxide and metal with atmosphere, as well as effects due to unstable oxides. The influences of alloying elements further complicate the problem on commercially available metals. When solutions to these problems are known, it will be possible to predict the controlled atmosphere requirement for each brazing application. Meanwhile, recommendations of suppliers of materials and equipment should be sought when difficulties are encountered or development of maximum efficiency high production is justified.

Many combinations of base metals and brazing filler metals can be used for the efficient production of brazed joints using suitable standard controlled atmospheres in lieu of flux and without such precise control of the atmosphere as may seem to be implied by explaining potential sources of difficulty.

Bases for Choice of Atmosphere

On the basis of cost, the atmosphere listed nearest the top of Table 4.2 for recommended application to the specific base metals and filler metal involved in the braze should be selected. The footnotes associated with the table are important, and these points are elaborated in the discussion of Composition of Atmospheres for Brazing. When base metals susceptible to carburization or decarburization (see chapters 15, 16, and 18) are to be heated in the atmosphere to temperatures at which these reactions take place, an atmosphere compatible to this requirement must be chosen. In applications involving metals that may be embrittled (see chapters 2 and 14) by hydrogen or that may combine chemically with hydrogen (e.g., tantalum), the amount of hydrogen present in the atmosphere must be kept below a suitable maximum value.

Probably the most widely used (on a tonnage basis) controlled atmosphere brazing is the use of BCu on steel with AWS Nos. 1, 2, 3, 5, or 6 atmospheres, depending upon the composition and requirements imposed upon the steel.

REFERENCES

1. Federal Specification O-F-499, Flux, Brazing (Silver Alloy, Low Melting)
2. Bredzs, N., and Tennenhouse, C. C. 1970. Metal-metal oxide-hydrogen atmosphere chart for brazing or bright metal processing. *Welding Journal* 49 (5), Research Supplement, pp. 189-s to 193-s.

CHAPTER 5

DRAFTING ROOM PRACTICES

As with any other industrial fabricating process, it is important that brazed joints be properly engineered. It is the function of the engineer to design the joint so that it will withstand the service conditions to which the finished fabrication will be exposed.

At the same time, the joint should be so disposed that it can be made and inspected without inconvenience and, as far as possible, with existing shop facilities. For this reason, close cooperation between shop and engineering personnel cannot be overemphasized.

The designer can best convey his design to the fabricating department by use of the latest edition of the AWS A2.4 Symbols for Welding and Nondestructive Testing. These symbols provide a means for identifying the brazing procedure to be used, as shown in Figs. 5.1 through 5.5. The type of brazing process to be used is indicated by the letter in the tail of the symbol. Table 5.1 shows the letter designation symbols for each of the brazing processes. If necessary, supplementary sketches and notes should be included on the design drawings to show details of the joint preparation and configuration.

It is strongly recommended that the designer familiarize himself with the other chapters in this manual so that he will fully appreciate and give due consideration in his design to such matters as assembly, filler metal feed, and the selection of the brazing and postbrazing processes.

DESIGN PARAMETERS

The designer and draftsman must define as many of the design parameters of the brazed joint as practical in order to assure obtaining the desired service properties and life of the joint and the completed brazement.

The following items list the specific variables of the brazement that should be evaluated.

1. Joint design: (a) joint clearance at room temperature, at brazing temperature; (b) physical shape of members, i.e., stress concentration points, base metal fillets
2. Base metal(s), i.e., specifications, chemistry, or others
3. Brazing filler metals (see AWS A5.8 or chapter 3) type and method of placement
4. Brazing atmosphere or flux (see Table 4.1, page 58 and Table 4.2, pages 62-63)
5. Brazing process and process variables: (a) brazing processes, see Table 5.1; (b) brazing process variables, i.e., temperature, time at heat, heat and cooling slopes, if required
6. Prebraze cleaning, i.e., oxide removal, grease and oil removal, prebraze clean-up cycle in furnace, out-gassing vacuum cycle, etc.
7. Postbraze cleaning, i.e., flux removal
8. Postbraze heat treatment, i.e., tempering, annealing, hardening heat treatment, solution treatment, aging, etc.
9. Inspection method, i.e., type of test(s), test requirements, frequency, test limits, qualification requirements, etc.
10. Special precautions, techniques, and other special information and requirements

Note that all the brazement parameters and instructions may be included or referenced on the assembly blueprint used during brazing so that adequate information is available to the brazing engineer and shop, i.e., base metal specification or chemistry, joint dimensions or clearance, etc. Joints should be accessible for cleaning, inspection, and repair.

INSPECTION

Not only is it important for the design engineer to consider the inspection requirements of each joint and the ease with which it may be inspected, but the inspection method and limits must be indicated or referenced on the blueprint to assure the brazement quality meets its design requirements.

The AWS A2.4, Symbols for Welding and Nondestructive Testing should be referred to for the nondestructive testing symbols (NDT) to be used on the blueprint.

Table 5.2 shows the basic testing symbols and the elements of the testing symbol. For complete information and instructions on the NDT symbols, see publication AWS A2.4.

QUALIFICATION OF THE BRAZING PROCESS AND BRAZING OPERATOR (BRAZER)

In applications where critical quality control is required, such as nuclear brazements, aerospace brazements, boiler and pressure vessel brazements, etc., the design engineer may wish to incorporate on the blueprint a reference to qualify the brazing process or brazing operator (brazer) to AWS B3.0, Standard Qualification Procedure or to ASME Boiler and Pressure Vessel Code Section IX.

Table 5.1—Brazing process letter designation symbols

Process	Symbol
Infrared brazing	IRB
Torch brazing	TB
Furnace brazing	FB
Induction brazing	IB
Resistance brazing	RB
Dip brazing	DB
Twin-carbon arc brazing*	TCAB
Block brazing*	BB
Flow brazing*	FLB

*These definitions have been relegated to obsolete or nonpreferred status due to present usage.

Table 5.2—Nondestructive testing symbols

Basic symbols

Type of test	Symbol
Visual	VT
*Dye Penetrant	DPT
*Fluorescent Penetrant	FPT
†Magnetic Particle	MT
†Eddy Current	ET
Ultrasonic	UT
Leak	LT
Proof	PRT
Radiographic	RT

Elements of the testing symbol

The assembled testing symbol consists of the following elements. Such of these elements as are necessary shall be used and shall have standard locations with respect to each other as shown below:

- Reference Line
  - Arrow
  - Basic testing symbol
  - Test-all-around symbol
  - (N) Number of Tests
- Tail
  - Extent of test
  - Specification, process, or other reference

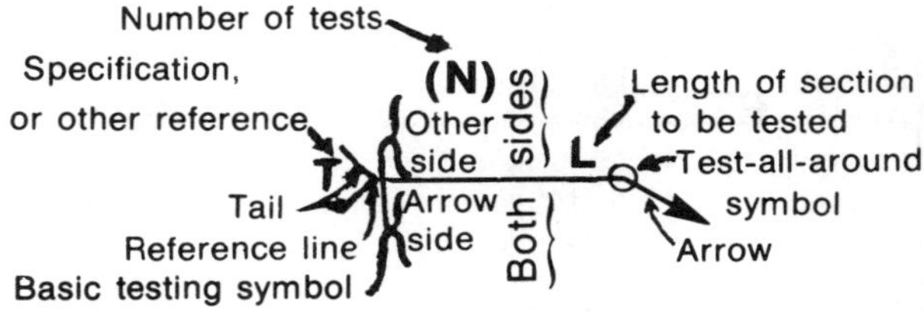

*Penetrant inspection (DPT, FPT) is generally not recommended because repair is difficult when contaminated with penetrant materials.

†Inspection Tests not generally applicable to brazements are the Magnetic Particle and Eddy Current Tests.

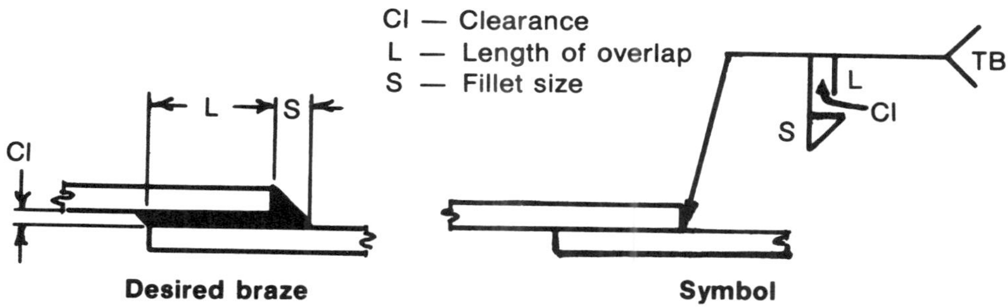

Fig. 5.1—Location of elements of the brazing symbol.

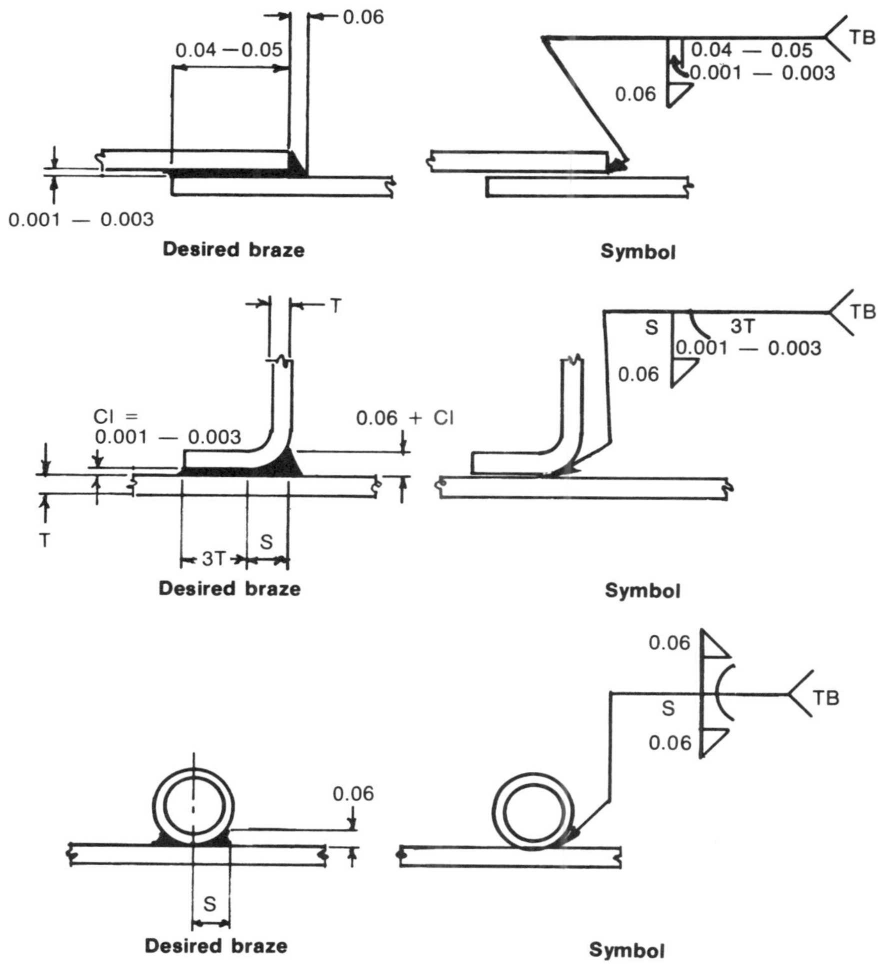

Fig. 5.2—Application of torch brazing symbols.

NOTE: The above symbols with fillet size should be used for brazements made with local heating methods only, e.g., torch, induction. Where high strength joints are desired and where furnace brazing is used, fillets should be fabricated in the base metal.

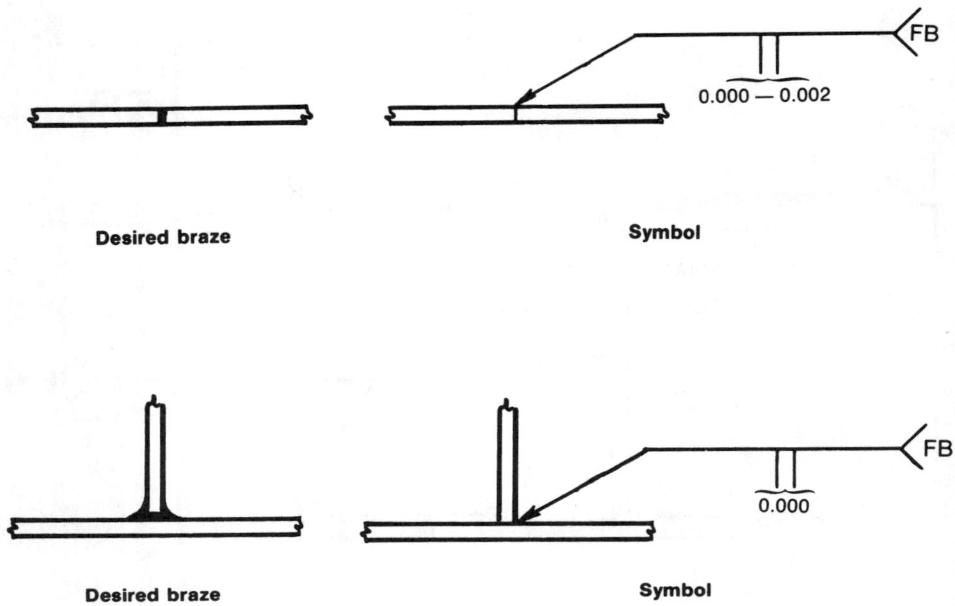

NOTE: No fillet symbol or dimensions should be used when furnace brazing unless for a specific application in a flat position or when special filler metals or techniques are used.

Fig. 5.3—Application of furnace brazing symbols.

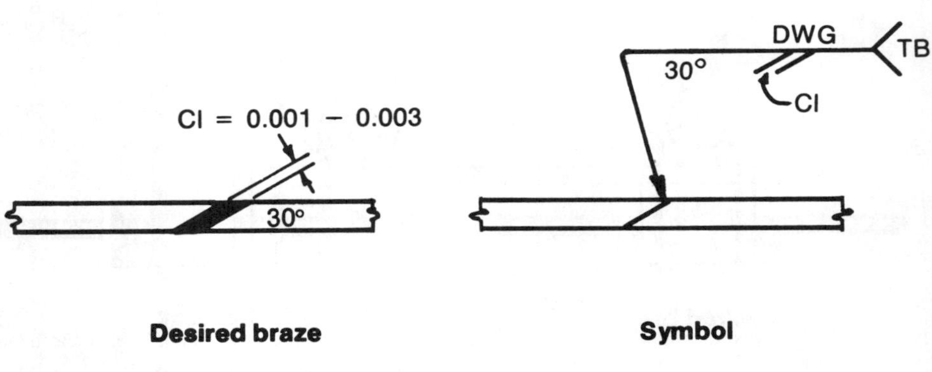

Fig. 5.4—Scarf joint.

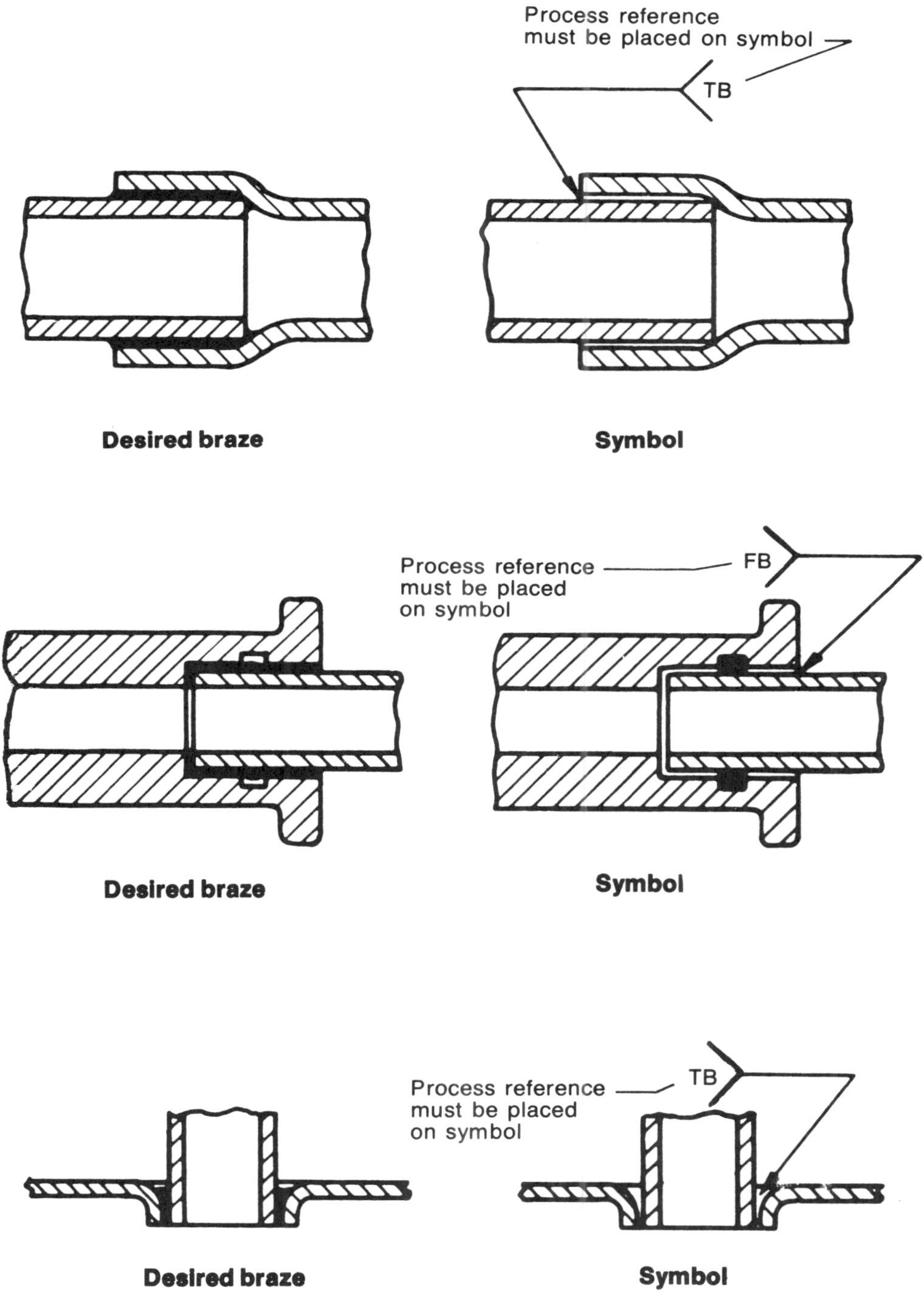

Fig. 5.5—Desired braze joint and application of brazing symbols.

CHAPTER 6

JOINT DESIGN

Many variables are to be considered in manufacturing a reliable mechanical design of the assembly. From the mechanical standpoint the design of a brazement is no different from the design of any other part. The rules applying to concentrated loads, stress risers, stress concentration, static loading, dynamic loading, etc., that apply to machined or other fabricated parts also apply to the brazed components.

In the past, it was considered that some types of brazed joints were not adequate for application in a brazement. Experience has shown that there is no such thing as a "bad" joint design. Today we know that if a joint design works in the application for which it is intended, it is a "good" joint design whether it is a very simple joint or a very sophisticated joint design. If the properly executed joint fails, it is evident that it is a bad application of a specific joint design and that it is necessary to redesign the joint to accommodate the service conditions to which it was subjected.

The design of a brazed joint does have specific requirements that must be met when adequate operating characteristics are to be achieved. Some of the more important factors are:

1. Composition of the base metal and filler metal (members to be brazed may be of similar or dissimilar materials)
2. Type and design of joint
3. Service requirements: mechanical performance, electrical conductivity, pressure tightness, corrosion resistance, and service temperature

Reference should be made to the appropriate sections in this chapter for information on filler metals and base metals.

SELECTION OF BASE METAL

In addition to the normal mechanical requirements of the base metal in the brazement the effect of the brazing cycle on the base metal and the final joint strength must be considered. Cold work strengthened base metals will be annealed and the joint strength reduced when the brazing process temperature and time are in the annealing range of the base metal being processed. "Hot-cold worked" heat resistant base metals can also be brazed; however, only the annealed physical properties will be available in the brazement. The brazing cycle by its very nature will usually anneal the cold worked base metal unless the brazing temperature is very low and the time at heat is very short. It is not practical to cold work the base metal after the brazing operation.

When it is essential to design a brazement having strength above the annealed properties of the base metal after the brazing operation, it will be necessary to specify a heat treatable base metal. The base metals can be of the oil quench type, the air quench type that can be brazed and hardened in the same or separate operation, or the precipitation hardening type in which the brazing cycle and solution treatment cycle may be combined; or the parts may be hardened and then brazed with a low temperature filler metal using short times at temperature in order to maintain the mechanical properties.

The strength of the base metal has a profound effect on the strength of the brazed joint; thus, this property must be clearly kept in mind when designing the joint for specific properties. Some base metals also are easier to braze than others, particularly by specific brazing processes. For example, the nickel base metal containing high titanium and/or aluminum additions will present special problems in furnace brazing. Thus, nickel plating is sometimes used as a barrier coating to prevent the oxidation of the titanium or aluminum and present a readily wettable surface to the brazing filler metal.

SELECTION OF FILLER METAL WITH RESPECT TO JOINT DESIGN

When designing a brazed joint for a specific service application, it is important to consider the properties and compatibility of the base metal and filler metal in the final brazed joint in the environment in which they will operate. There are almost as many filler metals as there are base metals; however, there are relatively few filler metals that find extensive usage. These filler metals are listed in chapter 3.

Brazing filler metals and their properties are covered more fully in the filler metal section.

The properties of the filler metal in the joint after brazing (and heat treating if required) must be taken into consideration when designing a joint for a given service temperature, some of the more important factors being strength at service temperature; galvanic corrosion under service conditions; corrosive media at service conditions; fatigue properties; flow of the filler metal through the joint clearance with base metal, etc.; base metal-filler metal interaction, and method of heating.

TYPES OF JOINTS

Several factors influence the selection of the type of joint to be used. These factors include brazing process to be used, fabrication techniques prior to brazing, number of items to be brazed, method of applying filler metal, and the ultimate service requirement of the joint.

The unit strength of the brazing filler metal may occasionally be higher than that of the base metal. In general, however, the strength of the filler metal is considered lower. The joint strength will vary according to the joint clearance, degree of filler metal-base metal interaction (diffusion and solution), and the presence of defects in the brazement.

There are basically two types of joints used in brazing operations — the lap joint and the butt joint. These joints are illustrated in Fig. 6.1.

In the lap joint, the area of overlap may be varied so that the joint will be as strong as the weaker member, even when using a lower strength filler metal or with the presence of small defects in the joint. An overlap at least three times the thickness of the thinner member will usually yield maximum joint efficiency. Lap joints are usually used because they offer the greatest possibility of obtaining joints with maximum efficiency and the greatest ease of fabrication. They do, however, have the disadvantage of increasing the metal thickness at the joint and of creating a stress concentration at each end of the joint where there is an abrupt change in cross section.

Butt joints are used where the thickness of the lap joint would be objectionable and where the strength of the completed joint will satisfactorily meet the requirements of the brazement in service. The strength of a properly executed butt joint may be sufficiently high that failures will occur in the base metal away from the joint, or it may fall far below the base metal strength. The joint strength will depend on the filler metal strength in the joint as compared to the base metal strength, the degree of brazing filler metal-base metal interaction during the brazing cycle, and the service requirements. Maximum efficiency may not be obtained with the butt joint when the filler metal in the joint is much weaker than the base metal. To obtain the greatest efficiency, the brazed joint should contain no defects (i.e., flux inclusions, voids, unbrazed areas, pores or porosity). The joint after brazing (and heat treatment, if required) should exhibit strength equal to or greater than the unit strength of the base metal. Another means of improving the joint strength of the butt joint is to utilize joint clearances as small as are compatible with the base metal and filler metal involved, as well as their compatibility with the brazing process to be used in making the brazement. Although the minimum joint clearance has been demonstrated to produce optimum joint strength, producing such clearances is sometimes considered impractical economically. However, with the current sophisticated metal working techniques, maintaining proper clearances is not a major problem. It is important to point out that if high quality and highly reliable brazements are to be manufactured it is imperative that clearances be controlled.

For our purposes T-joints and corner joints are regarded as butt joints. A variation of the butt joint is the scarf joint (Fig. 6.2). In this joint an

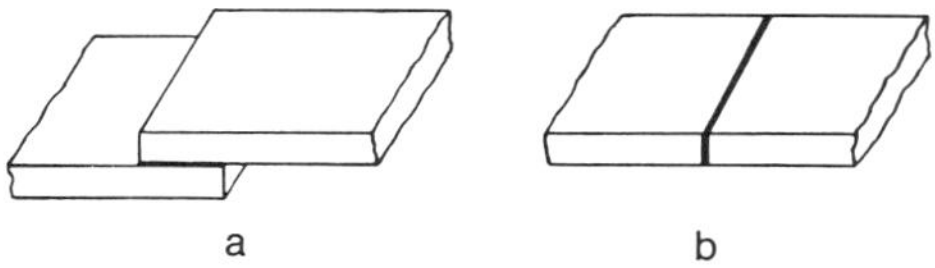

Fig. 6.1—Basic lap and butt joints for brazing. **a,** Lap joint. **b,** Butt joint.

Fig. 6.2—Typical scarf joint design.

attempt is made to increase the cross-sectional area of the joint without increase in thickness. There are a number of disadvantages to this joint which limit its use to any great extent. The sections are difficult to align and the joint is difficult to prepare, particularly in thin members. Since the joint is not perpendicular to the axis of tensile loading, the load-carrying capacity is similar to the lap shear joint rather than to the butt tensile joint. The design consideration of the percent of tensile plus the percent of shear value is not applicable to brazed joints.

JOINT CLEARANCE OR "FIT-UP"

Joint clearance generally has a potent effect on the mechanical performance of a brazed joint. This applies to all types of loading including static, fatigue, impact, etc., and is applicable to all joint designs. Joint clearance has several effects on mechanical performance and includes (1) the purely mechanical effect of restraint to plastic flow of the filler metal offered by the greater strength of the base metal, (2) the possibility of slag entrapment, (3) the possibility of voids, and (4) the relationship between joint thickness and capillary force which accounts for filler metal distribution.

Clearance is the distance between the faying surfaces of the joint. The clearance between members of similar metals is easily maintained in assemblies where parts are a press or shrink fit. In some parts it is necessary to use spacer wires, shims, prick punch marks, grit blasting, etc., to assure the proper clearance gap for optimum flow into the joint. Clearance between the brazed parts must be considered in terms of conditions at one specific instant, i.e., room temperature or brazing temperature. With similar metals of about equal mass, the room temperature clearance (before brazing) is a satisfactory guide. In brazing dissimilar metals, the one with the higher thermal expansion may tend to increase or decrease the clearance, depending on the relative positions and configurations of the base metals. Thus, when brazing dissimilar metals (or greatly differing masses of similar metals), consideration must be given to the clearance at brazing temperature and adjustments made in the room temperature clearance to achieve the desired clearance at brazing temperature. See further information under Effect of Dissimilar Base Metals.

The influence of joint clearance is illustrated in Fig. 6.3, which indicates joint shear strength for various joint clearances. Table 6.1 may be used as a guide for clearances when designing brazed joints for maximum strength.

Some specific clearance versus strength data for a given base metal-filler metal combination and brazing condition are contained in Fig. 6.4* and Fig. 6.5.* Figure 6.4* shows the *optimum* shear values obtained on a 0.5 in. (12.7 mm) round drill rod using pure silver. The rods were butt brazed by an induction heating method in a dry 10% hydrogen-90% nitrogen atmosphere. Testing was accomplished in a specially designed guillotine-type fixture. Figure 6.5* shows the curve produced with a butt brazed joint tested in tension. Although this curve shows the normal ascending joint strength with decreasing clearance at practical production clearance, it is of interest to note that the strength drops off when extremely small clearances were obtained. It must be pointed out that each individual design of test specimens will vary the indicated "joint strength" obtained; thus tests must be conducted on the final brazement in order to obtain load-carrying values for a specific design.

Variations in the process used, clearance, brazing temperature, fluxing agent (mineral flux or gas atmosphere), etc., will produce variations in the strength versus clearance curve. For example, a free-flowing filler metal brazed in a high quality atmosphere will adequately flow through a joint

*Note, the above data were obtained with nonstandard test specimens but do provide specific laboratory data only for the specific specimen design and test method as noted.

Table 6.1—Recommended joint clearance at brazing temperature

Filler metal AWS classification	Joint clearance, in.
BA1Si group	0.002-0.008 for length of lap less than 0.25 in.
	0.008-0.010 for length of lap greater than 0.25 in.
BCuP group	0.001-0.005 no flux and for flux brazing (mineral fluxes)
BAg group	0.002-0.005 flux brazing (mineral fluxes)
	0.000-0.002† atmosphere brazing (gas phase fluxes)
BAu group	0.002-0.005 flux brazing (mineral fluxes)
	0.000-0.002† atmosphere brazing (gas phase fluxes)
BCu group	0.000-0.002† atmosphere brazing (gas phase fluxes)
BCuZn group	0.002-0.005 flux brazing (mineral fluxes)
BMg	0.004-0.010 flux brazing (mineral fluxes)
BNi group	0.002-0.005 general applications (flux/atmosphere)
	0.000-0.002 free-flowing types, atmosphere brazing

*Clearance on the radius when rings, plugs, or tubular members are involved. On some applications it may be necessary to use the recommended clearance on the diameter to assure not having excessive clearance when all the clearance is on one side. An excessive clearance will produce voids. This is particularly true when brazing is accomplished in a high quality atmosphere (gas phase fluxing).

†For maximum strength a press fit of 0.001 in. per in. of diameter should be used.

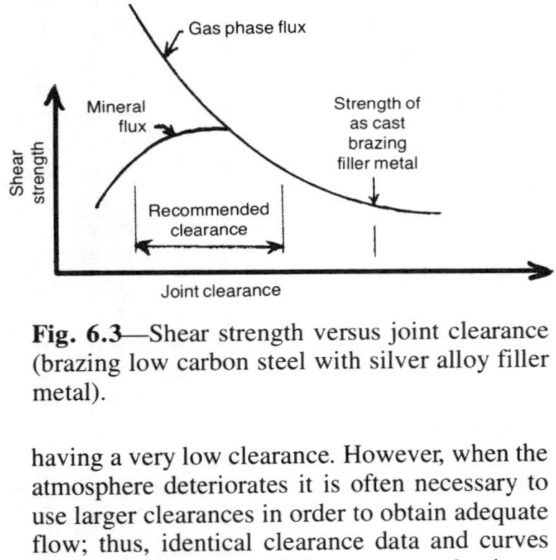

Fig. 6.3—Shear strength versus joint clearance (brazing low carbon steel with silver alloy filler metal).

having a very low clearance. However, when the atmosphere deteriorates it is often necessary to use larger clearances in order to obtain adequate flow; thus, identical clearance data and curves would not be expected from sources having a different quality atmosphere even though all other variables were identical. Figures 6.4 and 6.5 show the optimum strength values for given clearances when employing a specific set of materials and brazing variables. Thus, it will be necessary to evaluate the brazing variables of the materials combination along with the specific production brazing facility. The current method for joint evaluation (shear joints) may be found in AWS C3.2, Standard Method for Evaluating the Strength of Brazed Joints, which is printed at the end of this chapter.

Preplaced filler is filler metal put into the joint, such as between two plates. In this application the clearances noted in Table 6.1 generally do not apply. In applications where filler metal is preplaced the two members of the joint are so loaded that the joint clearance is reduced during the brazing operation. This reduction in clearance provides for filler metal to fill the voids in the normal roughness of the faying surfaces. In some applications additional filler metal is added by extending the filler metal shim out beyond the joint (note Fig. 6.22).

EFFECT OF MINERAL AND GAS PHASE FLUXES ON CLEARANCE

Flux is usually used in brazing. It may be mineral-type flux, gas atmosphere-type flux, or a combination of both the mineral flux and the gas atmosphere flux. The type of fluxing will have an important bearing on the clearance used to accomplish a given brazement.

Mineral fluxes melt at a temperature below the brazing filler metal and must flow into the joint to prepare it for the filler metal. The filler metal on melting must force out the flux as the filler metal is drawn into the joint by capillary action. When the clearance is too small, the mineral flux will be held in the joint so firmly that displacement by the liquid brazing filler metal may be difficult or

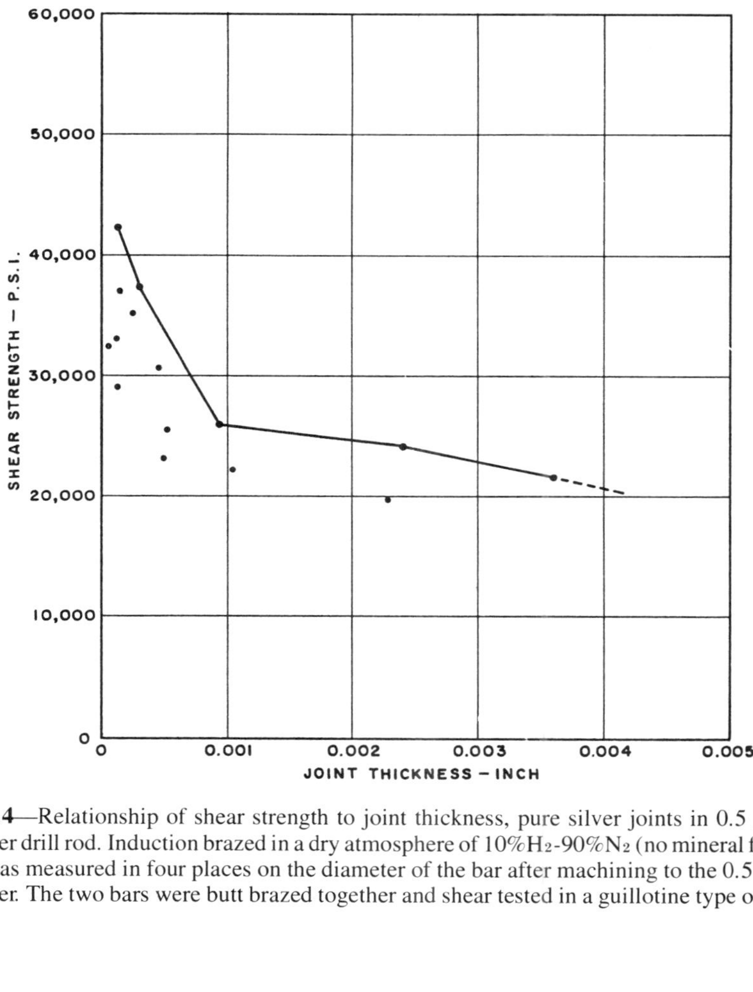

Fig. 6.4—Relationship of shear strength to joint thickness, pure silver joints in 0.5 in. (12.7 mm) diameter drill rod. Induction brazed in a dry atmosphere of 10%H_2-90%N_2 (no mineral flux used). The joint was measured in four places on the diameter of the bar after machining to the 0.5 in. (12.7 mm) diameter. The two bars were butt brazed together and shear tested in a guillotine type of fixture.

impossible. Thus joint defects may be produced. When the clearance is too large the liquid filler metal will travel around pockets of flux, thus giving rise to excessive flux inclusions.

Gas phase fluxes are atmospheres that have the same function as the mineral fluxes: combusted fuel gas, hydrogen, vacuum, argon, etc. (see chapter 4). Atmospheres also have their effect on the design of clearances for a specific brazement; it is essential to specify the brazing process on the blueprint when the clearance is determined so that they are compatible. The gas fluxes require lower clearances to obtain optimum strength, so the load-carrying capacity of the joint can be higher. In general, when atmosphere furnace brazing a joint in a vertical position, a free-flowing filler metal will flow out of a joint having a clearance in excess of 0.003 in. (0.08 mm). Larger clearances can be filled by using filler metals that have a wide liquidus to solidus range and using a brazing temperature between the liquidus and solidus. This technique is often helpful in experimental applications, but it is not recommended for production applications because it is hard to control and the joints have lower load-carrying capacity with the larger clearances. The "purity" of an atmosphere, as measured by the ability to dissociate the oxides of a specific base metal in the brazement, will also vary the clearance requirement. When the atmosphere rapidly and thoroughly dissociates the metal oxides of the base metal and free-flowing filler metal, the filler metal may be drawn out of a 0.002 in. (0.05 mm) clearance joint and onto the

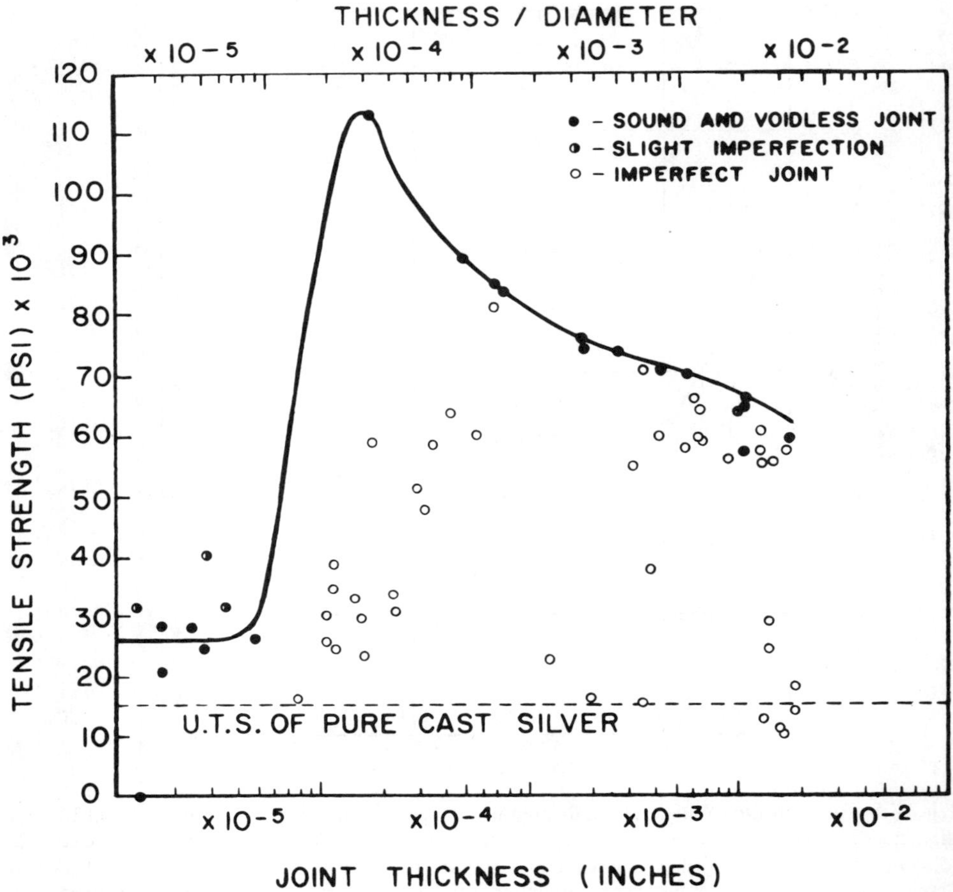

Fig. 6.5—Tensile strength of 0.5 in. (12.7 mm) diameter silver brazed butt joints in 4340 steel versus joint thickness.

base metal surface. In this case, a lower purity atmosphere or a lower clearance can be used. Lower clearance is recommended because it is easier to control in production operations, i.e., copper brazing filler metal on a low carbon steel in a −60 °F (−51 °C) dew point (or lower) pure dry hydrogen atmosphere and other high purity atmospheres where the flow of copper is excessive.

EFFECT OF SURFACE FINISH ON CLEARANCE

In general, the filler metal is drawn into the joint by capillary action. Thus, if the surface finish of the base metal is too smooth, the filler metal may not distribute itself throughout the entire joint. The resulting voids will lower the strength of the joint. To assure adequate flow of the filler metal throughout the joint, particularly when the clearance is zero or is a press fit, the faying surfaces of the joint should be roughened, preferably with a clean metallic grit compatible with the base metal. Care must be taken here not to contaminate the surfaces to be brazed with nonmetallic materials (ceramics, nonmetallic blasting materials, minerals from a water rinse bath, etc.) or other detrimental contaminants as this will seriously lower the joint strength.

Surfaces that are too rough will also lower the joint strength because only the points may braze or the average clearance is too large. A surface roughness of 30 to 80 μin. (0.7 to 2 μm) RMS is generally acceptable, but is not to be considered as optimum in all base metal-filler metal combinations, and tests must be conducted to assure optimum conditions for a specific brazement.

EFFECT OF BASE METAL-FILLER METAL INTERACTION ON CLEARANCE

The mutual solubility of many base metals and filler metals causes interaction to take place through solution of the base metal with the liquid filler metal and diffusion both in the liquid and solid states. This interaction has its effect on the clearance for a specific brazement. If the interaction is low, the clearances will in general be smaller; however, when the interaction is high, which is usually in the liquid phase, then the clearances required will have to be larger. This also affects the flow through the joint; thus, when the joint is long and interaction is large, the clearance should be increased. Since the aluminum base metal is brazed with a lower melting alloy of aluminum at a temperature close to the base metal melting point, interaction can be expected.

Aluminum is by no means the only filler metal that exhibits interaction, as all filler metals show varying degrees of interaction to different base metals — i.e., silver filler metals with copper base metal, copper filler metal with nickel and copper-nickel base metals, gold filler metals with nickel base metals, nickel filler metals with nickel and cobalt base metals, etc.

EFFECT OF BASE METALS ON CLEARANCE

The chemistry of base metals often includes one or more elements in varying quantities whose oxide is not easily dissociated in a specific atmosphere or by a specific mineral flux. Since the metal-metal oxide dissociation of a given base metal by a specific mineral flux or atmosphere is dependent on many factors (see chapter 4), it is important to match the proper clearance with the brazing process and flux or atmosphere. For example, in aluminum additions, the clearance can be small; but, when a small aluminum addition is incorporated, the clearance may have to be increased. When still larger aluminum additions are incorporated, brazing may not take place and the atmosphere must be made more active (low dew point or lower vacuum pressures), or the surface must be protected with a barrier coating, such as electrolytic nickel plating. The clearances would then require reappraisal to obtain optimum braze joint properties.

The coefficient of thermal expansion will, of course, have an effect on the joint clearance at brazing temperature when the base metals are dissimilar. Refer to the section on the Effect of Dissimilar Base Metals for additional information.

EFFECT OF BRAZING FILLER METAL ON CLEARANCE

Filler metals have varying degrees of flowability and different viscosities at various brazing temperatures and conditions. At one end of the scale there is a group of free-flowing filler metals and at the other end are the sluggish filler metals. The free-flowing filler metals generally require a lower clearance than the sluggish filler metals. Filler metals that have a single melting point, such as pure metals (i.e., copper), or eutectic melting filler metals and self-fluxing filler metals, will usually be free flowing, particularly when there is no or very little interaction with the base metal.

Variations in the quality of fluxes and atmospheres can enhance the free-flowing qualities of the filler metals or can result in no flow at all as fluxes may be of poor quality or become oxidized, or atmospheres may be of low relative purity. Thus the clearances may appear to be improper for a given set of brazing conditions when in fact the flux or atmosphere requires improved control.

EFFECT OF JOINT LENGTH AND GEOMETRY ON CLEARANCE

The joint length affects the clearance, particularly when there is interaction between the filler metal and the base metal. As the filler metal is drawn into a long joint, it may pick up enough base metal to freeze before it reaches the other end. The more interaction that exists with a specific base metal-filler metal combination under given brazing conditions, the larger the clearance must be as the joint becomes longer. This is only one of a number of reasons why it is

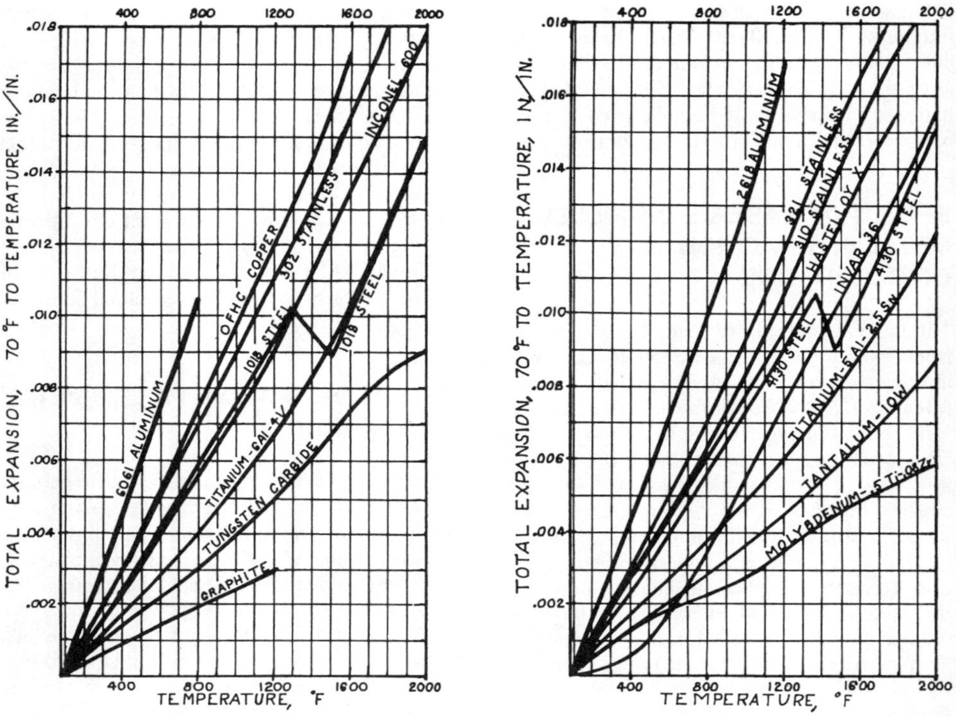

Fig. 6.6—Thermal expansion curve for some common materials (data from NASA publication).

important to make the joint length as short as possible, consistent with an optimum joint strength.

EFFECT OF DISSIMILAR BASE METALS

When designing a joint where dissimilar base metals are involved, it will be necessary to calculate the joint clearance at the brazing temperature to be used. Figure 6.6 shows the expansion with temperature. The thermal coefficient of expansion for each member and the brazing temperature must be taken into consideration to assure the desired mechanical strength in the final joint (Fig. 6.7). With high differential thermal expansion between two details, the brazing filler metal must be strong enough to resist fracture and the base metal must yield during cooling. It is important to remember that some residual stress will remain in the final brazement as a result of joining at the brazing temperature and subsequently cooling to room temperature. Thermal cycling of such a brazement during its service life will also stress the joint area which may or may not shorten the service life of the brazement. Whenever possible, the brazement should be designed so that any residual stresses do not add to the stress imposed during service. Few problems will be more difficult and will require more ingenuity during the brazing operation. One example of differential expansion is the copper brazing of a 20 in. (508 mm) outside diameter carbon steel ring with BCu on the inside of a 300-series stainless steel ring. If the rings were machined for 0.000 in. clearance at room temperature, we would expect the joint to open up to 0.100 to 0.150 in. (2.54 to 3.81 mm) on the diameter or 0.050 to 0.075 in. (0.27 to 1.91 mm) clearance per side. With this clearance no filler metal would flow into the joint by capillary action, and filler metal flowing into the joint by gravity would flow out the other side. If such a joint were made, it would be of poor quality. One solution of this problem would be to employ a 300-series

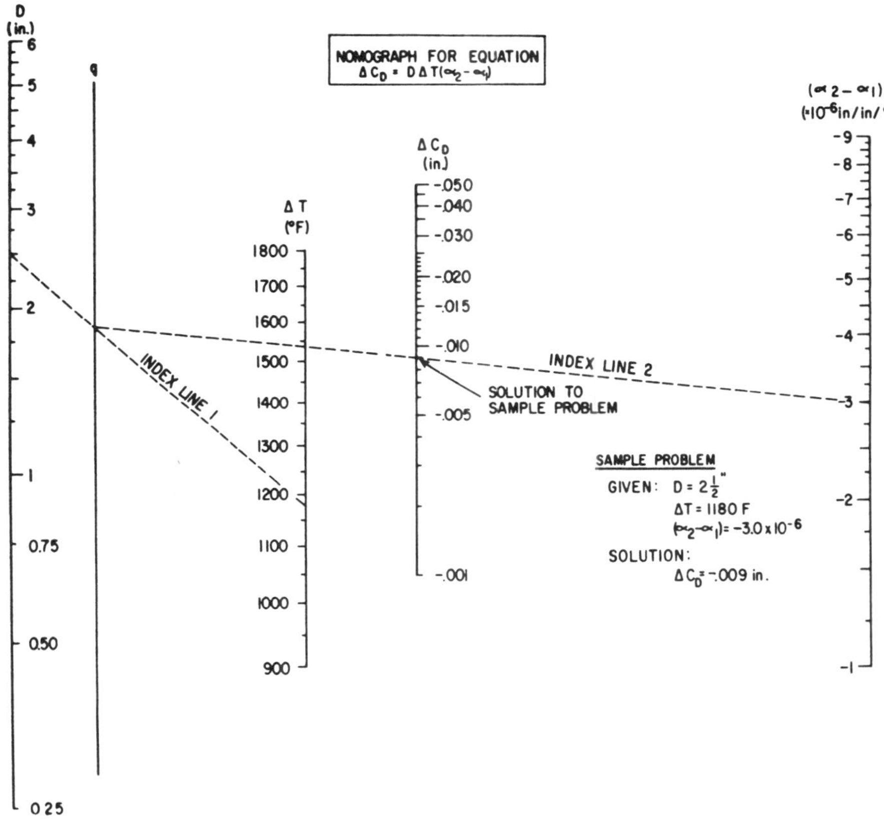

NOTES:

1. This nomograph gives change in diameter caused by heating. Clearance to promote brazing filler metal flow must be provided at brazing temperature.
2. D = nominal diameter of joint, in.
   ΔC_D = change in clearance, in.
   ΔT = brazing temperature minus room temperature, °F
   α_1 = mean coefficient of thermal expansion, male member, in./in./°F
   α_2 = mean coefficient of thermal expansion, female member, in./in./°F
3. This nomograph assumes a case where α_1 exceeds α_2; thus, scale value for $(\alpha_2 - \alpha_1)$ is negative. Resultant values for ΔC_D are therefore also negative, signifying that the joint gap reduces upon heating. Where $(\alpha_2 - \alpha_1)$ is positive, values of ΔC_D are read as positive, signifying enlargement of the joint gap upon heating.

Fig. 6.7—Nomograph for finding the change in diametrical clearance in joints of dissimilar metals for a variety of brazing situations.

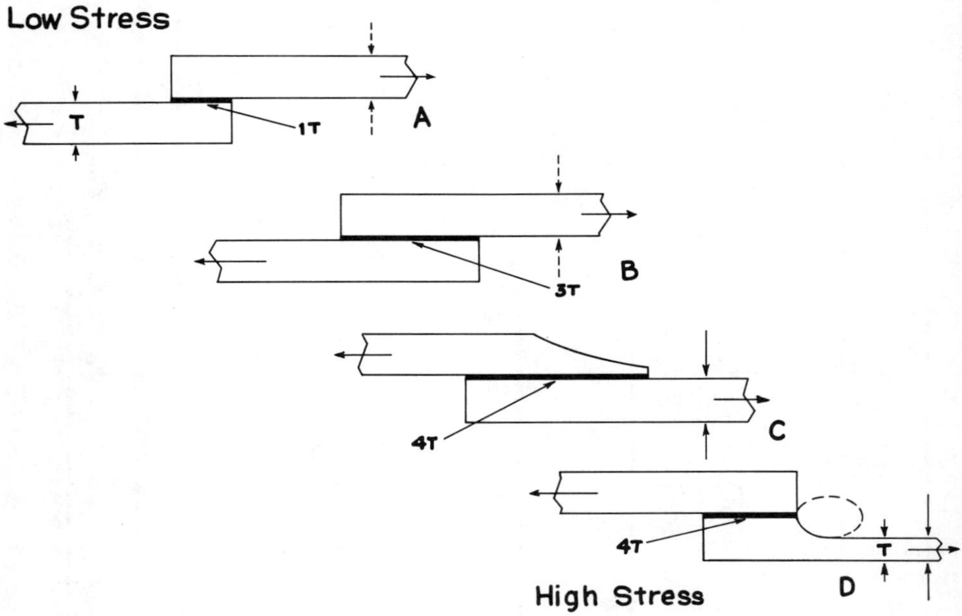

Fig. 6.8—Lap joint designs to be used at low and high stress. Flexure of right member in **c** and **d** will distribute load through base metal.

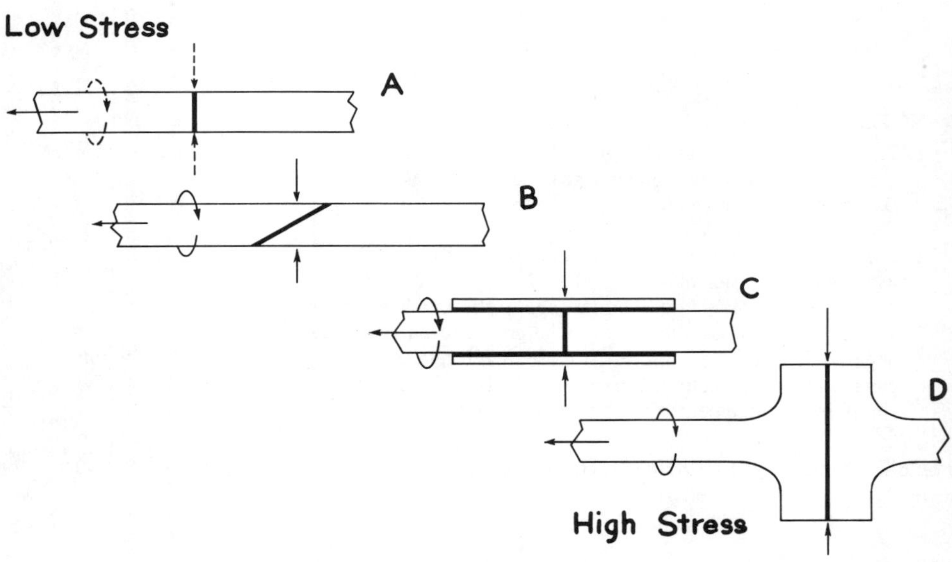

Fig. 6.9—Butt joint designs and modifications to increase load-carrying capacity of joint under high stress and dynamic loading.

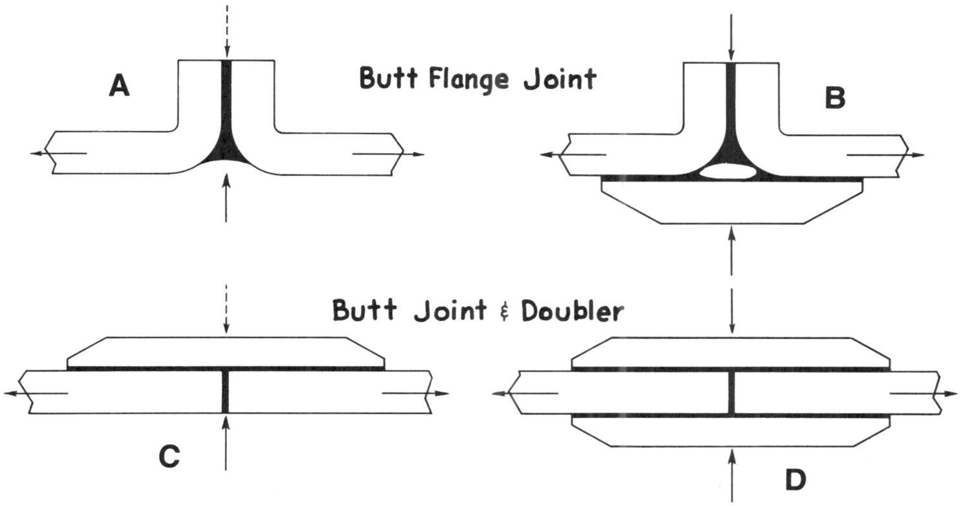

Fig. 6.10—Joint designs useful with sheet metal brazements. Note that loading in sketch **a** cannot be symmetrical.

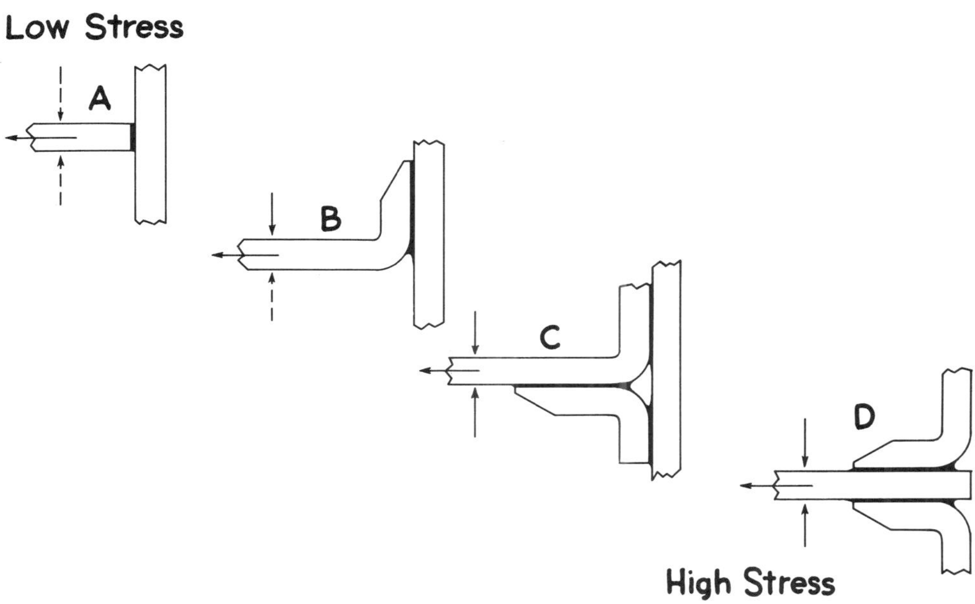

Fig. 6.11—Joint designs for sheet metal brazements, including increasing ability to carry higher static and dynamic loading.

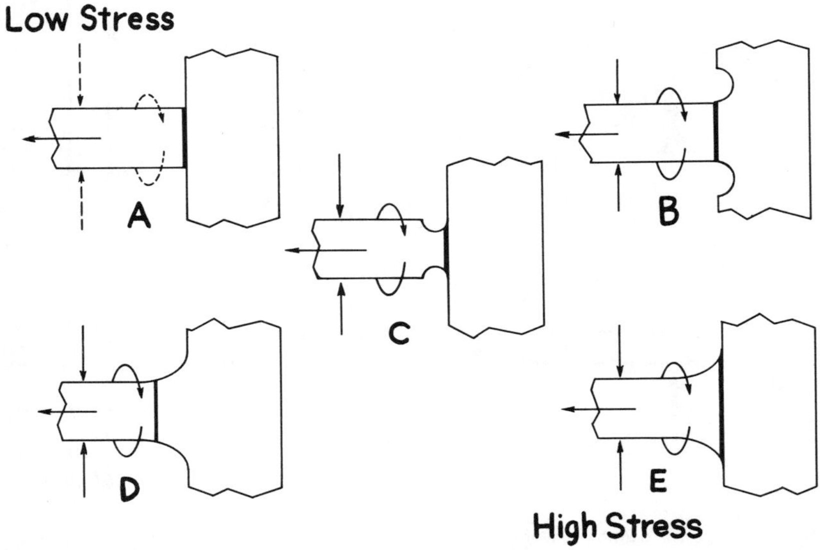

Fig. 6.12—Plate or hub and shaft joint designs for rotating turbine wheels. The joints are in tension due to the centrifugal loading of the rotating wheel.

stainless backup ring inside the carbon steel ring to stretch the carbon steel ring at the same rate as the outer stainless ring, thus maintaining the clearance preset at room temperature.

In a few specific cases the base metals, such as carbides, do not possess sufficient ductility, and it is necessary to use a soft ductile base metal spacer, such as nickel or copper, which will yield during the cooling cycle to prevent high stress between two high strength base metals; i.e., tungsten carbide brazed to a nickel base or cobalt base stainless steel that is heat and corrosion resistant. Problems can also be expected when brazing low ductility materials, such as carbides and ceramics, to heat treatable base metals requiring a fast to moderate quenching rate and base metals that have a volume increase as a result of transformation. This problem is particularly acute when the volume change is at the lower temperature ranges (600 °F [316 °C] and below).

STRESS DISTRIBUTION

It can be stated that a properly designed and executed high strength brazement will fail in the base metal. In brazements where joints may be lightly loaded it may fail in the base metal. In brazements where joints may be lightly loaded it may be more economical to use more simplified joint designs which may break in the brazed joint if overstressed in testing or in service.

In general the loading of a brazed joint demands the same design consideration given any other joint or change in base metal cross section. Brazed joints have a few specific requirements that are usually only apparent under conditions of high dynamic stresses or of high static stresses. It is well known that the base metal itself can best withstand high stresses and dynamic loading; thus a good brazement design will incorporate a joint design that will remove any high stress concentration from the edges of the braze joint and will distribute the stresses into the base metal. Numerous examples of this will be noted in the joint design sketches, Figs. 6.8 through 6.17.

A fillet of brazing filler metal is often erroneously incorporated by the designer as a method of eliminating stress concentrations. This is a very poor brazing design consideration, as it leaves much to chance and it is seldom possible to have the brazing filler metal form the desired fillet size and contour. Large fillets of brazing filler metal are of a cast structure although the filler metal in a

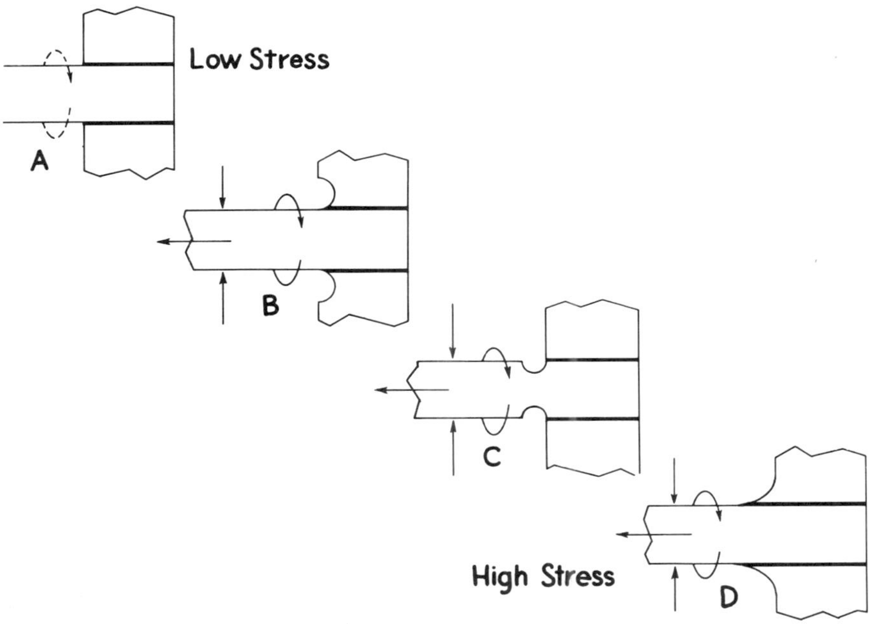

Fig. 6.13—Sheet and plate or hub and shaft joint designs showing methods of removing stress concentration from edge of brazed joint. Note that though grooves indicated in **b** and **c** may appear to weaken member, they actually increase overall brazement strength.

properly brazed joint will not be. When the fillets become too large, shrinkage or pipe may be present and will act as a stress riser rather than a stress reducer. Fillets of brazing filler metal, while they can be beneficial under some limited conditions, are a poor crutch in a highly stressed design and should not be used to replace a base metal fillet required to produce an optimum joint design with stress adequately distributed across a base metal fillet.

A common pitfall in the use of a brazed joint operating with high static stresses is to disregard the flexibility of the design and assume the brazement to be a rigid member for purposes of stress analysis. Thus a brazed joint that is calculated to satisfactorily operate with a load of 130 000 psi (910 MPa) in the base metal may fail when loaded to 5000 psi (35 kPa) in the base metal due to the excessive stress concentration that is imposed by the bending of one member (Fig. 6.16a). Here again, it is imperative to move the stress into the base metal by using a base metal fillet, or else to increase the rigidity of the assembly in the vicinity of the joint.

PLACEMENT OF BRAZING FILLER METAL

Before one designs a brazed joint, it is necessary to determine the brazing process to be used and the manner in which the filler metal will be introduced to the joint. In most manually brazed joints, the filler metal is simply face fed.

Furnace brazing and high production brazing operations, however, employ preplacement of the filler metal and may also incorporate the use of automatic dispensing equipment.

Figures 6.18 through 6.21 illustrate a number of ways in which the filler metal may be preplaced. Brazing filler metal may take the form of wire, shims, strip, powder; or it may be sprayed, plated, or vacuum deposited. Preplaced filler metal is most commonly used in the form of wire or strips. When the base metal is grooved to accept the preplaced filler metal, the groove should be cut in the heavier section (Figs. 6.18 and 6.19). When designing for strength, it is essential that the groove area be subtracted from the joint area. This step is necessary since the

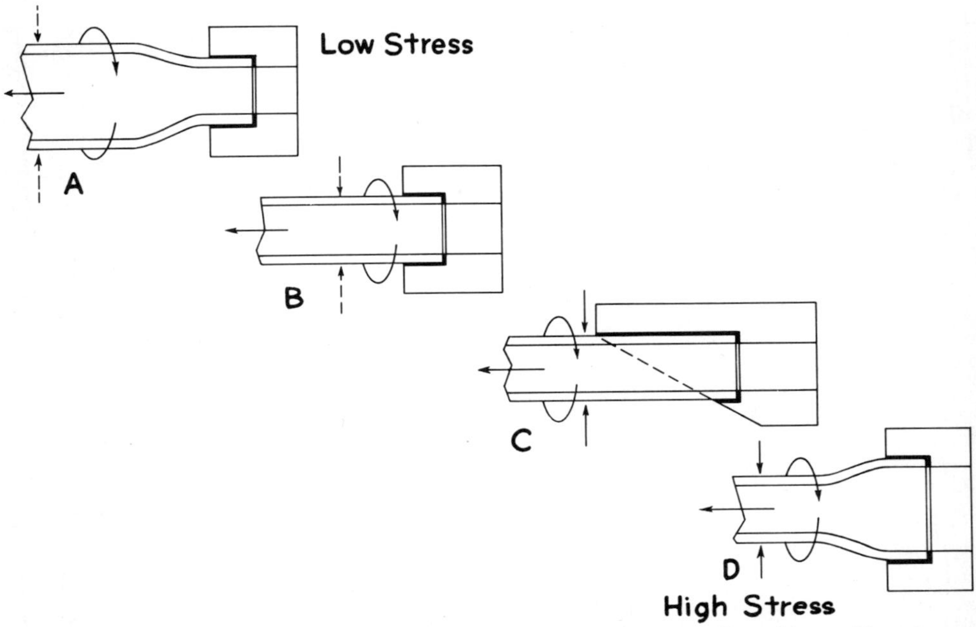

Fig. 6.14—Joint designs for tube and fittings. Reducing tube sizing as indicated in **a** will produce a serviceable brazement for low stresses; however, multiple stress concentrations preclude its use in high stressed dynamic loading. Expanding tube as indicated in **d** will produce best overall strength. Exact increase in diameter, contour of radii, and angle of expanded tube (as in **c**) will have to be developed for optimum service life with a specific design requirement.

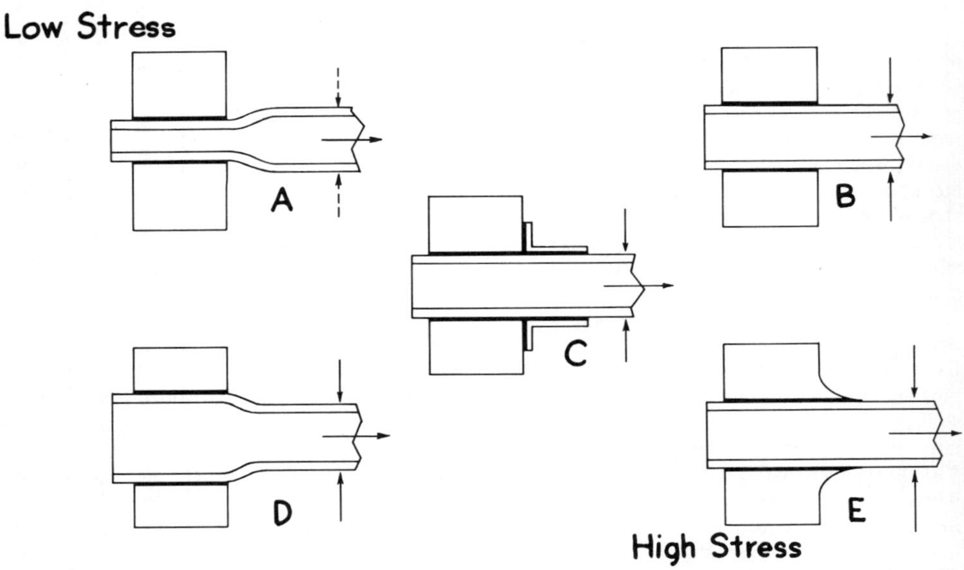

Fig. 6.15—Designs for tube and header joints in a heat exchanger. Pulsations, sonic stresses, and high velocity air across tubes can vibrate tubes to destruction. In **a** and **b** maximum stress is at header; tube will thus fail in fatigue at this point. Designs **c, d,** and **e** distribute stresses, thus increasing allowable loading and service life.

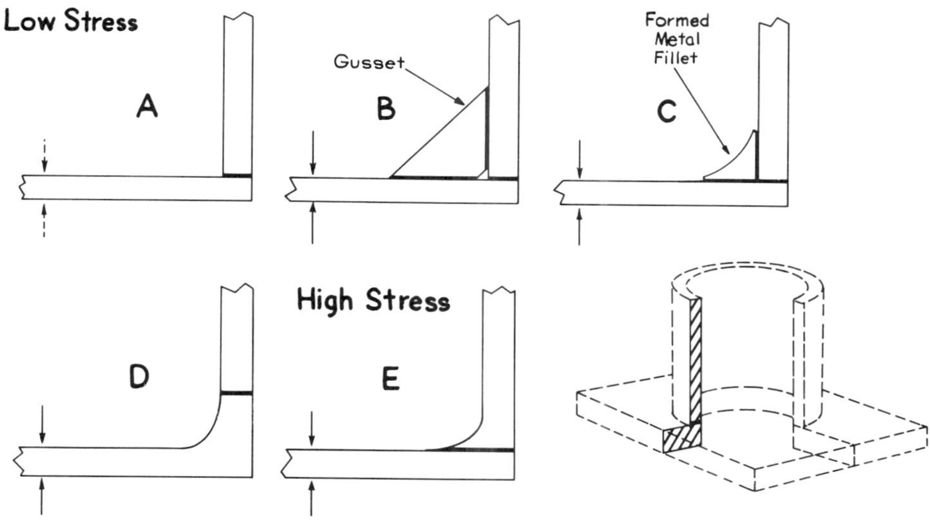

Fig. 6.16—Tube and diaphragm section of an actual part where diaphragm was hydraulically loaded. When designed as in **a,** premature failure resulted; the redesign stiffened the assembly to remove high stresses at one side of joint. Fillets, such as in **d** and **e,** will also improve serviceability.

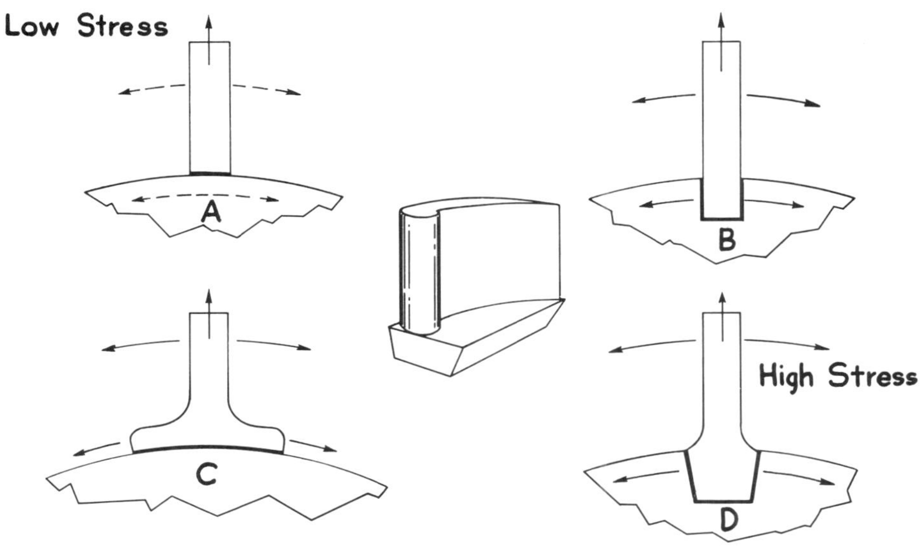

Fig. 6.17—Brazed turbine blade attachments for thin blading. Long, thin blades should not be attached as indicated in **a;** however, short, squat bladings have been adequately brazed and operated using this design. Design **d** has proven serviceable for high vibrating and centrifugal stresses.

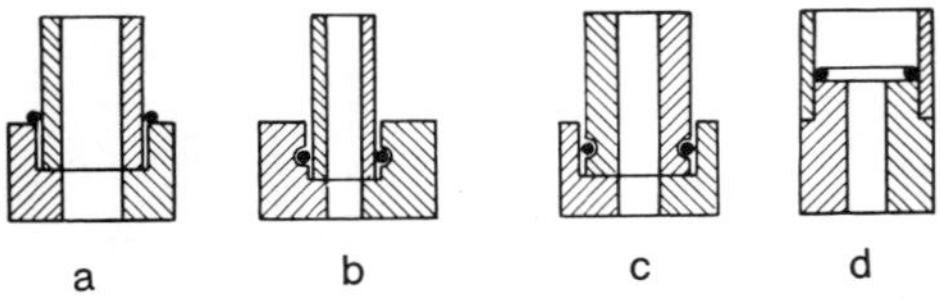

Fig. 6.18—Methods of preplacing brazing filler metal in wire form.

brazing filler metal flows from the groove to the joint interfaces.

Powdered filler metals can be applied in any of the locations indicated in Fig. 6.18, and are available in suspension mixtures with neutral binders or binders containing fluxes. Powdered filler metals are also applied to the joint area and wet down with binders or can be premixed with the binder and applied to the joint. Note that the density of the powder is less than the solid wire or shim, and therefore the groove volume must be larger since the powder is 50 to 70% of the solid metal volume.

In applications where there is to be a large difference in cross section of the members to be brazed and where there is a degree of mutual solubility between the filler metal and the base metal, undercutting of the thin member can be anticipated. This usually takes place because the filler metal and the thinner member reach the filler metal melting temperature well before the heavy member of the joint; thus, the filler metal and the thinner member continue to increase in temperature. With all the liquid filler metal in contact with the thinner member and continuing to increase in temperature, erosion and undercutting may be expected. Figure 6.19 shows two methods of preventing the liquid filler metal from reaching the tube wall until the heavier member reaches the brazing temperature. If solution does occur in the trepanned section (bottom two sketches, Fig. 6.19), or on the shield ferrule (top sketch, Fig. 6.19) it is of little or no concern because the properties of the tube are unaffected.

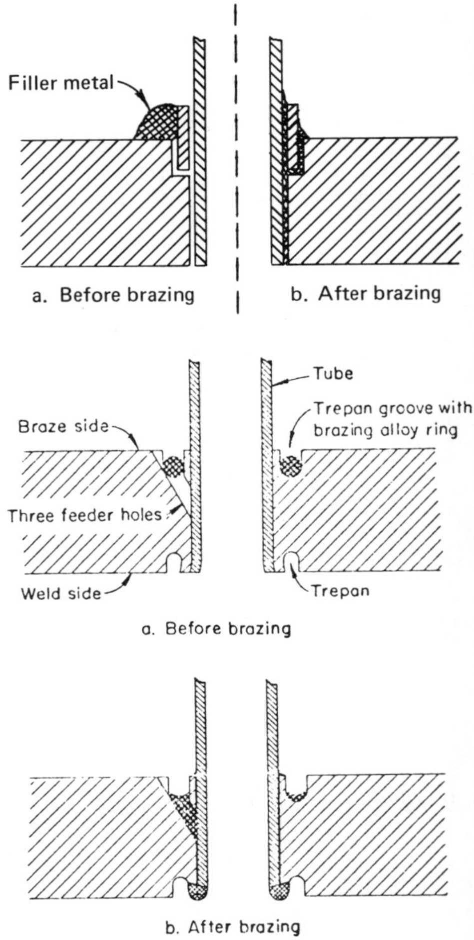

Fig. 6.19—Braze joint design and filler metal feeding for tube and header in nuclear application. Note that the filler metal is located away from the tube wall, and that the filler metal cannot flow into the joint until the entire joint is up to heat and will accept the filler metal. This is very helpful where base metal-filler metal interaction occurs.

Where preplaced shims affecting fit are used, the sections being brazed should be free to move with respect to one another so that the excess filler metal and flux will be forced out of the joint by the application of pressure, either deliberately or by the expansion of the joint elements during the brazing cycle.

The designer should also consider subsequent inspection needs when designing a brazed joint. He can help in this matter, for example, by providing for the placement of filler metal inside the assembly so that it will flow outward and thus provide visible proof of filler metal flow and complete joint penetration. In many joints the filler metal cannot be placed in the joint. To assure adequate flow through the joint with proper provision for inspection, the filler metal should be placed on one side of the joint and

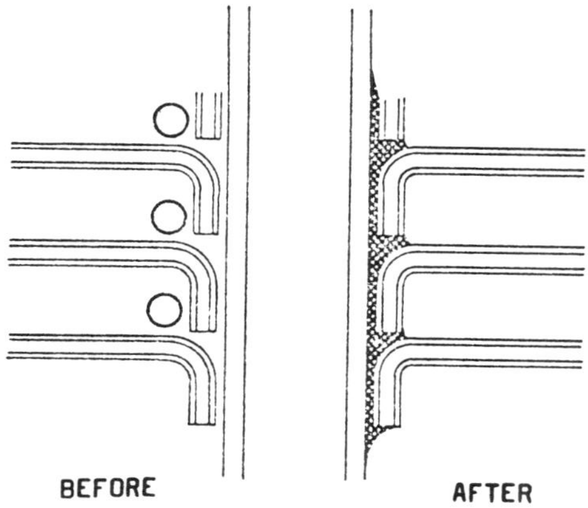

Fig. 6.20—Joint design for bonding fins to tubes for heat exchanger. Fins can be solid metal or clad, i.e., copper-stainless clad for high temperature applications.

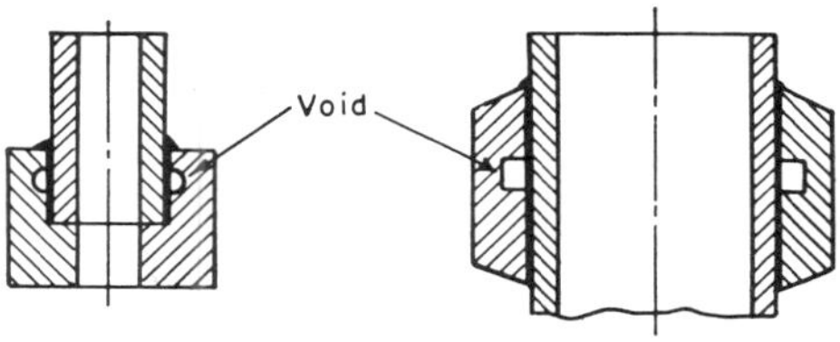

Fig. 6.21—Brazed joints with grooves for preplacement of filler metal.

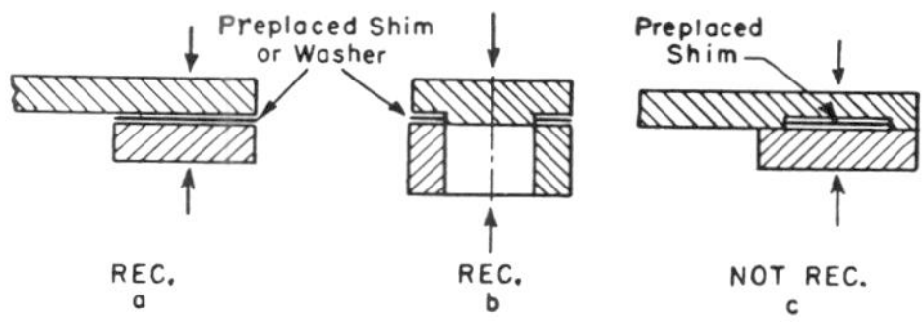

Fig. 6.22—Preplacement of brazing filler metal in shim form.

allowed to flow through the joint. Inspection for flow can then be made on the side of the joint opposite the filler metal.

In some applications, perforations or holes provide an excellent method of inspection. This is particularly true of long sheet metal joints and some tube and fitting joints. On large doublers, larger holes can be used as lightening holes, and filler metal can be applied in this area.

PROPERTIES OF JOINT VERSUS DESIGN

The mechanical properties of the joint will depend to a large extent on the design of the joint. Although an unlimited number of joint designs is possible, a few of the more commonly used ones are shown in Figs. 6.8 through 6.17. It will be noted that a given joint design, with changes in joint cross section and base metal contour, can be a very simple one usable with lower joint loading, or it can be a highly sophisticated joint design employable with high joint loading of both static and dynamic types. A good example of this is seen in Fig. 6.8.

Specific mechanical property data for specific joint applications are not included in this chapter since such data are dependent on joint geometry. Test methods may vary considerably for each joint design. It is recommended that preliminary data be obtained using AWS C3.2, Standard Method for Evaluating the Strength of Brazed Joints. Testing of simulated joints and the final brazement should be carried out under service conditions to assure adequate service life of the joint design.

Laboratory testing of brazing filler metals and base metal combination as well as other brazing variables should also be done using AWS C3.2.

In specific cases it will be necessary to use other test specimens to obtain additional data for structural members. Data on fatigue (using nonstandard specimens) for silver brazed butt joint specimens are shown in Fig. 6.23; for nickel brazed lap joints see Fig. 6.24.

ELECTRICAL CONDUCTIVITY

The principal factor to be considered in the design of an electrical joint is electrical conductivity. The joint, when properly designed, should not add appreciable electrical resistance to the circuit.

Brazing filler metals in general have very low electrical conductivity compared to copper. As an example, BCuP-5 filler metal has approximately 10% the conductivity of copper and BAg-6 filler metal 24.4% the conductivity of copper. However, a brazed joint will not add any appreciable resistance to the circuit. The shorter path through the brazing filler metal, as compared to the longer path through the conductor, results in only a negligible increase in the total resistance in the circuit.

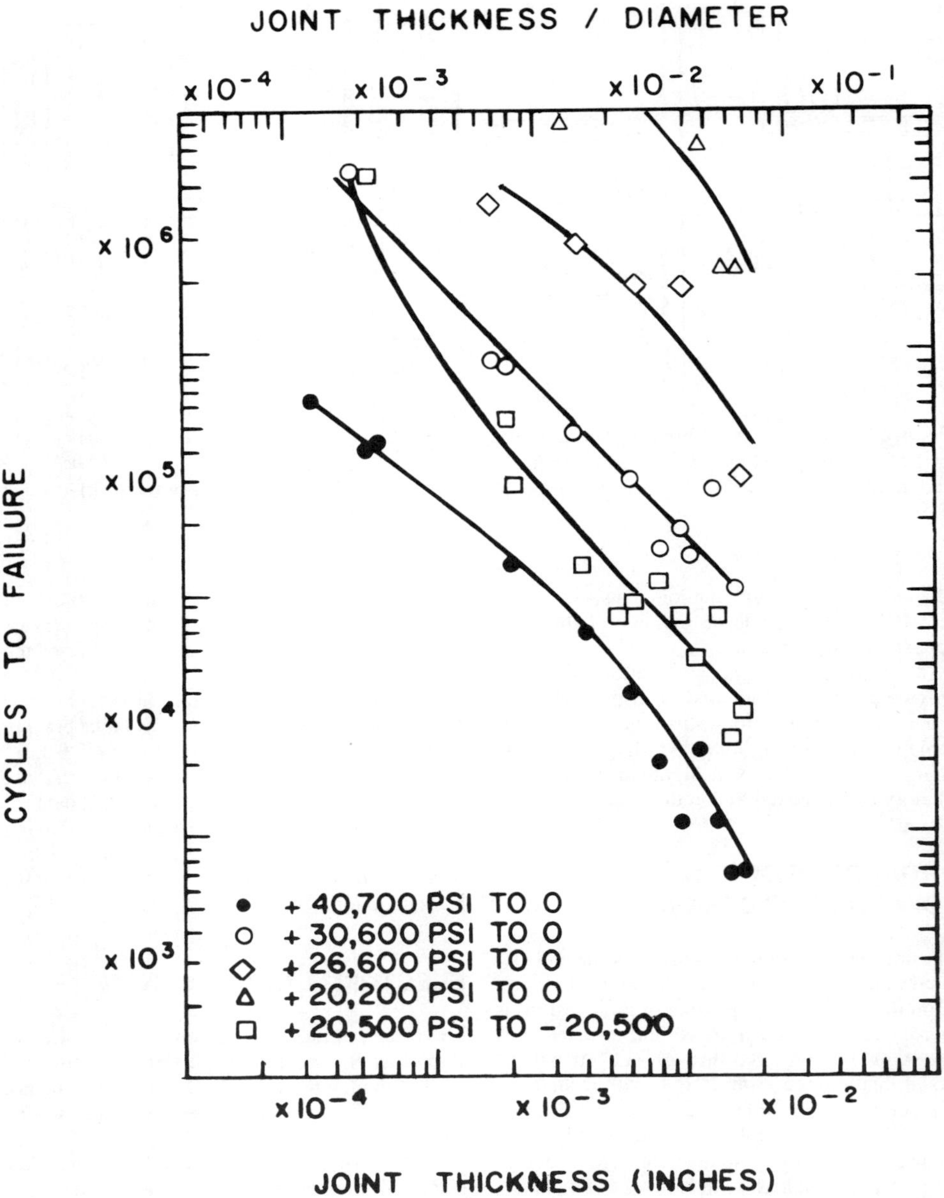

Fig. 6.23—Fatigue properties of 0.25 in. (6.35mm) diameter silver brazed butt joints in 4340 steel.

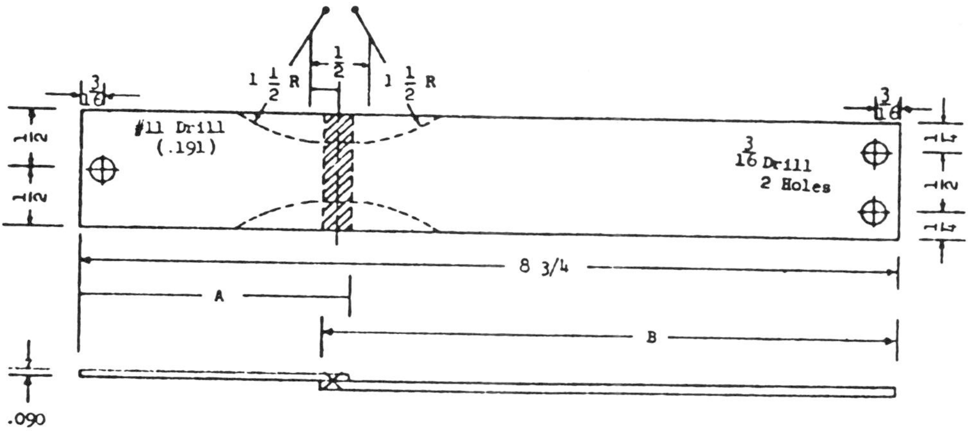

X Indicates tack weld

//// Indicates brazed joint

Dotted area indicates area milled away after brazing

A	B	Overlap
3¼	5¾	¼
3³/₁₆	5¹¹/₁₆	⅛

Fig. 6.24a—Fatigue data for AISI sheet metal brazed with BNi-1 and for the base metal processed through the brazing cycle. Single lap joint fatigue specimen.

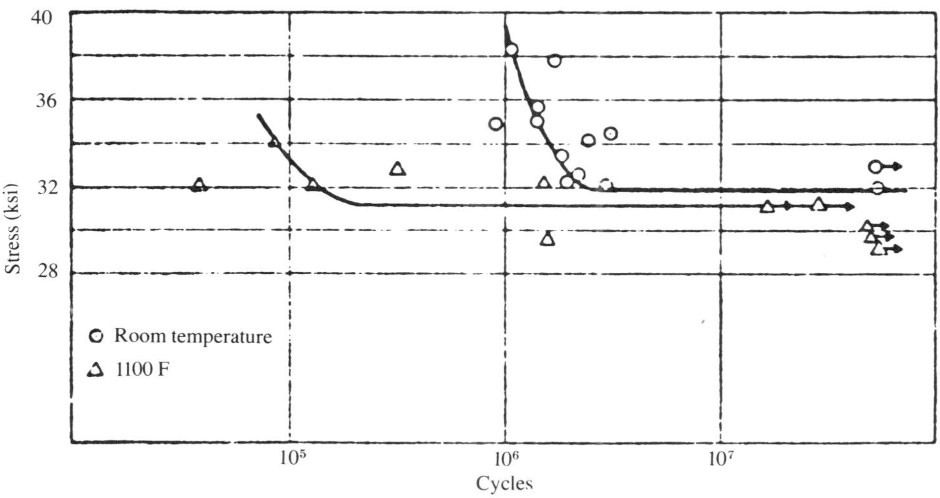

Fig. 6.24b—Fatigue data for AISI sheet metal brazed with BNi-1 and for the base metal processed through the brazing cycle. The effect of temperature on the fatigue strength of AISI 347 stainless steel sheet.

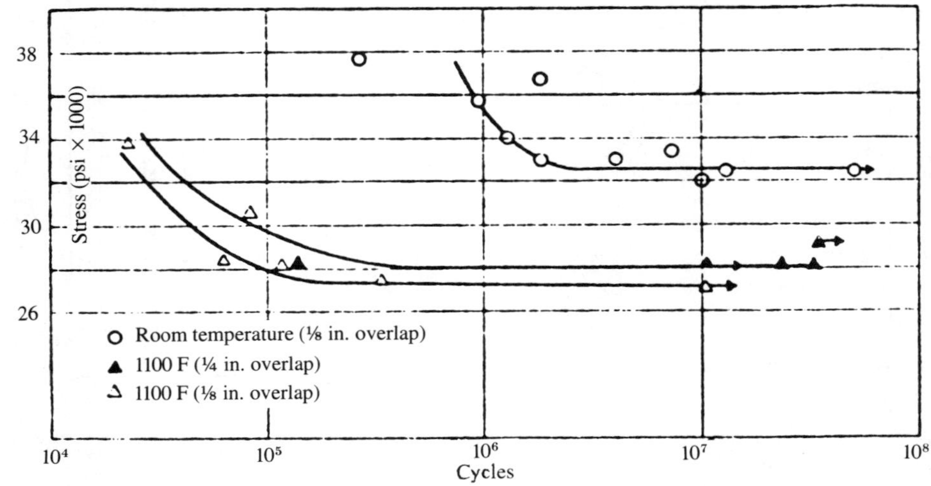

Fig. 6.24c — Fatigue data for AISI sheet metal brazed with BNi-1 and for the base metal processed through the brazing cycle. The effect of temperature and joint overlap on the fatigue strength of single lap joints of AISI 347, brazed with BNi-1 filler metal.

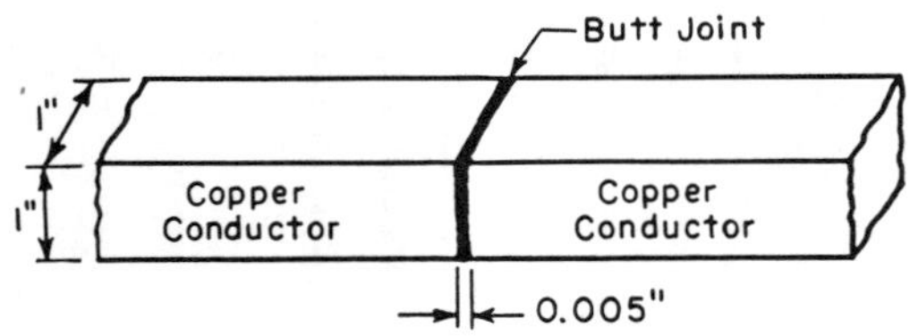

Fig. 6.25—Brazed joint in an electrical conductor.

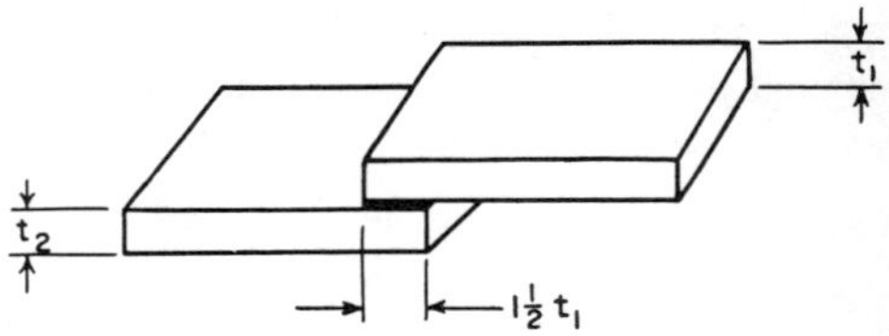

Fig. 6.26—Optimum lap for joints in electrical conductors.

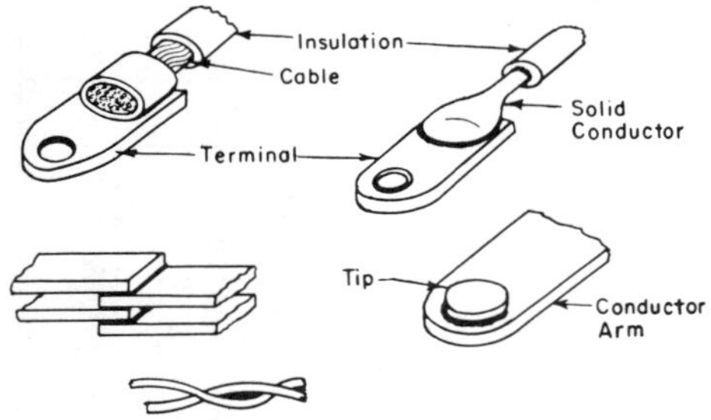

Fig. 6.27—Typical joints for brazing.

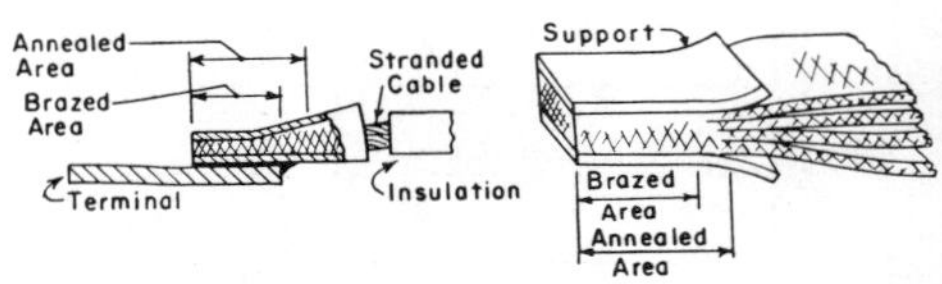

Fig. 6.28—Recommended brazed joints.

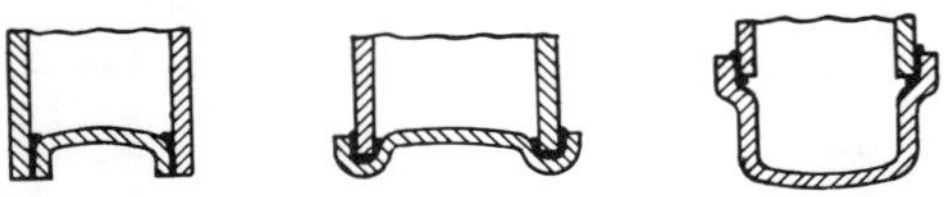

Fig. 6.29—Typical brazed joints for pressure-tight containers.

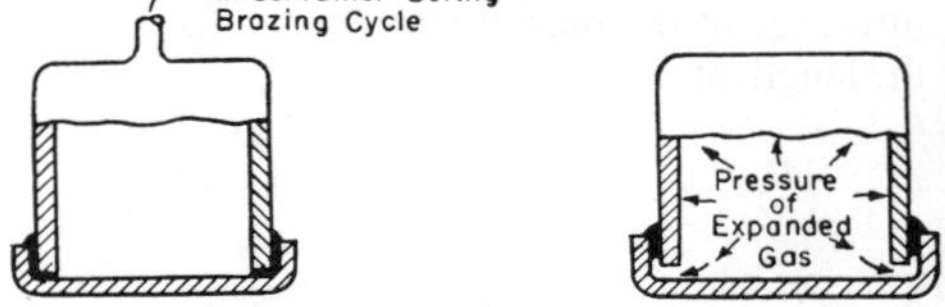

Fig. 6.30—Venting of solid joints during brazing.

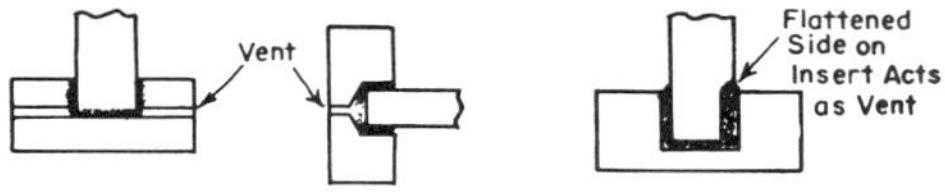

Fig. 6.31—Venting of solid joints during brazing.

As an illustration, consider the extreme case shown in Fig. 6.25. This figure shows a copper conductor of one square inch cross section containing a joint brazed with BCuP-5 brazing filler metal using an 0.005 in. (0.127 mm) joint clearance.

R_{Cu} = Resistance of copper
$= 6.79 \times 10^{-7}$ ohms/in./in.^{2} IACS at 20 °C
R_{FM} = Resistance of filler metal
$$= \left(\frac{1}{C_{FM}}\right) R_{Cu}$$
Where C_{FM} = conductivity of filler metal
= approximately 10 percent for BCuP-5.
Therefore,
$$R_{FM} = \left(\frac{1}{0.10}\right) 6.79 \times 10^{-7} = 67.9 \times 10^{-7} \text{ ohms/in./in.}^2$$
l = Joint clearance = Electrical path
= 0.005 in. for BCuP-5 through joint
Let L = Length of copper conductor + length brazed butt joint
(assume $L = 1.0$ in.)
Then total resistance $= R_T$
$$= (L - l) R_{Cu} + 1 R_{FM}$$
$$R_T = (0.995)\ 6.79 \times 10^{-7} + (0.005)\ 67.9 \times 10^{-7}$$
$$R_T = 7.10 \times 10^{-7} \text{ ohms/in./in.}^2$$
Percent resistance raised by butt joint in 1 in. of
$$\text{conductor} = \left(\frac{R_T - R_{Cu}}{R_{Cu}}\right) \times 100 = 4\tfrac{1}{2}\%$$

As the length of the conductor increases, the influence of the brazed joint decreases. Thus, if the length of conductor were 2 in. instead of 1 in., the increase in electrical resistance would be 2.25%.

From a practical standpoint it must be recognized that there will be a certain number of voids which will cut the effective area of the electrical path. For this reason lap joints are recommended where design will permit.

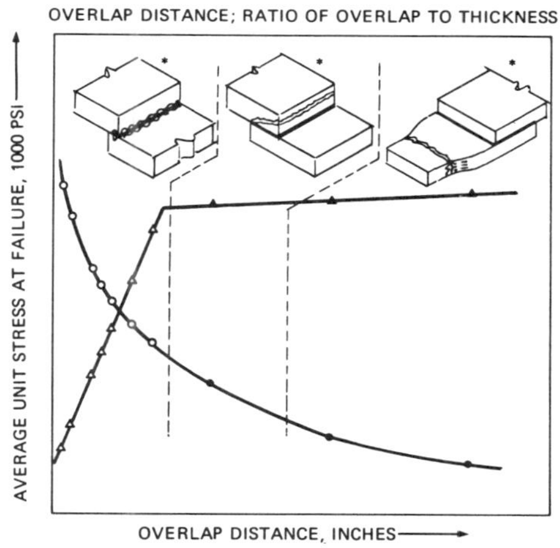

Fig. 6.32—Average unit shear stress (circles) and average unit tensile strength (triangles) as functions of overlap distance. (Open symbols represent failure in the filler metal; filled symbols represent failure in the base metal.)

A lap length of 1.5 times the thickness of the thinner member of the joint will give a resistance approximately equal to the same length in solid copper (Fig. 6.26). This is a rule of thumb which has been used extensively with good results. A longer lap may be used as necessary for convenience or for some special reason.

Other joint designs for electrical applications are shown in Fig. 6.27. In applications where leads must have flexibility to prevent breakage, designs such as shown in Fig. 6.28 may be used.

PRESSURE TIGHTNESS (PRESSURE OR VACUUM)

Joints in brazements required to be pressure tight should be of the lap (shear) type wherever possible. This type of joint provides the large braze area (interface) with less chance of leakage through the joint (Fig. 6.29).

The joint clearance is an important factor in obtaining the best quality pressure-tight joint. It is generally best to use the lower side of the clearance range (Table 6.1), since a joint having fewer inclusions and open passageways can be obtained. Most problems result from excessive clearances, poor flowing filler metal-base metal combinations, and poor brazing technique.

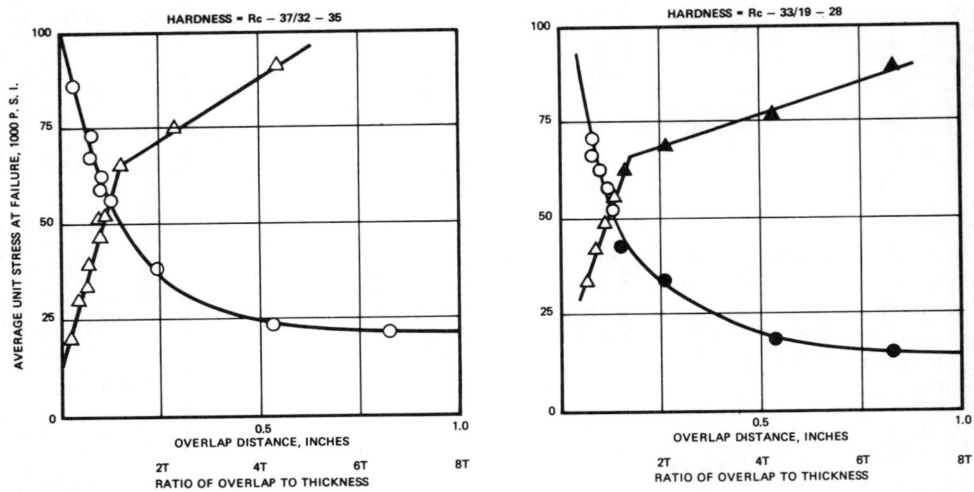

Fig. 6.33—Shear test data employing AWS Standard Method. Base metal AISI 410 stainless steel; filler metal: BNi-1; brazing method: furnace, 2150°F(1177°C), 30 min. pure dry hydrogen.

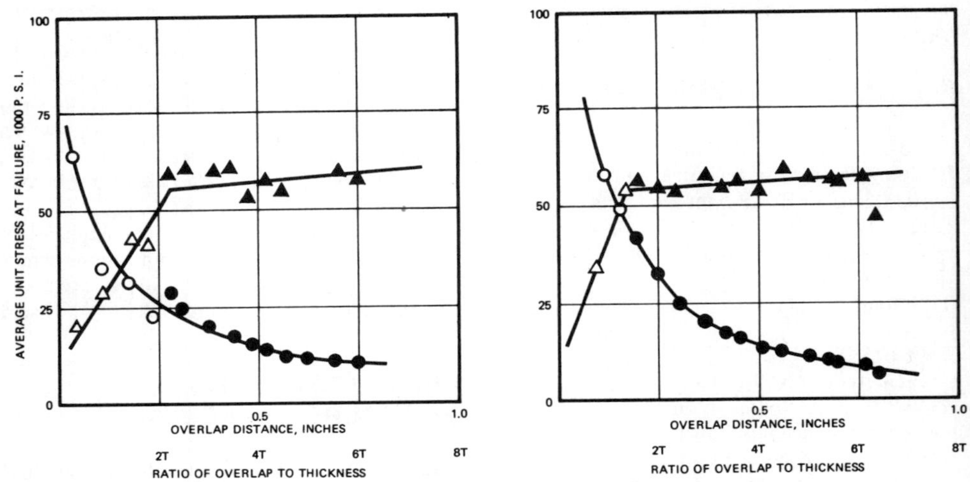

Fig. 6.34—Shear test data employing AWS Standard Method. Base metal: SAE 1010 mild steel; filler metal: BAg-1; brazing method: torch.

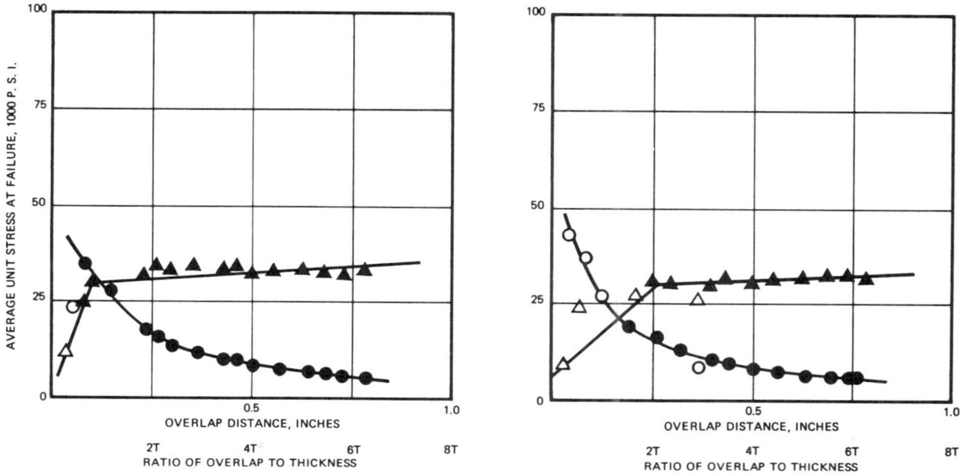

Fig. 6.35—Shear test data employing AWS Standard Method. Base metal: copper; filler metal: BAg-1; brazing method: torch.

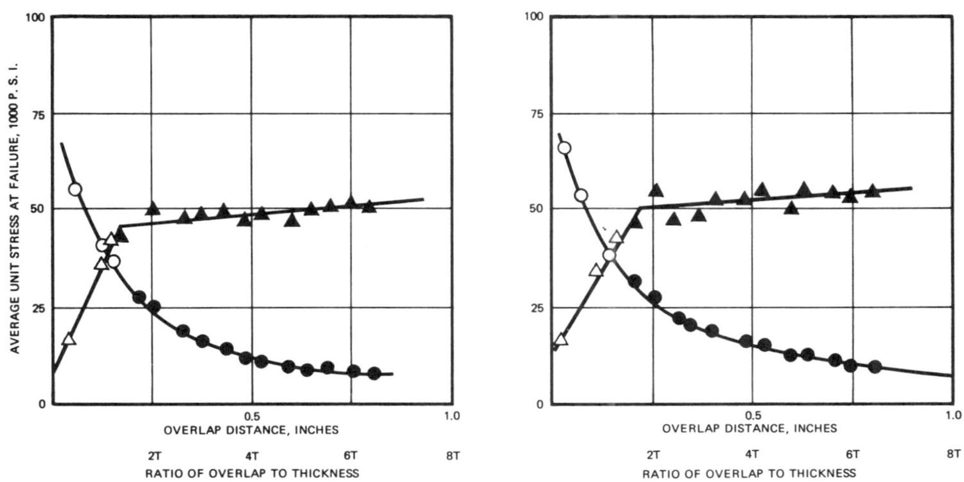

Fig. 6.36—Shear test data employing AWS Standard Method. Base metal: SAE 1010 mild steel; filler metal BCu-1; brazing method: furnace, 2050 °F (1121 °C), 30 min, pure dry hydrogen.

High pressure brazements are currently being fabricated satisfactorily when proper attention is paid to joint design and properties of the filler metals and base metals. The principal factor other than strength to be considered in the design of pressure- or vacuum-tight assemblies is the importance of proper venting. The heat from the brazing operation expands the air or gases within the closed assembly so rapidly that unless the assembly is well vented (open to the atmosphere) it is likely to be forced apart. At the same time, forces may act on the filler metal entering and flowing through the joint that tend to minimize the effect of capillary action (Fig. 6.30). If the unvented brazement survives the expanding gases as the part is heated to the brazing temperature, the cooling of the brazement to the solidus temperature of the filler metal (where the filler metal is completely solid) will usually provide sufficient forces acting on the filler metal in the joint to pull the fillets all to the inside. Air passages will be left through the joint area from the outside directly to the inside of the unvented brazement.

A similar problem exists when brazing in a vacuum atmosphere where desorption (outgassing) on increasing temperature or absorption on decreasing temperature can result in leakage through the joint if venting is not provided. This problem is independent of the ones caused on pumping down to a vacuum when no venting is provided.

Dead-end holes may be considered as small pressure containers. Figure 6.31 illustrates a means of designing joints with vents to eliminate the buildup of pressure during the brazing cycle. Similarly, any brazement that is designed with closed segments or areas can be subject to problems just mentioned. Thus, it is essential to have adequate venting of all enclosed containers or passages to assure good quality joints.

TESTING OF BRAZED JOINTS FOR DESIGN DATA

The importance of standardizing a method for evaluating the strength of brazed joints cannot be overemphasized. Different designs of test specimens will result in different sets of data. It is readily apparent in Fig. 6.32 that the "apparent joint strength" at a very low overlap distance (low T overlap where the overlap in T = overlap distance ÷ thickness of thinner sheet member) is high in comparison to the long overlap. Thus, two laboratories that each test a single overlap distance may be testing at opposite ends of the curve, and as a result come up with different conclusions. It is therefore highly desirable to use the entire usable overlap range of the curve to obtain adequate data.

The most sensitive portion of the curve is the low overlap portion where the apparent joint strength (average unit stress) is the highest. At this same portion of the curve the most scatter (or variation in results) will occur, as the joint is very sensitive to all the factors affecting the brazing of the joint.

The designer, on the other hand, is looking for the load-carrying capacity of the joint and thus is most interested in the right-hand portion of the base metal curve. The brazement should be designed, where possible, so that the failure occurs in the base metal and without an excessive overlap. Figures 6.33, 6.34, 6.35, and 6.36 give data on specific materials obtained using the standard method. The two curves for each combination show the results obtained by two different laboratories.

For further information refer to AWS C3.2, Standard Method For Evaluating the Strength of Brazed Joints, which follows.

STANDARD METHOD FOR EVALUATING THE STRENGTH OF BRAZED JOINTS

1. INTRODUCTION

This publication describes a standard method of evaluating the strength of brazed joints. The method of performing the test is presented as well as the method of representing the data. The method will be especially valuable in standardizing practices in various organizations so that their results can be compared with greater validity in such areas as filler metal development, process development, and study of process parameters. It may be of some assistance to designers provided due caution is exercised in applying the results.

2. SPECIMEN PREPARATION, BRAZING AND POSTBRAZING PROCEDURES

The procedure for making and brazing the specimens shall be carried out according to the following instructions:

2.1 General Instructions

2.1.1 Sheet nominally ⅛ in. thick shall be utilized.

2.1.2 For each set of conditions at least 10 specimens at various overlap distances shall be brazed and tested. The following overlap distances shall be included:

Overlap as Ratio of Overlap to Thickness	Actual Overlap in Inches
¼	0.031
⅜	0.047
½	0.062
¾	0.094
1	0.125
2	0.25
4	0.5
6	0.75

Overlap distances shall be chosen so that approximately 2/3 of all the specimens tested shall fail in the filler metal, unless this ratio cannot be achieved within the dimensional specifications for the specimens. It will probably be necessary to braze a few specimens over a wide range of overlap distances, and determine from the location of failure the proper overlap distances for the remainder of the tests.

2.1.3 All specimens shall be made as uniformly and as accurately as possible.

2.2 Specimen Preparation Procedure

2.2.1 Blanks for brazing shall be nominally 5 in. x 1⅛ in. rectangles.

2.2.2 On the submitted blanks, drill a hole in one end and grind the other end square as shown in Figure 1.

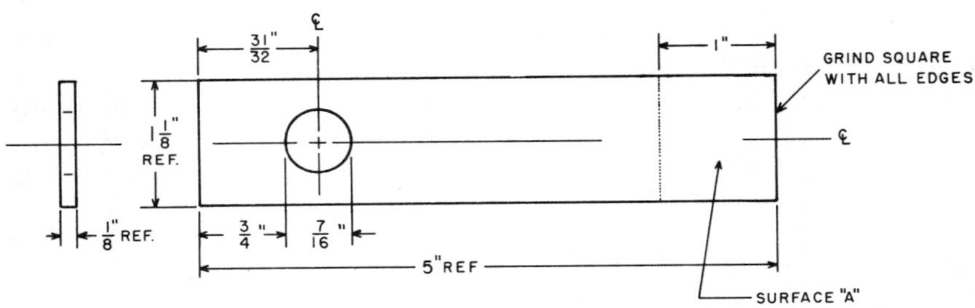

NOTES

1. Surface "A" must be flat and corners sharp. No rounding of edges on mating surfaces.
2. Surface "A" to be grit blasted, one side only.
3. Tolerances: ± 1/32".

Fig. 1 – Single-lap shear test bar leg.

2.2.3 Roughen the mating surfaces of the blanks by grit blasting with a clean cast iron grit (not sand or alumina) of approximately 60 mesh or by acid etching. State which is used in data sheet.

2.2.4 Specimens to be torch brazed should be fluxed on the faying surfaces.

2.2.5 Tack weld blanks together lightly by gas tungsten-arc welding with contact clearance. Tack welds should be made on each side of the test specimen either at the center of the lap or at each end as shown in Figure 2. Use fixture similar to that shown in Figure 3. Alignment in all directions must be held closely.

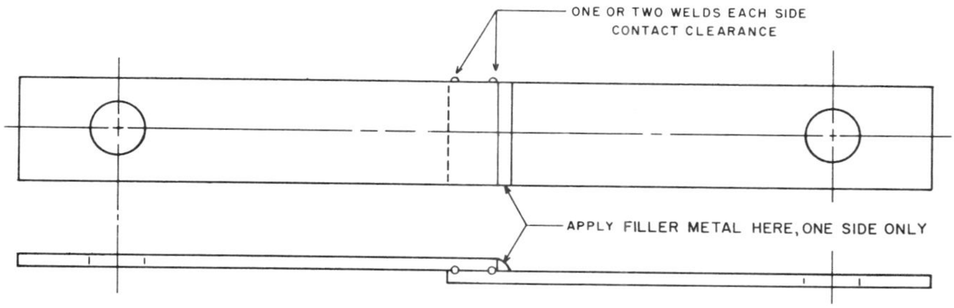

NOTE
Apply filler metal in powder or wire form where shown, and secure with cement. Quantity of filler metal to vary with overlap.

Fig. 2 – Method of applying filler metal.

2.2.6 On furnace-brazed specimens apply filler metal as shown in Figure 2. Hold filler metal in place with a cement. Amount of filler metal shall vary with the different overlaps.

2.3 Brazing Procedure

Specimens shall be brazed with the heat source and procedures under study. The best practice possible shall be utilized. When flux has been utilized, it should be removed after brazing with an appropriate cleaning procedure.

2.4 Postbrazing Procedure

2.4.1 After brazing, all specimens should be ground with a hand grinder or belt sander to remove surplus filler metal. Fillets should be filed, ground, or machine to 1/64 in. max. radius if necessary.

2.4.2 Three hardness readings shall be taken near the gage section from the portion to be machined away.

2.4.3 Edges of specimen shall be machined to produce a gage area as shown in Figure 4.

3. RECORDING OF DATA

After specimens are brazed, details of the specimen preparation and brazing operations should be accurately recorded on a data sheet such as Table I.

TABLE I — DATA SHEET FOR DETAILS OF SINGLE-LAP SPECIMEN PREPARATION AND BRAZING

Base metal__________ Date__________

Base metal condition and hardness before brazing__________

Filler metal__________

Surface preparation method__________

Joint clearance: Before brazing__________ After brazing__________

Hardness after brazing__________

Furnace-brazed specimens

Type of furnace__________

Type of retort (bell, tube, muffle)__________ Size__________

Seal geometry and type of seal__________

Flow rate (retort vol. per hr)__________

Atmosphere__________ Dew point__________ Where taken__________

Vacuum: Pressure__________ Rate of decay, microns/min__________

Brazing temperature__________ How measured__________

Heating time, minutes sol. to liq.__________ Liq. to braze temp.__________

Time at temperature__________

Cooling time, min., Brazing temp. to sol.__________

Torch-brazed specimens

Type of flux__________

Type, size of torch__________ Tip size__________

Length of flame__________ Oxid. or reducing flame__________

Approximate brazing time__________

Other heating methods. Describe and give pertinent details__________

____

____

Postbraze heat treatment. Describe and give hardness after heat treatment__________

____

Remarks __________

____

____

Company__________ Engineer__________

Fig. 3 — Fixture for tack welding single-lap shear specimen.

4. MECHANICAL TESTING

4.1 Prior to testing, all specimens shall be measured for overlap distance, width, and thickness (A, w, and t, respectively, on Figure 4).

4.2 All specimens shall be tension tested. The breaking load and the location of failure, whether through the filler metal or through the base metal, shall be noted.

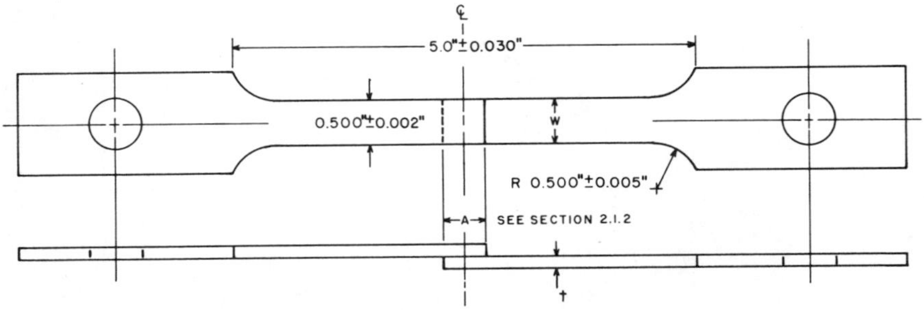

Fig. 4 — Finished single-lap shear test specimen.

5. PROCESSING AND PRESENTATION OF DATA

5.1 For each specimen, compute average unit shear stress in the filler metal at failure and average unit tensile stress in the base metal at failure as follows (see Figure 4):

$$\text{Average unit shear stress} = \frac{\text{Breaking load}}{A \cdot w}$$

$$\text{Average unit tensile stress} = \frac{\text{Breaking load}}{w \cdot t}$$

5.2 Plots of average unit shear stress in the filler metal and average unit tensile stress in the base metal as functions of overlap distance shall be prepared as in Figure 5. Overlap distance shall be represented by two scales as shown, one in terms of overlap distance in inches and the other in terms of ratio of overlap distance to sheet thickness. The shear stress values shall be plotted as circles, and the tensile stress values as triangles. The location of failure is designated by whether or not the symbols are filled; open symbols represent failure through the filler metal, and filled symbols represent failure through the base metal.

5.3 An optional treatment is to determine the equations of the curves of average unit shear stress and average unit tensile stress versus overlap distance (see Figure 5). To do this, assume an exponential function for the shear stress of the form:

$$Y = aX^b$$

where Y is the average unit shear stress, X is the overlap, and a and b are the intercept and slope, respectively, on a log-log plot. For the

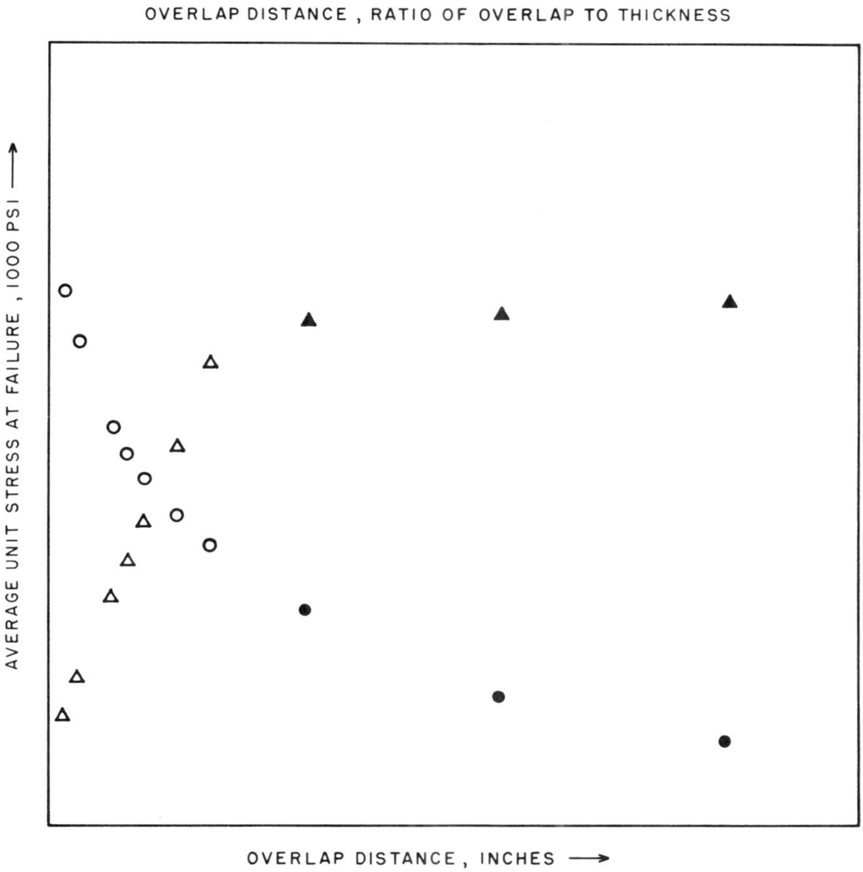

Fig. 5 — Average unit shear stress (circles) and average unit tensile stress (triangles) as functions of overlap distance. (Open symbols represent failure in the filler metal; filled symbols represent failure in the base metal.)

tensile stress, assume a two-straight-line curve, line 1 for low overlaps and line 2 for high overlaps. It has been determined that these equations best fit such data.

CHAPTER 7

PRECLEANING AND SURFACE PREPARATION

Clean, oxide-free surfaces are imperative to ensure sound brazed joints of uniform quality. Uniform capillary attraction may be obtained only when all grease, oil, dirt, and oxides have been removed from both the filler metal and the base metal before brazing. Although many fluxes have some cleaning effect, this is not the primary reason for their use, and complete reliance should not be placed upon them for this function.

Fluxes are used primarily to prevent the formation of oxides during brazing, to reduce the surface tension of the filler metal, and to form a protective covering of slag over the solidifying brazed joint. Their effectiveness in removing existing oxides is only slight and incidental (see chapter 4).

The length of time that cleaning remains effective depends on the metals involved, the atmospheric conditions, the amount of handling the parts may receive, the manner of storage, and similar factors. Therefore, it is recommended that brazing be done as soon as possible after the material has been cleaned.

Cleaning is commonly divided into two categories, chemical and mechanical. Chemical surface cleaning methods vary from simple manual immersions to complex multistage operations. Nine chemical cleaning methods commonly in use are:

1. Solvent cleaning: petroleum solvents or chlorinated hydrocarbons
2. Vapor degreasing: stabilized trichloroethylene or stabilized perchloroethylene
3. Alkaline cleaning: commercial mixtures of silicates, phosphates, carbonates, detergents, soaps, wetting agents, and in some cases hydroxides
4. Emulsion cleaning: mixtures of hydrocarbons, fatty acids, wetting agents, and surface activators
5. Electrolytic cleaning: anodic and cathodic
6. Acid cleaning: phosphate type acid cleaners
7. Acid pickling: sulfuric, nitric, and hydrochloric
8. Salt bath pickling: electrolytic and nonelectrolytic
9. Ultrasonic cleaning

The selection of the chemical cleaning agent depends upon the nature of the contaminant, the base metal, the surface condition, and the joint design. Many commercial products are available for use with the various cleaning methods by a dip or a spray process. Regardless of the cleaning agent or the method used, it is very important that all residue or surface film be removed from the cleaned parts by adequate rinsing to prevent the formation of other equally undesirable films on the faying surfaces.

Mechanical cleaning methods, such as grinding, filing, machining, blasting, and wire brushing, are also used to remove objectionable surface conditions and roughen faying surfaces in preparation for brazing. If a power-driven wire wheel is used, care should be exercised to prevent burnishing. Burnishing can result in surface oxide embedment, which interferes with the proper wetting of the base metal by the filler metal. When rolling, fine grinding, or lapping has produced a base metal surface that is too smooth, the filler metal may not effectively wet the faying surfaces. In this case, the parts can be roughened slightly by rubbing with 30 to 40 grit emery cloth, and better wetting will be obtained. Provision should be made to adjust and control joint clearances after machining and blasting operations. Kerosene or oils used in machining must be removed prior to brazing.

When faying surfaces of parts to be brazed are prepared by blasting techniques there are several factors that should be understood and considered. The purpose of blasting parts to be brazed is to remove any oxide film and to roughen the mating surfaces so that capillary attraction of the brazing filler metal will be increased. The blasting material must be clean and of a substance that will not leave a deposit on the surfaces to be joined that will restrict filler metal flow or impair brazing. The materials should be fragmented rather than spherical so that the blasted parts are lightly roughened rather than peened. The operation should be done in such a way that delicate parts are not distorted or otherwise harmed. Vapor blasting and similar wet blasting methods require care because of possible surface contamination.

Blasting materials, together with their advantages and disadvantages for use in preparing surfaces for brazing, are as follows:

1. Recommended blasting materials
    - A. Chilled cast iron fragmented shot or hardened steel fragmented shot
        1. Advantages
            - a. Roughened faying surface promotes filler metal flow
            - b. Residual iron on surface promotes filler metal flow
        2. Disadvantages
            - a. Contaminates surface with iron that may rust upon standing
    - B. Stainless steel grits or powders
        1. Advantages
            - a. Promotes filler metal flow on stainless steel
            - b. Leaves nonrusting residue
        2. Disadvantages
            - a. Does not clean or roughen surface as readily as fragmented iron or steel grit
2. Not recommended blasting materials and methods
    - A. Nonmetallic materials, such as alumina, zirconia, silica, silicon-carbide, and other similar materials. These materials, although excellent for cleaning and roughening, can be undesirable because they embed themselves in the surface and may retard filler metal flow.
    - B. Wet blasting methods. Vapor blasting and similar wet blasting methods are not generally recommended because of surface contamination with minerals from water, rust inhibitors, and refractory oxides.

For some applications parts to be brazed are precoated by electrodeposition, hot dip, or flame spraying methods. Precoatings are used to retain clean surfaces and prevent the formation of oxides on base metals that contain refractory alloy additions. When brazing dissimilar metals, precoatings reduce the tendency of a filler metal to wet and flow preferentially on one of the base metals. Occasionally, precoatings are used on refractory metals to prevent the rapid diffusion of filler metal constituents into the base metal and the subsequent formation of brittle intermetallic compounds. The selection of the precoating depends on the base metal, filler metal, and brazing technique, and the appropriate chapters should be consulted. For information on the cleaning of specific metals see the chapter covering the particular metal to be brazed.

The flow of filler metal may be restricted to the joint area and prevented from wetting certain areas of the base metal by using suitable stop-off materials. These materials are used to prevent the flow of brazing filler metal in holes, threads, and between mating surfaces that are not to be brazed. Many stop-off materials are commercially available. Manufacturers' literature should be consulted for the proper application of these materials.

When brazing is performed in an atmosphere without a flux, the cleanliness of the parts is of great importance. Although chemical fluxes are capable of handling a greater degree of residual oxides than those generated during the brazing cycle, atmospheres cannot handle these same conditions. Precleaning therefore must be more thorough, and the components in the clean condition must be preserved and protected. One of the more common techniques is to utilize a "clean room" for final handling and assembling of the cleaned parts.

Clean rooms are areas physically separated from the rest of the shop and provided with means for controlling the atmosphere and cleanliness within the area. In some instances a plastic "bubble" or tent, using low pressure circulating air, is used. In almost every instance the clean rooms are air-conditioned. The objective is to reduce or eliminate airborne contamination.

Workers in the clean room may be required to wear special lint-free clothing. Fresh clothing is usually provided daily. Special shoes may also be necesary. To avoid carrying outside contamination into the clean room, it may be advisable to provide dressing rooms between the shop area and the clean room where street clothes are changed for the clean room clothes. Workers in clean rooms may be required to wear hair coverings. This final step represents a high degree of control in cleanliness and is not necessary in every case.

One of the most important factors concerning clothing used in clean rooms is the protection of the parts from contamination by handling with bare hands. Cleaned parts can become so contaminated by bare hands that the flow of brazing filler metal is impaired or prevented even in a satisfactory atmosphere. Therefore, workers in a clean room should be required to wear gloves. White, lint-free, cotton gloves (known as an inspector's-type glove, of reversible gun pattern) have proven most satisfactory. These gloves are changed several times a day depending upon the amount of soil they accumulate. In some operations, the cotton gloves are further supplemented by a pair of nylon gloves worn over the cotton.

Surgeon's rubber gloves have also been used for clean room work. Although they provide protection, it has been found that in working with metals, particularly thin sheet metals with thicknesses in the range of 0.0005 to 0.002 in., (0.013-0.05 mm) the rubber glove may be slit by the sheet metal. The worker may not be aware of this and, in consequence, a large amount of work may be contaminated. The cotton gloves have proven more practical.

Clean rooms have been widely used as final assembly areas for honeycomb brazed assemblies and brazed stainless steel heat exchangers used in connection with jet engine fuel systems. They are also used as assembly areas for brazing electronic components, such as vacuum tubes and solid state devices.

CHAPTER 8

ASSEMBLY AND FIXTURING

Components to be joined by brazing must be assembled in a fixed position relative to each other which must be maintained throughout the brazing cycle. Surfaces to be joined must be properly spaced to preserve joint clearance at room and brazing temperatures in order that the brazing filler metal may fill the joint and achieve maximum properties. The method of assembly and fixturing for brazing therefore can be critical. The cost of assembly tooling and fixturing, plus added processing times for brazing of fixtured assemblies, makes it advantageous to design parts that are easily assembled and self-fixturing during brazing.

Assembly of parts for brazing depends on the brazing technique to be used, the materials being joined, and the configuration of individual details in a brazed component. Sheet metal structures often are assembled by intermittent resistance welding, capacitor-discharge tack welding, gas tungsten arc welding, riveting, strapping with thin foils, etc. Assembling by welding methods requires the use of a flux or a good atmosphere to prevent oxidation of welded areas, which could inhibit braze filler metal flow. Provisions are sometimes made for the portion of the structure that is welded or strapped to be removed by machining after brazing. Aluminum components to be brazed are normally assembled through the use of tabs and slots that are interlocked at assembly.

Cylindrical parts, tubing, and solid members are assembled for brazing in a variety of methods. The assembly method used should, if possible, produce a uniform clearance that can be maintained throughout the brazing operation. Some of the methods employed for assembling cylindrical parts are: staking, expanding or flaring, spinning or swaging, knurling, and dimpling.

More complex structures, as found in the aerospace industry, usually require elaborate assembly methods. First, a prerequisite is that the components in a structure being brazed must have an acceptable braze joint clearance at brazing temperature consistent with the brazing filler metal, base metal, and brazing process to be used. See Table 6.1 for recommended joint clearances. Assembly to assure such clearances being obtained is accomplished using shims — wire, ribbon, or screen materials compatible with base materials at the joint. Sometimes dimpling, pinning, and crimping are used to locate or fit details within a component, as with turbine blades requiring the use of sophisticated assembly fixturing.

Assembly for brazing must be performed, in many instances, taking into account that air passages, cooling holes, etc., must be free of brazing filler metal upon completion of the brazing operation. Therefore a variety of braze stop-off methods and filler metal placement techniques are used to prevent the unwanted flow of the brazing filler metal into areas of the braze assembly.

Honeycomb structures, sandwich type and open face, are assembled by selective resistance welding the honeycomb to the brazing substrate, with or without preplaced brazing alloy at the joint. Expansion rings are also used for fixturing for open-faced honeycomb.

Some examples of methods used to assemble

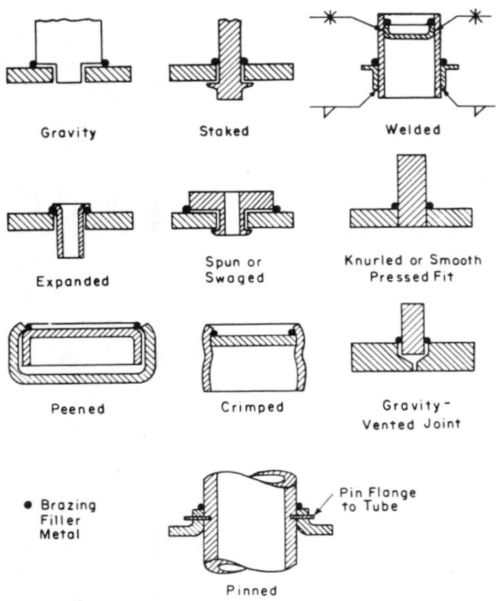

Fig. 8.1—Typical self-jigging brazing assemblies.

parts for brazing without the use of auxiliary fixtures are shown in Fig. 8.1. In these cases, fixturing for brazing is accomplished by placing the assembled part on a plate, rack, or tray and subjecting the part to the prescribed brazing cycle. In some cases, however, self-fixtured parts may require the use of auxiliary fixtures due to the complexity of the part, location of braze joints, and mass distribution. Auxiliary fixturing may consist of clamps, springs, support blocks, channels, rings, etc., to keep parts in proper attitude for brazing. Mild steel, stainless steel, machinable ceramics, and graphite are often used for auxiliary fixturing.

Brazing of complex parts often results in the need for complex fixtures — to locate part details, to maintain surface contours, and in special cases to creep form the part to dimensional requirements during the brazing cycle. Creep forming consists of applying external loading to the braze fixture to make contained components conform to the contour of the fixture.

Selecting the fixture best suited for a particular brazing application requires considering, among other things, the choice of fixture material. The choice must be based on (1) the brazing method to be used, either furnace, induction, torch, dip, etc., and (2) the base materials in the assembly and the required brazing temperature and atmosphere. In vacuum brazing, the only limitations on fixture material are that they be stable at brazing temperature, possess compatible thermal expansion properties, and contain no materials that could outgas and contaminate the brazing atmosphere. Whenever metal fixtures are used for brazing, fixture areas contacting the part being brazed should be adequately protected to prevent brazing the part to the fixture. Brazing fixture materials must be selected that do not interact with the brazing atmosphere to produce undesirable reactions detrimental to braze flow or to the mechanical properties of the base material. For example, carbon or graphite will react with the water vapor in a hydrogen atmosphere to produce methane gas.

Induction brazing often requires that fixturing be located in the proximity of the inductor. When this situation exists, the choice of materials for the fixture is limited to those materials that remain relatively unaffected by the induced field, such as transite, fiberglass, and ceramics. With a few exceptions, depending on the base metals and frequency of the inductor, the use of metals readily heated by the induced field should be avoided.

In salt bath brazing, fixtures when used should be free of moisture and be positive in their holding action because the buoyancy effect of the molten salt must be considered. Also, the fixture material must not react with the salt bath. Mass effect, due to the fast cooling rates associated with salt bath brazing, should be considered and components and fixture preheated if necessary.

Due to the high thermal expansion of aluminum and magnesium, fixtures for these materials must be designed with spring or some other type of relief. Nickel base superalloys and stainless steels are generally used for spring materials. Nickel base superalloys are preferred because they will retain more of their spring characteristics at brazing temperature and have better corrosion resistance. Fixtures when required for torch brazing must not interfere with the torch flame, operator's view, or application of the brazing filler metal.

As applications requiring braze fixturing vary widely, it is impossible to be very specific about designs. However, basic rules to follow in metallic fixture selection are as follows:

1. Keep fixtures as simple as possible. Use thin sections consistent with rigidity and durability required.

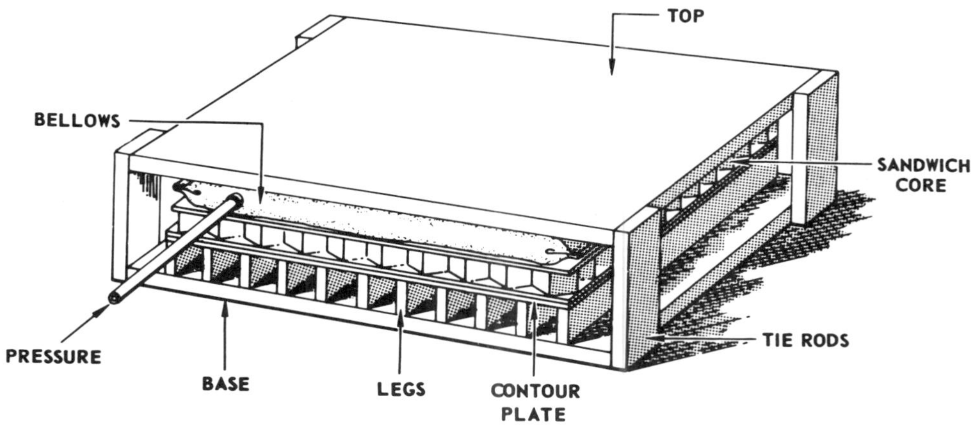

Fig. 8.2—Pressurized bellows fixture.

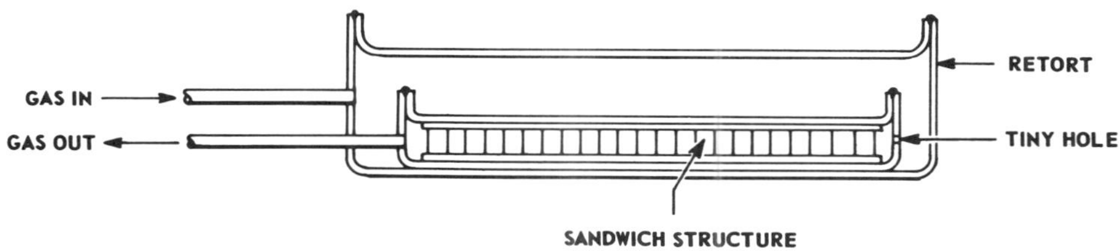

Fig. 8.3—Pressure differential fixturing.

2. Avoid use of bolts or screws in fixtures as they tend to relax upon heating and often pressure weld in place.

3. Springs or clamps must be designed to withstand brazing temperatures if used in the heated area, otherwise relaxation will take place in the brazing operation.

4. Fabricate fixtures to a consistent uniform thickness. Avoid heavy sections.

5. In most cases, avoid the use of dissimilar materials where differences in thermal expansion could affect assembly dimensions. An exception in this case is where expansion differences are an advantage.

6. Fixtures should be subjected to the brazing environment and temperatures before they are used for brazing hardware to ensure stability and relieve stresses.

7. Fixtures should be designed for easy inspection; dimensionally inspect fixtures often.

8. Nickel base superalloys containing aluminum and titanium develop protective oxides upon heating in air which helps to prevent joining the parts to fixtures.

9. Avoid use of fixtures made of materials that might react at elevated temperatures with the assembly being brazed while in intimate contact. For example, never use nickel base alloy fixtures for brazing titanium or vice versa, since the nickel titanium eutectic forms at 1729 °F (943 °C), which is lower than brazing temperatures normally used for titanium or nickel base alloys.

10. Fixturing can be made to apply localized pressure to brazed components through use of flexible refractory blanket-type pads, steel balls, or pins.

Brazing of assemblies, such as contours, which require that close tolerances be maintained dictates the need for very stable fixturing. Metallic fixturing normally lacks the stability required for repeated use; therefore, for holding close tolerances, ceramic fixturing is often used. Castable ceramics such as fused silica are excellent; they possess superior dimensional stability and can be cast to close tolerances without machining. Primary drawbacks are their poor thermal conductivity, creating extended heat-up and cool-down times, and porosity, which makes fixtures hygroscopic when not in use. Usually a bake-out cycle to remove moisture is required before use. Alumina is also used for braze fixturing, but it is difficult to fabricate to close tolerances, is hard to machine, and has poor thermal shock properties.

Carbon (graphite) is used for braze fixturing of honeycomb components and contoured sheet metal assemblies. Carbon has good thermal conductivity and dimensional stability but also has drawbacks: low abrasion resistance, diffusion at elevated temperatures into some materials being brazed (carburization), and reaction with water vapor in hydrogen atmospheres to produce contaminants. Silicon carbide is also used for braze fixturing, but it has lower thermal conductivity than carbon and the same drawbacks.

Fixturing for brazing honeycomb sandwich structures is usually based on pressure differentials or external pressurization. Fixturing using pressure differentials involves using thin alloy steel welded retort bags that are evacuated. External pressurization involves the use of a thin metal bellows placed between the assembly and a rigid backup structure. During the brazing cycle, the bellows is inflated to a few pounds per square inch. Figures 8.2 and 8.3 show these fixturing methods.

SUMMARY

The assembly of parts for brazing is very important since brazing is often used to reduce costs and improve quality. Assembly methods vary with the brazing process to be used, the materials being joined, and the complexity of the part. Designs which are self-locating for brazing upon assembly are preferable. In many cases, however, some form of auxiliary tooling is required to locate details within components prior to brazing.

The fixturing of parts for brazing can be as simple as a clamp or spring or as intricate as a multicomponent superalloy structure costing hundreds of dollars. The type of fixturing selected must be compatible with the brazing method employed, with base materials in the brazed assembly, with dimensional requirements for the finished assembly, and with brazing temperature and atmosphere used. In any case, the fixturing used should be clean, as simple as possible, subjected to a simulated brazing cycle before use with parts, and metallurgically compatible with the base metals being brazed.

CHAPTER 9

TECHNIQUE OF BRAZING

SECTION I
FUNDAMENTALS

A basic knowledge of the many variables affecting the quality of a brazed joint is of fundamental importance in developing a technique for brazing. These factors include:

1. Selection of base metals
2. Joint design
3. Surface preparation
4. Jigging
5. Selection of flux or atmosphere
6. Use of correct filler metal
7. Method of heating

Correct brazing techniques will assure satisfactory results when combined with sound engineering. It is advisable to heat the joint to brazing temperature as quickly as possible consistent with uniform heating of assemblies, but care should be exercised to avoid localized overheating.

BRAZING PROCESSES AND FUNDAMENTALS

Torch Brazing

The proper adjustment of the torch flame is essential for satisfactory results in torch brazing. (See Fig. 9.1.) Generally, a slightly reducing flame is desirable. One exception is in the case of undeoxidized copper, where a neutral or slightly oxidizing flame is employed to prevent rupture of the copper during heating. Reaction of the gases in a reducing flame will result in small pockets of steam forming inside the metal from reduction of the internal oxides.

The adjustment of the oxyacetylene flame is relatively simple and may be maintained by observing the characteristics of the flame. The proper adjustment of flames of other types of fuel gases is more difficult since no marked visual change in flame characteristics takes place. The adjustment can be made and controlled by the use of flowmeters. The outer envelope of the flame, not the inner cone, should be applied to the work. This technique is different from that used in gas welding where heat is concentrated in the specific area to be welded. Torch preheating of adjacent members will depend upon the size, shape, and thermal conductivity of the metals and the type of joint. Improper heating, primarily due to poor manipulation of the torch, may cause cracking, excessive interaction, or oxidation of the brazing filler metals, as well as incomplete joint penetration.

In addition to proper heating to offset high thermal conductivity of joint members and stress cracking due to thermal shock, it is important that the joint be brought to a uniform temperature within the brazing temperature range so that the filler metal will flow freely and fill the joint. Overheating must be avoided; this is best accomplished by using an active flux with a melting temperature not too far below the brazing temperature range of the filler metal. The melting of the flux can then serve as an indicator of the approach to the proper brazing temperature. If the brazing filler metal is to be face fed, particular

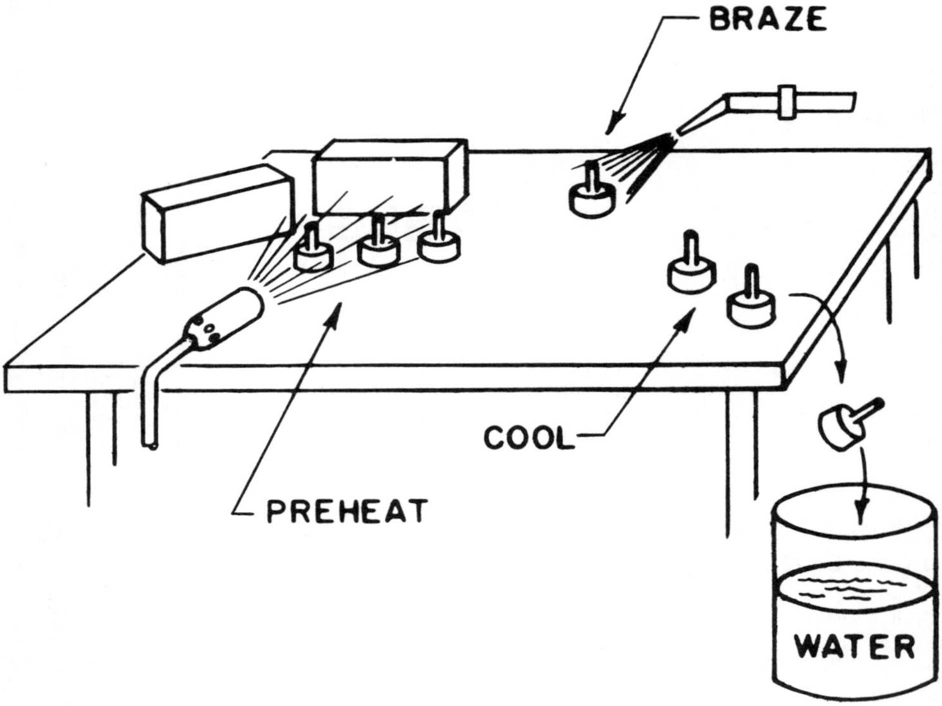

Fig. 9.1—Torch brazing small parts. Parts are first preheated using gas-air torch, and then brazed using an oxyacetylene flame and face fed filler metal. They are allowed to air-cool to approximately 800 °F (427 °C) to allow the filler metal to solidify and are finally quenched to remove brazing flux.

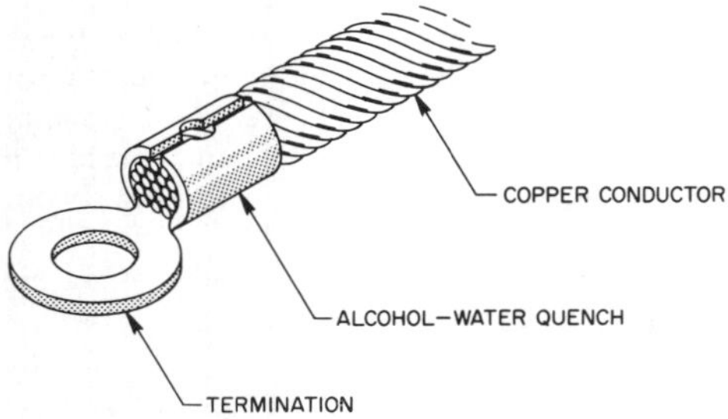

Fig. 9.2—Copper lug brazed to flexible copper conductor using BCuP filler metal. To preserve the bright finish on the copper, the assembly is dipped while hot into an 80% alcohol-20% water solution.

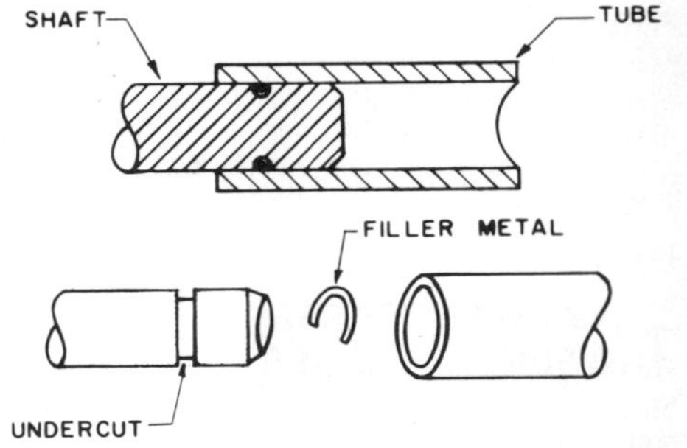

Fig. 9.3—Enclosed filler metal in the form of a snap ring is placed in a groove of the shaft prior to brazing.

attention should be paid to the appearance of the flux. As soon as the flux is completely fluid, the filler metal is touched to the joint; the filler metal is applied until it flows completely through the joint, and heating is then stopped. This technique allows the molten flux to act as a temperature guide, and the part itself supplies the heat to cause the filler metal to melt and flow. It is poor practice to apply the flame directly to the filler metal.

Maintaining a uniform temperature is advisable at the joint. To accomplish this, auxiliary or multiple tip or multiflame torches may be necessary. In mechanized operations, the parts can be moved, oscillated, or rotated as they pass through the heating zone. Alternatively, the flames are moved around the parts. In some operations, torches or burners are located to heat the parts over the entire joint area. In any case, the purpose is to obtain and maintain a uniform metal temperature.

The filler metal is most commonly face fed in manual torch brazing operations. However, in mechanized operations where higher production is desired, proper design will permit the use of preplaced filler metal in the form of rings, washers, shims, paste, or powder. This preplaced filler metal should be positioned with respect to the flame to avoid premature melting.

One interesting example of a postbraze technique used in torch brazing is illustrated in Fig. 9.2. This figure describes the use of dilute alcohol as a quench to improve the surface finish on brazed copper.

Furnace Brazing

Furnace brazing is used extensively, and furnace manufacturers build furnaces specifically for this method of metal joining. The process is not limited to any one base metal nor to any one brazing filler metal, but is governed by a combination of factors.

Equipment — The brazing furnace may be electric, gas, or oil fired. The type of furnace selected must be of suitable construction to provide proper control of temperature, time, and atmosphere. Suitably calibrated controls for both temperature and atmosphere for the service intended are required. In conveyor-type furnaces, the speed through the furnace must be controlled because this governs the time at temperature for a particular brazing operation.

Design of Joints — Joints should be specifically designed for furnace brazing. If possible, the brazing filler metal should be preplaced at the joint or adjacent to the heaviest member. This will minimize the tendency of the light gauge filler metal to melt and flow before the parts to be brazed have reached the brazing temperature. Figure 9.3 illustrates this design.

Since the entire assembly is raised to the brazing temperature, specific consideration must be given to metal expansion to obtain correct joint clearances at brazing temperature. This differs from localized heating methods where only the joint and immediately adjacent areas are heated.

Assemblies involving lightweight sections to be joined to heavy sections will often braze more uniformly if the lighter section is shielded from the intense radiation of the furnace chamber. Such shielding will allow the heavy section to come to proper temperature at approximately the same time as the thin section and result in better quality joints.

For furnace brazing it is desirable to design the assemblies to be self-jigging. The economies available through the use of high production furnace brazing are made possible to a degree by this feature. Figures 9.4 and 9.5 are examples of assemblies designed to take advantage of the self-jigging feature.

Jigs are sometimes used to align the parts for brazing. Since jigs will introduce additional weight into the furnace and must be heated to the brazing temperature, they should be kept as light as possible. Jigging material must, of course, be able to withstand repeated heating without serious loss of shape, etc.

Filler Metals and Fluxes — Filler metal is preplaced in the form of wire, shims, washers, rings, paste, or powder. As mentioned in chapter 4, certain fluxes have shorter life spans at brazing temperature than others, and this characteristic, together with the time of the brazing cycle, should be considered in the selection of a flux. Even when controlled atmospheres are used, it is sometimes advisable to include a small amount of flux in the joint area. This flux addition prevents oxidation that may occur before the assembly reaches the temperature at which the atmosphere is effective. Care must be taken in the selection of the flux to avoid undue damage to the furnace. Volatile fluorides from the fluxes shorten the life of furnace heating elements and wall materials.

Brazing Procedure — The parts to be brazed must be properly cleaned. Flux, if necessary, is then applied to the parts with the preplaced filler metal. No flux should be used in vacuum atmosphere. Protective atmosphere, neutral (N_2), reducing (H_2), or vacuum, must be controlled.

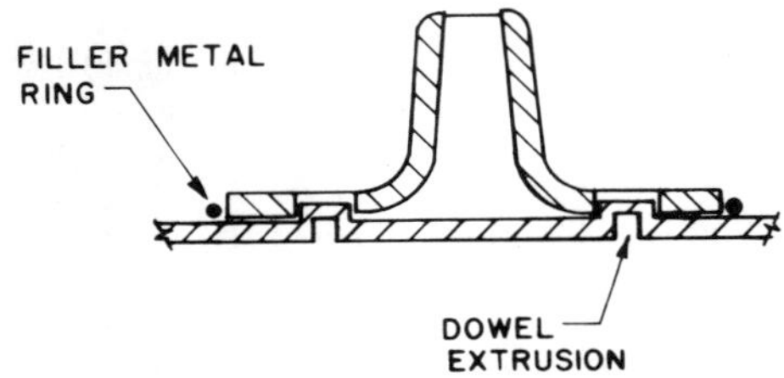

Fig. 9.4—Self-fixturing for furnace brazing utilizes extrusions on one part and matching holes on the other. Filler metal is a preplaced ring.

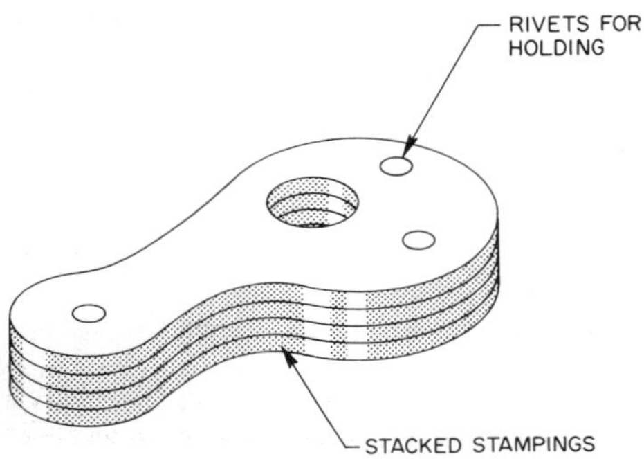

Fig. 9.5—Laminated steel stampings furnace brazed to form a cam. Rivets or pins are used to maintain proper positioning. In some cases tack welding at two or more places on the edges of the stampings has proven equally effective. Filler metal is applied by dipping the positioned assembly into a BCu paste or laying cut lengths of copper at edge of laminations.

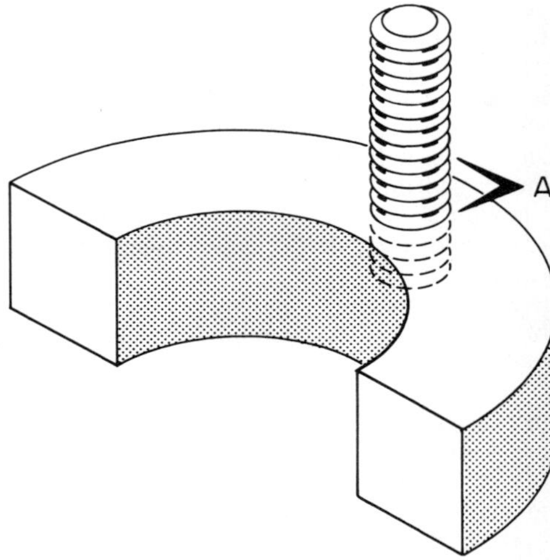

Fig. 9.6—Brazing stop-off applied with a brush to area **a** of the threaded stud suppresses the flow of filler metal up the threads during brazing.

Gases must be dried to a dew point prior to introduction into the furnace. A dew point of −20 °F (−6.7 °C) may be suitable for brazing steel parts with BCu in exothermic atmosphere. For stainless steel a dew point of −75 °F (−58 °C) may be required for brazing in hydrogen. For brazing titanium a vacuum atmosphere at pressures of 10^{-5} torr (0.0015 Pa) may be required. In each case the concentration of oxygen, which would suppress adequate wetting of the parent metal by the filler metal, is adjusted to accomplish the results required.

Time at temperature measurements in furnace brazing are of utmost importance, because in most cases no observations can be made when the filler metal begins to flow. The placement of the thermocouples is of prime importance. The control thermocouple, unless it is actually placed inside the furnace and adjacent to the parts being brazed, will not always indicate actual brazing temperatures. For this reason, it is good practice to use two or more couples. The first thermocouple controls furnace temperature. Another thermocouple placed inside the furnace and in the brazing area will indicate actual temperature in the area of the part. Where very careful control of the heating rate is desired, it is necessary to attach a couple to the actual part being brazed.

When a controlled atmosphere is used in brazing, the part is usually cooled in the same atmosphere to avoid oxidation or exposure to air. In all cases it is necessary to prevent relative movement of the parts until the filler metal has solidified and the joint strength is strong enough to be moved.

In furnace brazing it is frequently desirable to restrict the flow of the filler metal from threaded areas, holes, or fixture contact points. A typical illustration of the use of stop-off materials is shown in Fig. 9.6. In this example the threads on the stud tend to draw the filler metal by capillary attraction, resulting in excess filler metal at the root of the threads. A number of stop-off materials are commercially available. Most of these contain oxides of aluminum, magnesium, or titanium suspended in an organic binder. They are applied by brush or spray. In some cases a strip of masking tape will facilitate closer control when painting stop-off on a part.

Atmospheres — Most furnace brazing is conducted in an atmosphere* designed to prevent oxidation of the parts during the brazing cycle. The principle gases used are:

EXOTHERMIC OR ENDOTHERMIC GASES: Chiefly made by the controlled combustion of natural or synthetic gases with air to form a mixture essentially composed of nitrogen, hydrogen, methane or ethane, carbon dioxide, carbon monoxide, and water vapor.

DISSOCIATED AMMONIA: A gas evolved by heating pure ammonia gas under controlled conditions to produce a product consisting of 75% hydrogen and 25% nitrogen with traces of water vapor.

HYDROGEN: This atmosphere is widely used in the brazing of stainless and heat resistant steel. Generators are available for production at the plant site, and hydrogen is also available liquified in cryogenic containers. Generally cylinders of hydrogen are most economical where consumption is low. In some cases, hydrogen must be dried to reduce the level of residual water vapor.

INERT GASES: Argon is widely used in special brazing applications for furnace brazing. An example is the brazing of honeycomb panels for aircraft utilizing an atmosphere of argon in the sealed "bag" which contains the panel during brazing. Argon is available only in tanks or cylinders, either as a gas or a liquid. Helium, although quite suitable for a brazing atmosphere, is not widely used because of its limited availability. It is a light gas and inert to all metals.

OTHER GASES: Nitrogen blends are useful for brazing where their reducing properties are sufficient for the base and filler metals being used. Nitrogen blends produced by generators may be used for brazing where they have sufficient reducing properties to suppress oxide formation on the base metals.

VACUUM: The use of vacuum brazing furnaces is widely accepted today. Various types of equipment are available both for hot wall (the vacuum is maintained inside a heavy walled retort only, and the heating elements are outside the retort, under partial pressure utilizing a separate pumping system) and cold wall (the heating elements are located within the retort, which remains relatively cool during operation). Vacuum brazing is often the preferred method when working with refractory and reactive metals and heat resistance metals containing elements of titanium and/or aluminum, which are difficult to reduce in other atmosphere brazing operations. Caution must be observed in selection of the filler metal for vacuum brazing (see chapter 4).

*See chapter 4 for further references on the use of atmospheres.

Induction Brazing

Careful consideration must be given to the equipment available, metals to be joined, joint design, induction coil design, filler metal temperatures, fluxing, and jigging. Each of these factors is important in itself and in conjunction with the others. By balancing all the factors properly, a wide range of base metals can be induction brazed.

The selection of induction brazing versus other methods usually is based on the assumption that it is a distinct advantage to heat only a local area of the part immediately adjacent to the joint. This assumption arises from a number of considerations all of which make induction heating attractive:

1. Heating some other part of the assembly may be undesirable (a hardened part might be softened by additional heating)
2. Heating the entire assembly may be uneconomical or undesirable (brazing a cap on a long tube)
3. Selective heating of the joint may prove advantageous in preventing distortion from overheating, or annealing of cold worked areas
4. Very rapid heating of the joint area may be desirable (time may be important and speed of heating may yield a lower cost, all other factors being equal)

Equipment — Induction brazing equipment is of three main types, namely, motor generator, spark gap, and vacuum tube. For classification purposes, it can be said that motor generators are low frequency (up to 10 000 Hz), spark gap units are medium frequency (20 000 to 300 000 Hz), and vacuum tube oscillators are medium-to-high frequency (200 000 to 5 000 000 Hz). Manufacturers of this equipment produce standard power sources in various electrical capacities and couple these power units to worktables designed for general or specific applications. The selection of the proper capacity power unit is dependent on the part to be brazed and the desired production.

An integral part of the induction heating equipment is the inductor coil (or block), called the work coil. Most work coils are constructed simply and can be fabricated by the user. The coils are constructed of copper, usually in the form of tubing, but may be made of a block of copper that has been machined to allow for an internal water passage. It is essential that the work coil be cooled during the brazing cycle, usually by means of circulating water. Small tube units of 0.75 kW capacity may have the work coils air-cooled.

The work coil must be tailored to the heating requirements of the part for maximum efficiency. It is possible to use oversized, general-purpose coils for more than one joint design, within the limits of power available. The number of turns of copper tubing or the thickness of the copper block is dependent on the extent of the heat zone desired. Coil design must allow for corner effect in rectangular parts and surface irregularities in the heat zone.

The gap between the work coil and the part to be brazed is known as the couple. By varying the couple from close (small gap) to loose (wide gap) over different portions of the same part, variations in the amount of heat input can be obtained.

This is very important where a change of section brings both heavy and light masses into the joint area. Both sections can be brought to the brazing temperature uniformly by forming the work coil with the proper coupling for each section. Multiple work coils are available in which several units are brazed at one time.

Induction brazing with high frequency power sources is typified by skin heating, and an allowance must be made for adequate conduction of heat to the area to be brazed. Lower frequency power sources will result in deeper heating and are often used in brazing heavy sections. Sufficient time for the filler metal to flow properly and form good fillets must be provided. It is usually better practice to use multiple work coils utilizing the full capacity of the machine and heat over a reasonable time, such as 30 to 60 s. However, relatively small assemblies can be brazed with heating times of 10 s or less. Larger assemblies may require extensive time.

Where exceptionally large parts are to be heated, an advantage is to construct a shading coil that only partially encloses the work. A shading coil is essentially round and is bent in a manner similar to the configuration of the part. The part is then rotated slowly past the coil. This technique has been successful in the brazing of bottoms to large tanks, as shown in Fig. 9.7.

Often, because of the small nature of a part, it is difficult to prevent overheating of other areas away from the brazed joint. One technique involves the use of a shield as shown in Fig. 9.8. The shield interferes with the transfer of high-frequency energy to the part from the induction coil in the area beneath the shield, keeping it from becoming heated, and confining the heat to the immediate joint area.

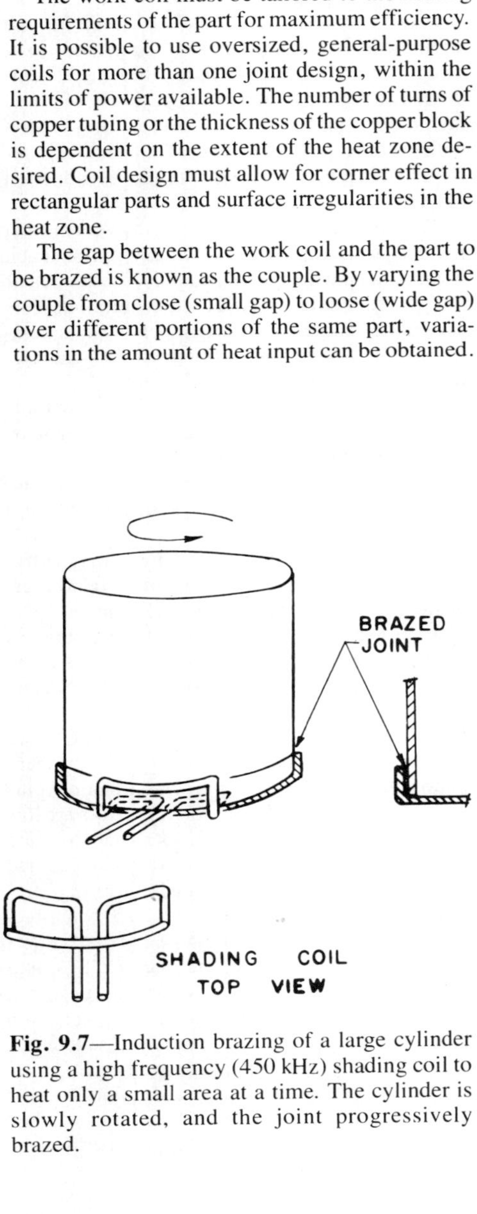

Fig. 9.7—Induction brazing of a large cylinder using a high frequency (450 kHz) shading coil to heat only a small area at a time. The cylinder is slowly rotated, and the joint progressively brazed.

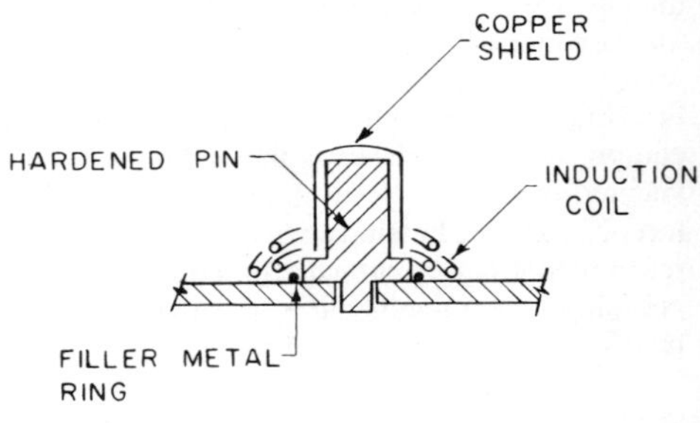

Fig. 9.8—Copper shield blocks effect of induction heating coil during brazing and prevents excessive softening of previously hardened pin.

Design of Joints — Since induction brazing involves localized heating, the design should take advantage of this factor. Subassemblies can be built up by step brazing or by using two separate brazes, using the same filler metal, provided the heated zone does not extend to the previously brazed joint. It is also possible to braze assemblies which contain parts or areas that would be adversely affected by the brazing temperature.

Very close attention must be given to the proper joint clearance with induction brazing because with this process the heat may be induced in only one component of the joint, causing it to heat faster and thereby expand more rapidly than the other. This uneven expansion could produce a very undesirable uneven heat condition during the brazing cycle and should be compensated for in the initial joint clearance. For example, if the joint clearance increases during the brazing cycle it should be compensated for by reducing the initial clearance of the cold parts.

Filler Metals, Fluxes, and Atmospheres — The filler metals are usually preplaced in the form of wire, rings, washers, shims, powder, and paste. Where flux is required because of the filler metals or base metals used, the primary consideration for selection is the brazing temperature of the filler metal. Induction brazing, involving a rapid heating method, minimizes the problem of time at temperature. Induction brazing with an active or protective atmosphere can be performed satisfactorily. Where a protective atmosphere is used, a small amount of flux must sometimes be inserted in the joint area to protect against oxidation until the atmosphere reaches its effective temperature (see Fig. 9.14).

Materials for Jigs and Fixtures — One of the most frequently misunderstood aspects of induction heating is the design of the holding fixtures for the parts being brazed. The use of steel fixtures should be avoided since they are magnetic and heat rapidly inside an induction coil. The best induction fixture materials are nonmagnetic in the area close to the work coil where the fixture touches the part. Ceramic materials are suitable. Ceramics as used for induction heating can be grouped into three main types: (1) natural occurring mineral, (2) manufactured materials in cast form, and (3) manufactured materials in other forms.

The principal types of natural materials usually used for fixtures are called steatites and are available widely from many sources. These materials are easily machined in the natural state. Later they are fired at high heat and change their molecular structure, becoming good insulators with excellent mechanical properties at high temperatures. There is shrinkage during the firing, but this may be controlled. Manufactured materials are also available as castable cements or mixtures of cements and other high temperature materials. They are cast, finished, and then fired at a high temperature to develop their full physical properties. Other materials frequently used are electrical refractories, such as those shapes commonly in use for insulating heating elements in stoves and heaters. These are available from standard electrical suppliers.

Brazing Procedure — The parts should be properly cleaned, fluxed if necessary, and assembled with the filler metal preplaced. This work may be done in advance, but a note of caution is warranted. This type of brazing involves a considerable amount of handling, and there is a possibility that the flux will flake off when it has dried. Completely prepared parts are placed in the work coil in the locating fixture. This fixture should automatically establish and uniformly maintain the amount of electrical coupling between the coil and parts. Automatic timing devices are almost a necessity for economical operation: they permit the brazing operator to establish the time cycle and keep it uniform from piece to piece. Where it is desirable to forcibly cool within the work coil, the timers can be preset to turn on a cooling medium, such as an air blast or water spray, and control its cycle. A water quench should be used only after the braze has cooled sufficiently to develop enough strength to hold the parts together. In most cases this necessitates cooling the part to a temperature below 800 °F (427 °C) before quenching.

Resistance Brazing

Design of Joints — Resistance brazing is most applicable to joints which have relatively simple configuration. It is difficult to obtain uniform current distribution, and therefore uniform heating, if the area to be brazed is large or discontinuous or is very much longer in one dimension. Joints to be resistance brazed should have their parts so disposed that pressure may be applied to them without causing distortion in the parts at brazing temperature. Wherever possible, the parts should be designed to be self-nesting, to eliminate the need for dimensional features in the fixtures. Parts should also be free to move as the filler metal melts and flows.

Brazing Equipment — One common source of

current for resistance brazing is a stepdown transformer whose secondary circuit can furnish sufficient current at low voltage (2-25 V). The current will range from about 50 amperes for small, delicate jobs to many thousands of amperes for larger jobs. Commercial equipment is available for resistance brazing. Spot welding machines of all sizes may be altered slightly for use in resistance brazing, as less pressure and a longer time of current flow is necessary than normally used for spot welding.

Electrodes for resistance brazing are made of high resistance electrical conductors, such as carbon or graphite blocks, tungsten or molybdenum rods or inserts, or even steel in some instances. The heat for brazing is mainly generated in the electrodes and flows to the work by conduction. It is generally unsatisfactory to attempt to use the resistance of the workpieces alone as a source of heat. The current requirements are high, as in spot welding, and it is difficult to allow time for the filler metal to flow without overheating.

Carbon or graphite blocks or rods are most frequently used since brazing filler metals will not wet them as they will metals. Various grades of these materials are available with ranges of resistance and hardness which permit adaption to various jobs. The carbon or graphite is easily worked to the desired shape. These parts, however, must be considered expendable since they waste away quite rapidly.

The pressure used by the spot welding machine, clamps, pliers, or other means must be sufficient to maintain good electrical contact and to hold the pieces firmly together as the filler metal melts. Too high a pressure may crack the carbon electrodes. The pressure must be maintained during the time of current flow, after the current is shut off, and until the joint solidifies sufficiently to permit moving the assembly. The time of current flow will vary from about one second for small, delicate work to several minutes for larger work. This time is usually controlled manually by the operator, who determines when the part is brazed by the temperature and extent of filler metal flow. In general, most resistance brazing operations are manually controlled because the effective resistance of the joints between the contacts and the parts is constantly changing due to the erosion and burning away of the contact points, yielding widely varying current values and resulting temperatures.

Filler Metals and Fluxes — Brazing filler metal in the form of preplaced wire, shims, washers, rings, powder, or paste is used. In a few instances, face feeding is possible. For copper and copper alloys, the BCuP filler metals are most satisfactory since they are self-fluxing. The BAg filler metals may be used, but a flux or atmosphere is necessary. A wet flux is usually applied as a very thin mixture just before the assembly is placed in the brazing fixture. Dry flux is a good insulator and will not permit sufficient current to flow. In some instances, powdered filler metal is mixed with a thinned flux to assure electrical contact and initiate current flow. Each part is dipped in this mixture prior to placing the part in the brazing fixture.

Brazing Procedure — The parts to be brazed must be clean. The parts, brazing filler metal, and flux are assembled and placed in the fixture and pressure applied. As current flows, the electrodes become heated, frequently to incandescence, and the flux and filler metal melt and flow. The current should be adjusted to obtain uniform and rapid heating in the parts. If too much current is used, the electrodes will become too hot and deteriorate. There is also danger of oxidizing or melting the work. If too little current is employed, the time of brazing will be excessive. The current can be adjusted by varying the voltage of the transformer and/or using more of the electrodes or higher resistance electrodes in the circuit. A little experimenting with electrode composition, geometry, and voltage will give the best combination of rapid heating with reasonable electrode life. The pressure should never be interrupted during current flow or arcing will occur and the assembly will frequently be spoiled. The pressure must be maintained after the current is cut off to permit the brazing filler metal to solidify and gain sufficient strength before it is moved. Quenching the parts from an elevated temperature will help flux removal. The control of pressure, current, and time must be mastered for resistance brazing. For example, in Fig. 9.9 too much pressure will prevent the filler metal from entering the joint if face fed. On the other hand, insufficient pressure will not produce a good joint if the filler metal is preplaced as in the illustration. Quenching should only be done after the assembly has cooled sufficiently to permit the braze to hold the parts together. An exception might be the brazing of insulated conductors: in this case, it is advisable to quench the parts rapidly while they are still in the electrodes to prevent overheating of the adjacent insulation. Water cooled clamps also prevent damage to the insulation.

Applications — Resistance brazing is limited in its application. Two important uses are:

1. Joining of large copper leads to transformers in which hand-held tongs are employed using graphite electrodes. The filler metal used usually falls in the BCuP class.
2. Joining electrical contacts to switch components. A modified spot welder is suitable. Electrodes may be copper/tungsten, or tungsten, or molybdenum. Filler metals are generally supplied as clad material on the back side of the electrical contact and are of the BAg series, depending on a specific application.

Dip Brazing

Two methods of dip brazing are: molten metal bath dip brazing and molten chemical (flux) bath dip brazing.

Molten Metal Bath Dip Brazing — This method is usually limited to brazing small assemblies, such as wire connections or metal strips. A crucible, usually made of graphite or ceramics, is heated externally to the required temperature to maintain the brazing filler metal in fluid form. A cover of flux may be maintained over the molten filler metal. The size of the molten bath (crucible) and the heating method must be such that the immersion of parts to be brazed will not lower the temperature of the bath below that necessary for brazing. Parts to be brazed must be clean and are often protected with flux before bathing.

Molten Chemical (Flux) Bath Dip Brazing — This method of brazing requires a suitable container, metal or ceramic, for the flux and a method of heating the flux to maintain it at temperature. The container or flux pot may be heated by a gas flame or electrical resistance. By far the most popular method is to heat the bath by passage of an electric current through it. The resistance of the bath, once it is molten, is sufficient to provide a convenient method of heating. Suitable controls are usually provided, and the temperature is measured by a thermocouple immersed in the molten bath.

To start a bath heated by electric resistance the initial charge of flux must be melted outside the bath and slowly poured into the flux pot. The chilled salt may be remelted by putting a carbon block between the electrodes or melting a portion of the chilled salt with a torch to form a conductive liquid path between the electrodes. Chilled salts at room temperature will not carry an electric current. For this reason, once started, a molten flux bath is usually idled at a temperature to maintain the salts in the liquid state.

Parts must be cleaned and assembled and preferably held in jigs prior to immersion into the bath. Brazing filler metal is preplaced as enclosed rings, washers, slugs, dried alloy-flux paste mixtures, or as a cladding on the base metal. It is important that parts be dry and free of water of hydration since a violent explosion may occur when the parts are immersed. A certain amount of flux adheres to the assembly after brazing, and this must be drained off while the parts are hot. Any flux remaining must subsequently be removed by washing in water or by chemical means after cooling.

Production brazing with molten chemical fluxes usually involves the use of a preheating furnace to bring the parts and holding fixture to a temperature close to the melting point of the melted flux before dipping. In aluminum brazing an air furnace capable of bringing the assembly to approximately 1000 °F (538 °C) is quite suitable. Preheating prevents excessive freezing of the salt on a cold fixture, avoids possibility of explosions from dipping moist parts in the salt, and serves to speed up production and shorten the cycle.

Frequently it has been found advantageous to paint the joint areas with brazing flux prior to preheating. This assists in retarding oxidation of the parts in the joint area and produces more uniform deposition of the filler metal along the brazed joint. On small assemblies it may be possible to preflux by first warming the part, then dipping into diluted brazing flux to produce a uniform coating over the part prior to brazing.

Chemical bath dip brazing with silver filler metals is performed in a bath of neutral salts. A small amount of the usual brazing flux, such as AWS Type 3A, is applied at the joint with the preplaced brazing filler metal. After drying the moisture from the salt in a preheating cycle, the work is immersed into the molten salt for heating and brazing. On the other hand, aluminum is brazed in a molten chemical bath consisting of active salts which serve as a brazing flux and which make it unnecessary to apply additional fluxes to the work before immersion.

Figure 9.10 illustrates an aluminum-cored heat exchanger that has been prepared for dip brazing. The assembly is carefully jigged to ensure tight metal-to-metal contact between the parts.

Infrared Brazing

High intensity quartz incandescent lamps with suitable reflectors and, sometimes, parabolic

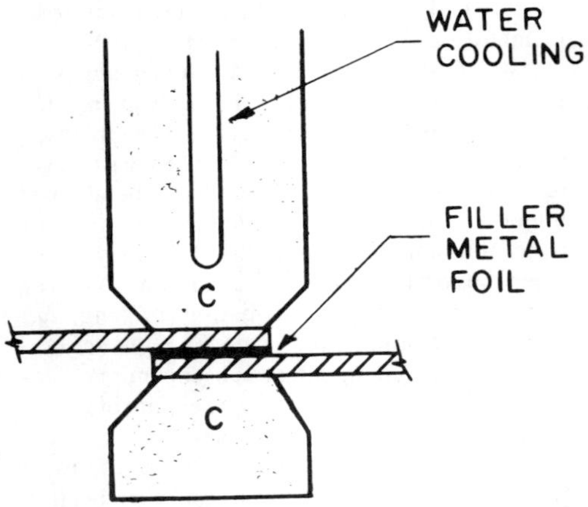

Fig. 9.9—Resistance brazing copper electric motor coil leads. The carbon blocks **c** are heated by passing low voltage alternating current through them and the work, which is held between them. The BCuP filler metal may be prepared or face fed.

Fig. 9.10—Dip brazing fixture for heat exchanger.

focusing features can present a unique source of energy for brazing. Honeycomb panels have been successfully brazed using opposed banks of lamps. Spot brazing can be done using parabolic reflectors on small parts. Split, clamshell reflectors have been used with some success in making tube joints under closely controlled conditions for marine piping. Gas infrared heaters are occasionally used to preheat assemblies prior to brazing.

Infrared brazing can concentrate large volumes of heat in small areas. This high concentration of heat makes the process desirable when applicable.

BRAZING PROCEDURES

Step Brazing

Step brazing makes use of the different brazing temperature ranges of related types of brazing filler metals. By means of this procedure, it is possible to braze one section of an assembly using the filler metal with the higher brazing temperature range. After brazing, any supplemental operations may be performed followed by another brazing operation using a filler metal with a lower brazing temperature range. The brazing filler metals are selected so that the temperature used for the second braze does not impair the braze made at the higher temperature. Step brazing involving multiple sequences has also been used.

The brazing process selected may be any of those mentioned, but furnace, induction, and resistance brazing have proven the most successful. Figure 9.11 illustrates a typical step brazing operation. In this case it was less expensive to prepare the shaft and hub assembly by atmosphere furnace brazing. The higher temperatures used in step one, of course, are unsuited for brazing the brass bushing in place.

One interesting feature of step brazing is often overlooked and not recognized as step brazing. This is the phenomenon of a brazed joint having a higher remelt temperature than its original braze temperature. A typical situation exists when brazing brass with BAg filler metals. Zinc in the brass diffuses into the brazed joint, producing an alloy of higher melting point than the original filler metal. Brass joints furnace brazed with BAg filler metal in which the heating is very long illustrate the phenomenon, known as "Christmas Treeing,"

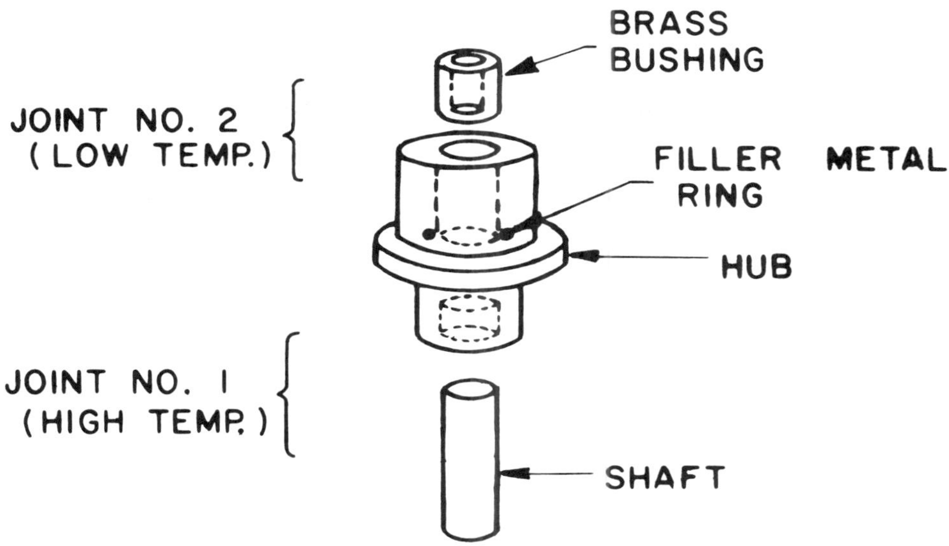

Fig. 9.11—Step brazed assembly. Steel shaft is dip coated with BCu paste, fitted into hub, and then furnace brazed. The brass bushing is later induction brazed to the hub with BAg-1.

in which zinc diffuses from the brass and causes the filler metal to take on a rough, dendritic appearance outside the joint.

Brazing in a Retort

In the discussion on furnace brazing it was assumed that the atmosphere used is introduced directly into the furnace chamber. In many instances the design of the furnace will not permit maintenance of the required atmosphere. Specifically, when hydrogen is used and the dew point of the gas must be below, for instance, −40 °F (−40 °C), the leakage of air into the furnace may not permit this, mainly because contamination of air from outside the furnace dilutes the atmosphere too rapidly.

When such is the case, it is common to put a lining, called a retort, in the furnace. The retort is sealed carefully from the outside air by welding, or employing a gasket seal, to exclude contaminating gas and preserve the low dew point desired. It is filled completely with an inert or reducing atmosphere and then heated by conventional furnace heat — electric, oil, or gas. Since retorts are exposed to high temperatures they are usually made of heat resistant alloys to resist oxidation, enabling them to be used numerous times. Where retorts are to be used only a few times and the initial cost is more of a factor, however, they may be made of mild steel.

Most brazing using a BNi filler metal or chromium-bearing base metals, or both, is conducted in a retort to preserve a low dew point atmosphere. Continuous mesh belt furnaces using hydrogen employ the same principle, except that the retort is open at each end to allow the mesh belt to enter and leave.

One type of brazing retort heated by a conventional box furnace is illustrated in Fig. 9.12. The furnace is not shown, but the heating area inside the furnace is illustrated by the dotted line enclosure. The door of the furnace is left in the elevated position and the remaining open areas around the retort are temporarily filled with firebrick while the charge is being heated.

Figure 9.13 is a typical all-welded retort in which the cover is welded in place after the charge is inserted in the retort to ensure a very low dew point. Note that the atmosphere is brought in from the rear of the retort for two reasons: it preheats the gas before it enters the work area, and it ensures the good sweeping action from rear to front which is necessary to remove the last of the purging gas used prior to introduction of the brazing atmosphere.

Figure 9.14 is a combination of induction heat-

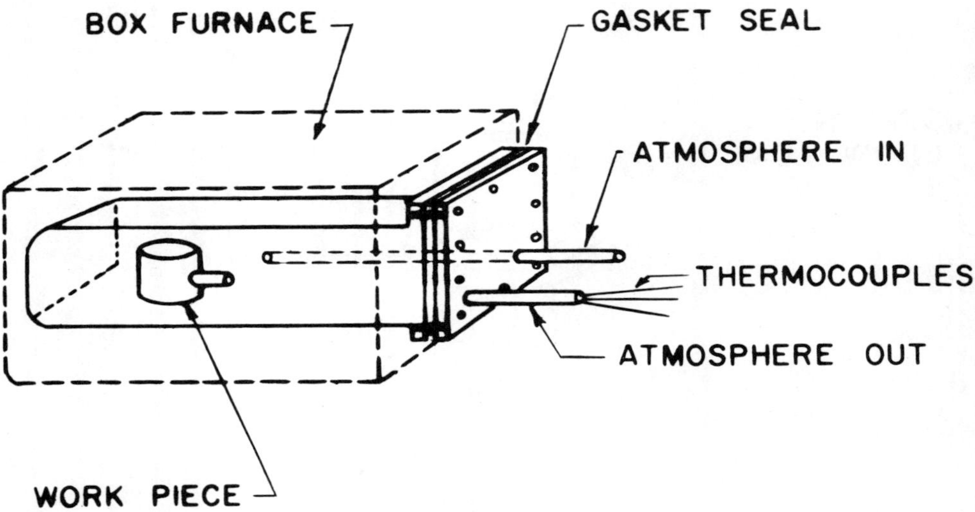

Fig. 9.12—Brazing retort placed in box furnace with loading door located outside heat zone. Gasket seal between cover and flange on retort is bolted in place and prevents gas leakage.

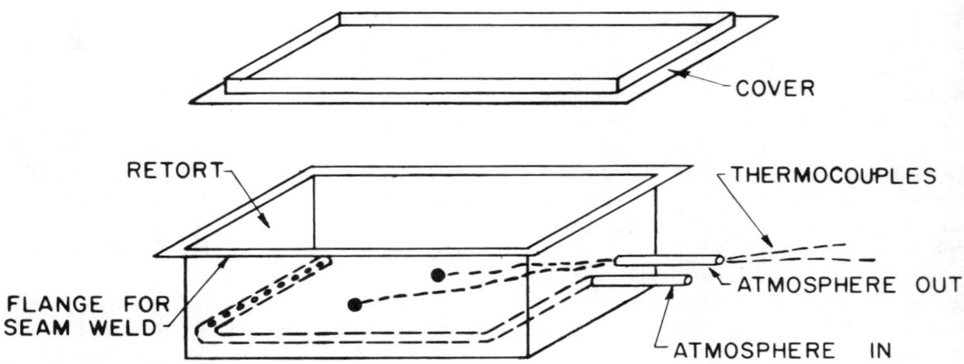

Fig. 9.13—Welded retort for brazing showing the arrangement of atmosphere inlet and outlet pipes. Thermocouple leads are passed through the outlet pipes. Not shown are potentiometers for temperature measurement, and valves, gages, and flow meters for controlling atmosphere.

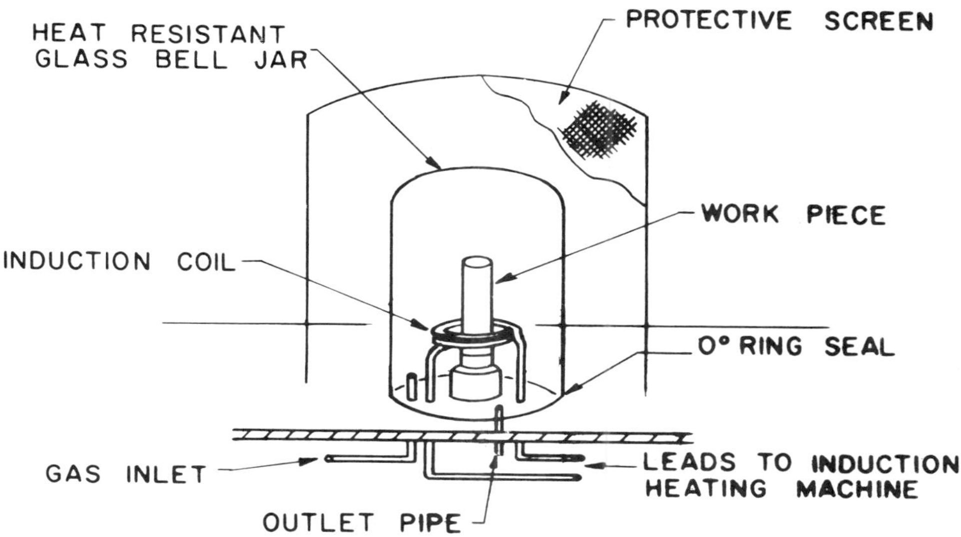

Fig. 9.14—Retort brazing using heat resistant glass bell jar and induction coils. The atmosphere is contained in the jar. A purge period allows the air to be displaced before heating begins.

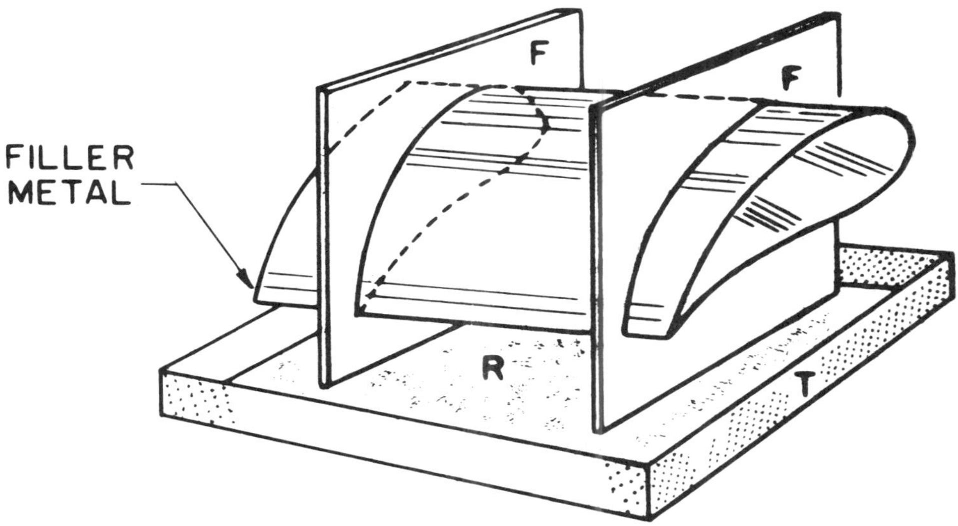

Fig. 9.15—Vacuum furnace brazing of hollow air foil sections at trailing edge. Fixtures **f** are die stamped from stainless steel sheet that has been painted with a stop-off before use. The fixture rests on a ceramic cloth **r** on a tray **t**. Nickel filler metal is prepared inside the trailing edge prior to brazing.

ing and a retort; in this case a glass bell jar contains the atmosphere. Such an arrangement is particularly advantageous in small scale production and developmental work since the operator can see the brazing operation as it progresses.

FURNACE BRAZING WITH VACUUM ATMOSPHERE

The design of the joint for vacuum brazing does not differ from the design used in furnace brazing, and the process is essentially pollution free because noxious fluxes are not used. Vacuum brazing is done in two major types of equipment, hot wall and cold wall. In hot wall equipment, the vacuum chamber is heated externally. Some of the units are built to be heated only after evacuation; others are maintained at even, high temperatures throughout the operating cycle. The disadvantages of the hot wall furnaces depend on the type. Units designed to be heated and cooled while evacuated require substantial time for heating and cooling of the heavy wall retort. The maintained high temperature equipment is usually inefficient from a space viewpoint because only a small segment of the hot zone provides the even radiation source needed to maintain low thermal differentials within the workpiece. Figure 9.15 is an example of a vacuum brazed, air foil section.

In some of the larger installations, the hot wall vacuum furnaces have a separate vacuum in a space between the outside of the work chamber and the heating elements of the furnace, commonly referred to as a double-pumped vacuum system. The purpose of this type is to permit the use of thinner retort wall materials, since the work chamber is not subjected to high pressure at working temperature. In addition, the heating elements are protected from oxidation, usually resulting in longer life.

Cold wall furnaces involve the suspension of the resistance heating elements within the vacuum chamber. Suitable radiation shields are placed between the heating elements and the furnace wall to concentrate the radiant heat on the work and prevent heat losses to the furnace walls. Heating rates are consequently very rapid.

In the search for higher temperatures and faster heating rates, the use of induction heating combined with vacuum has shown some benefit. Induction generator leads, brought through vacuum chamber walls, surround a graphite cylinder, called a "susceptor," which is heated by the induction field. The theoretical top temperature attainable is graphite's sublimation point.

The techniques of vacuum brazing, well known in the laboratory for a number of years, emerged as a production tool for space hardware. Now, vacuum brazed aluminum heat exchangers are common in automotive and home air-conditioned systems. Some of the stainless steel alloys, as well as the more active metals, show decidedly poor results when brazed in gaseous atmospheres. Up to a certain point dry hydrogen, argon, or helium atmospheres will perform satisfactorily, but when brazing conditions require that the contaminants be reduced to an absolute minimum, vacuum heating is definitely indicated. For high volume production applications, such as those commonly associated with aluminum heat exchanger brazing, the low costs available with vacuum make it especially attractive. Base metals containing more than 1% of titanium and aluminum, such as A-286, PH15-7Mo, Inconel X-750, etc., may under certain conditions show better wetting of the base metal in a vacuum than in other atmospheres.

Whether or not a clean surface is achieved depends upon the type and level of oxide or nitride formation on the surface of the alloy. If low vapor pressure oxides are present, brazing temperatures may not be high enough or vacuum high enough to remove or dissociate them. Under these conditions, either a higher vacuum or mechanisms to break the surface film are required. High temperature alloys containing titanium and aluminum are examples. Aluminum alloys depend on thermal crazing and magnesium vapors to diminish the influence of persistent aluminum oxide surface layers. Normally, brazing pressures are in the range of 10^{-5}–10^{-1} torr (0.0015–13.5 Pa).

Solid and liquid brazing fluxes are not used in a vacuum. In a few cases gaseous fluxes have been used with some success — boron tri-flouride is an example. Its disadvantage is that it is a noxious air pollutant. The effect of a flux would be to aid in cleaning the braze surface and to prevent further contamination under imperfect vacuum conditions. Another method of reducing contamination in vacuum is by "gettering." Lithium, magnesium, sodium, potassium, calcium, titanium, and barium can be vaporized in the chamber to reduce the volume of oxides and nitrides present in the vacuum atmosphere. These materials may condense on the chamber walls. Their disadvantage is that most of them will either react with the workload or form a coating on the wall when exposed to atmosphere moisture. For most materials best results are obtained by exercising due care with regard to cleanliness of materials to be vacuum brazed and then using nothing but a high vacuum. Aluminum is one exception. Magnesium contained in the brazing filler metal does double duty; as it vaporizes, it tends to disrupt the aluminum oxide at the aluminum boundary layer and later as a vapor in the vacuum chamber reacts with oxygen and oxides (mainly water) to reduce or eliminate formation of additional aluminum oxide surface films that could inhibit good wetting and capillary flow.

SECTION 2

APPLICATIONS OF BRAZING TECHNIQUES

Once the design of the brazement is completed and the hardware manufactured, the first brazed joints must be made to prove the process. If at this stage a satisfactory joint is not immediately attainable, the real trouble shooting begins. Sometimes small details escape the engineer responsible for this preliminary brazing. Recognizing these details and correcting for them may be the difference between success or failure. This section deals with practical solutions. Most of the information, although not new, is new to this book and will, it is hoped, be helpful. This practical information is passed along by those experienced in the field. Although not completely comprehensive in nature, it may suggest ways to avoid difficulties which result in defective brazing.

PREBRAZE PREPARATION

Controlling the Gap for Proper Filler Metal Flow

CLEARANCE BETWEEN MATING PARTS. The fluidity of filler metals above their liquidus temperature and within their melting range varies widely depending on chemical composition. The more fluid filler metals, as a rule, require narrow gaps, whereas those less fluid can tolerate wider gaps. It is extremely important in brazing that the proper gap be maintained at brazing temperature for best results.

DIFFERENTIAL EXPANSION. Brazing of metals with different coefficients of expansion can often create problems. Clearances should be calculated at braze metal flow points, and appropriate allowances should be incorporated into the design before the assembly is committed to production. When this differential is very large, such as brazing copper to steel, some assemblies may actually be committed to the heating operation in a manner which may dictate a press fit, or a rather wide gap, at room temperature. Such circumstances may dictate the heating method required to successfully braze the assembly. Furnace brazing, which tends to bring both components to temperature simultaneously, may not be indicated, and the choice may be torch or induction brazing.

USE OF PUNCH MARKS OR POPPET (Fig. 9.16). In assembly of many components, part of or all of the faying surfaces may be positioned so as to prevent flow of the filler metal to the entire braze joint. One example would be a round shaft brazed into a round ring, for the unsupported

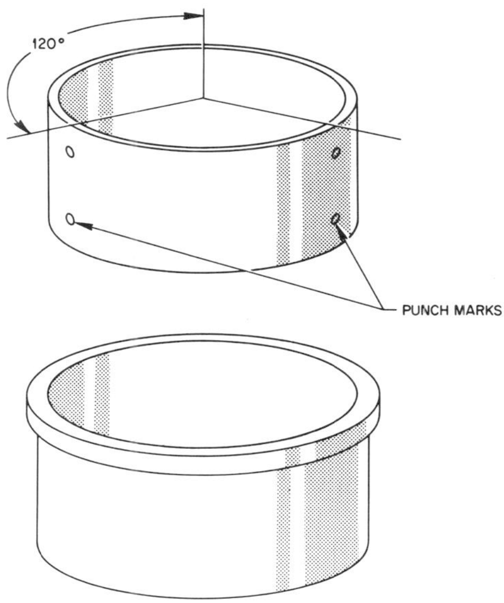

Fig. 9.16—Use of punch marks or poppet. Three sets of double punch marks spaced equally on the outside diameter of the upper member will aid in controlling the gap after assembly into the detail below, resulting in more uniform distribution of filler metal during brazing.

shaft does not usually maintain uniform clearance on the circumference. The parts tend to be off-center at assembly, and in some cases parts of the faying surfaces may actually touch each other in one area and have wide clearance in other areas. Out-of-round conditions are often present in a manufacturing process; burrs, likewise, will tend to prevent concentric alignment. Proper use of punch marks prior to assembly will often correct this problem. For such an operation three sets of punch marks at 120 degrees at appropriate distances along the joint area will suffice. The punch marks should be to a depth sufficient to raise metal on either member so as to cause minor contact points during assembly. The same condition applies to flat faying surfaces where punch marks will prevent narrowing of the braze gap in critical areas.

DIAMOND AND PLAIN KNURLS ON ASSEMBLY (Fig. 9.17). Diamond and plain knurls applied to parts, particularly screw machine assemblies, add little to the cost of the component. They do, however, ensure reasonably accurate centering of a round component into a hole, etc. Statistics indicate that in any metal manufacturing operation some components will be oversize and some undersize. Mismatch is not, then, a probability but almost a certainty on a predictable quantity of the total. The knurl anticipates this, and the extent of the knurl must be calculated so that assembly is still possible even under conditions where the smallest opening is mated with the largest shaft size.

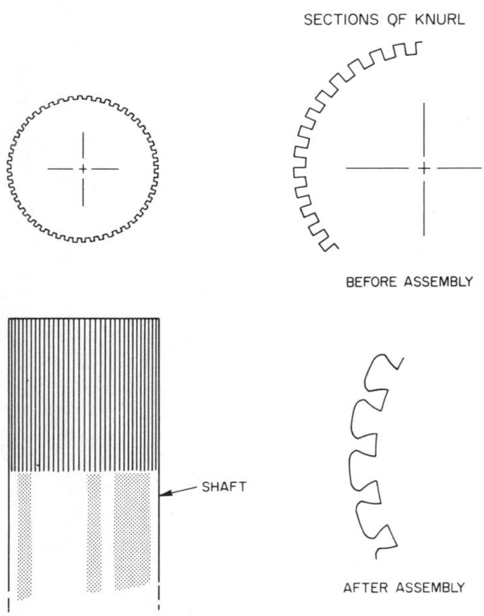

Fig. 9.17—The plain knurl used on brazed assemblies. Knurled end of shaft provides alignment of male part to hole in assembly after press-fitting prior to brazing. Knurl diameter is larger than hole of mating part by a few thousandths of an inch. The brazing filler metal flows into the space between the knurl marks.

In furnace brazing with copper filler metals, knurls are used to make possible self-locating components which require no fixturing. The same practice also applies to filler metals used in torch and induction heating, such as the BAg series of silver-bearing filler metals, the emphasis being on centering of the components properly rather than on holding the components for the braze operation.

USE OF SPACERS BETWEEN MATING PARTS. Shim stock spacers can be prepositioned between mating parts in strategic areas to accomplish control of the gap. The use of capacitor discharge welding to hold these shims is often indicated and will prevent movement of the shim during assembly. Fine wires, likewise, may replace the shim under certain circumstances, and will work as well.

Compensating for Excessive Clearance Between Parts

TOO MUCH CLEARANCE BETWEEN MATING PARTS. Excessive clearance for proper brazing occurs just as often in manufacturing operations as does inadequate clearance. This "wide gap" situation may be corrected in several ways.

SELECTION OF FILLER METAL WITH GAP-FILLING PROPERTIES. There are many examples of use of this technique in industry. In furnace brazing of steel components, copper filler metal, which is very fluid, may not uniformly fill the joint area. Selection of a copper filler metal containing nickel in small proportions (3 to 8%) will aid in bridging the gap during brazing. Likewise the BCuZn series will perform well in furnace brazing of assemblies which could not be assembled by normal copper furnace brazing because of the excessive gaps. In the lower melting filler metal series, the substitution of less fluid filler metals in the BAg series will permit bridging a wider gap. The addition of nickel in the

BAg-3 filler metal improves gap-filling properties. The addition of pure nickel or nickel-silicon powder to conventional BNi filler metals used in the brazing of stainless steel components in hydrogen furnace brazing or vacuum furnace brazing, will also improve the gap-filling abilities. In the latter case, the simultaneous alloying of the filler metal to the nickel powder and the consequent rise of the melting temperature results in a sluggish behavior of the filler metal at temperature, and an ability to bridge the excessive clearance present in the assembly.

ELECTROPLATING ONE MEMBER TO REDUCE JOINT CLEARANCE. This means of correction, although seldom used because of the high cost involved, is, however, a means to be considered as a rework of a male component that for one reason or another is undersize. The nature of electroplating usually prohibits use of this technique on female members since plating will not normally occur inside holes unless special conditions warrant placing an anode inside the hole to accomplish adequate throwing power in the plating bath.

WIRES AND SHIM STOCK USED TO FILL WIDE GAP AREAS. Here again, to resort to such operations to accomplish proper braze clearance is not often warranted because of the additional cost. Two examples might be:

1. A formed wire ring may be located in the joint area prior to brazing. Brazing then proceeds normally, the ring becoming a part of the final brazement.

2. Fine wires are often wrapped around poor fitting joint areas and tack welded in place prior to brazing if the assembly is not of simple geometry as described above. Cut shim stock accomplishes the same end, correcting for mismatch at the assembly and making possible satisfactory brazing. The presence of such additional material in the joint area may or may not be acceptable depending on the end use of the finished product.

POWDER METAL COMPONENT BRAZING, FERROUS PARTS. Whether one considers the steel powder metal component as a candidate for excessive clearance, or as a unique case of its own, the problem of brazing these components into completed hardware is a real one. Regardless of the filler metal selected, the wicking of this material into the porous sintered compact is a factor that often makes brazing extremely difficult. The filler metal often ends up being flushed into the porous component with not enough left to accomplish joining at the faying surfaces. Properties of the powder compact may be seriously altered by this procedure, which in this case results in a secondary liquid phase (siltering) operation not contemplated at the design stage. One answer to such a problem is the creation of a special powder metal compact braze preform, which is made up at the time the powder metal compact is made for specific use in the brazing operation.

Composition of these preforms is proprietary, but generally they follow the following theoretical approach. The addition of a higher melting component (iron, manganese, etc.) preform made by compaction places small nuclei of a second metal in strategic locations throughout the matrix so that at the moment the braze preform is melted a secondary realloying takes place which raises the melting point of the filler metal and restricts its fluidity and penetration of the base metal powder compact. In this manner penetration may be restricted to a satisfactory depth. Note that in this special case the brazing preform is placed between the mating parts, not outside the joint area, and becomes part of the completed assembly.

Most efforts in solving this brazing problem by this means have been confined to steel parts joined by copper furnace brazing. The same technique might logically apply to nonferrous components, but the state of the art at this writing has not been fully explored.

Restricting the Flow of Filler Metal. To assure a strong, neat appearing, and economical joint, filler metals should be restricted to the joint area and the fillet should be kept to a minimum size.

EFFECT OF EXCESSIVE FILLER METALS. Obviously the best way to avoid excessive filler metal in the immediate joint area is to use the proper quantity in the first place. An excess often results in erosion of the base metal by solution of the base metal in the filler metal. The ready solution of the copper alloys in filler metals high in silver and the solution of ferrous base metals in the BNi series are examples.

THE USE OF STOP-OFF MATERIALS. Chapter 21 deals with the brazing of refractory metals. Particular attention is called to the metal-metal oxide formation and reformation data. The materials most difficult to reduce as oxides are titania and magnesia; both require very low dew points in hydrogen or corresponding low pressure values in vacuum to reduce the metal constituents. These materials and others of the rare earth metals are the basis of stop-off materials. The objective is simply to coat the base metal adjacent to the area of the braze joint with an overlayer of these oxides prior to brazing. Stop-off materials using various bases, such as lacquers, varnishes, and gels, are now in wide use throughout the brazing industry.

Very little of these oxide deposits is required to accomplish the desired results. More often the tendency is to use too much. Nonsettling varieties are available which can be applied in minute streams through hypodermic needles under pressure. Removing the stop-off residues after brazing is costly, so diligence in application often pays off in reduced clean up labor. Furthermore, excessive quantities can interfere with other operations taking place in the joint area after brazing is complete. In brazing aluminum a pencil mark from a carbon pencil is sufficient to restrict the flow of BA1Si filler metal during brazing.

Atmosphere brazing accounts for the widest use of stop-off materials since it is here that there is little possibility of controlling filler metal flow after the parts have entered a brazing furnace. Torch and induction brazing occasionally require the use of a stop-off, but the effectiveness is somewhat reduced, particularly if fluxes are used in brazing. A compromise is often possible, however, and stop-offs will work even under these difficult circumstances.

An example of the wide use of stop-off is the brazing of threaded studs to any assembly by furnace brazing in atmosphere. The filler metal will immediately wet the area around the stud, and rise by capillary action up the threads, thus spoiling the fit of the fastener to a nut intended for use in a final assembly. A small amount of stop-off applied at the base of the threads prior to brazing will prevent this filler metal from rising in the thread roots. Small diameter holes in components are kept open in a similar manner by application of stop-off if the hole is in the immediate braze area and if exposed to filler metal that may wet the base metal and flow away from the joint area.

SELECTIVE ABRASIVE BLASTING. Areas where filler metal is objectionable may on occasion be abrasive blasted prior to assembly using aluminum oxide grit or titanium oxide, etc. Enough of the abrasive will often be impinged on the surface to act as a stop-off. Most generally accepted grit for abrasive blasting is one which might deposit a material acceptable to filler metal wetting. For example, silicon carbide grit or steel shot are often applied to help in wetting.

Care in the choice of abrasive blast materials is essential if proper results are to be obtained. Thorough cleaning of an abrasive blast system is necessary if the same system is to be used as a stop-off in one instance and to promote wetting in another. Enough metal or oxide may be removed in one operation to be objectional in a subsequent operation with different materials if the system is inadequately cleaned. It is often possible to experience this same phenomenon of dewetting when ground surfaces are present in which aluminum oxide wheels are used in the grinding. Such occurrences, although rare, can present a tough problem for the brazing engineer, particularly when the previous history of the type of grinding wheel being used is unknown.

Pretreatments to Ensure Adequate Flow and Wetting

PLATING OF COMPONENTS PRIOR TO BRAZING. Electroplating, when properly performed, using a metal that is easily wet by filler metals, will help ensure adequate flow of the filler in brazing. Copper plating of steel components, nickel plating of stainless steel parts, and refractory alloys, such as Inconel and Hastelloy, are processes frequently used. Caution should be observed, however; firm adhesion of the plating to the base metal is required, and plating done under inadequate control can result in surfaces that will peel off. The copper plating of tungsten carbide components to aid in furnace or induction brazing is a good example. Many carbides are now blends of titanium and other carbides in addition to tungsten carbide, which presents an additional burden on the plating process to secure good adhesion.

Electroless plated nickel will throw a good coating on many materials prior to brazing. When using this process, one should remember that the basic deposit may be as high as 12% phosphorus. This percentage can change during the braze cycle by combining with hydrogen during furnace brazing, in which case, with proper time and temperature, the phosphorus may pass off in the atmosphere gas as oxide of phosphorus, leaving a layer enriched in nickel. Also, when brazing with the BNi series, the deposited plate will rapidly combine with the filler metal to form an altered composition of the resulting filler metal-base metal diffusion layer.

Silver plating on copper parts is common in the electronic industry, and components properly prepositioned will join together without the use of any filler metal by the formation of the eutectic

72 Ag-28 Cu at the interface. This phase becomes liquid and proceeds to penetrate the base metal along the grain boundaries to form a brazed joint.

Zinc and cadmium plating are frequently encountered. Caution should be used with these materials since the heat of the brazing operation may alter the surface away from the brazed joint enough to lower expected corrosion resistance. Furthermore, both zinc and cadmium are volatile at braze temperatures and can present a health hazard to personnel (see chapter 29).

PRETREATMENTS IN REDUCING ATMOSPHERES. Another pretreatment to aid flow of the brazing filler metal involves heating in reducing (or vacuum) furnace atmospheres. An example of this practice, which is sometimes termed bright-annealing, is the cleaning of carbon steel parts in a controlled atmosphere at temperatures ranging from 1600 to 2100 °F (871 to 1149 °C). Stainless steel is frequently precleaned in hydrogen atmospheres or vacuum. By using a rather high vacuum (10^{-6} torr [0.001 Pa]), materials such as Inconel and high nickel alloys containing up to 4 to 5% combined titanium and aluminum can be rendered more wettable by pretreating at 2100 to 2200 °F (1149 to 1204 °C). Such an approach has little or no effect in precleaning of aluminum prior to brazing and is not recommended.

USE OF VOLATILE FLUXES IN BRAZING. The use of ammonium bifluoride during furnace brazing is an interesting example of cleaning to ensure adequate wetting. When surface films present on the components are sufficient to prevent adequate flow, incorporation of a water base fluoride wetting agent into the joint area prior to brazing will sometimes be useful. Nickel fluoride has been used in a similar manner. Using the BNi series of filler metals, it is sometimes helpful to pretreat the powdered filler metals with these volatile salts and allow the deposit to dry on the powder mixture. It should be remembered that precleaning of the filler metal itself may be advisable in some cases to prevent introduction of oxides to the joint area from this source.

Residues from this process are very small and difficult to detect even with silver nitrate washing of the joint area after brazing. It seems apparent that the bulk of the fluoride passes off as a gas during the brazing operation.

SPECIAL CONTAMINANTS: REMOVAL BEFORE BRAZING. Many unusual surface contaminants may appear on occasion to plague the brazing engineer. Basic in the solution to these problems is a thorough knowledge of the prior history of components received for brazing. A few of these contaminants are:

1. *Aluminum, titanium, and other refractory oxides* — Source may be deburring, vibratory finishing, grinding, belt sanding, etc.

2. *Graphite and molybdenum disulphide* — Common ingredients in deep drawing, impact extrusion, wire drawing compounds, these are excellent for protection of the tooling and for minimizing wear to these surfaces. Unfortunately, they can become well bonded to the metal parts being processed when these components are employed.

3. *Lead oxide* — Frequently used for proving dies prior to use in production, this material is, unfortunately, often present in proof parts submitted for prebrazing operations prior to production.

4. *Nitrogen* — Nitrided surfaces can be present on components prior to brazing. There may be severe wetting problems with the components in the braze area.

If any of these contaminants is suspected, it is important that sufficient testing be performed to determine the best course of action. A few component sets should be carefully treated to remove a few thousandths of an inch of surface metal in the braze area. Returning to pure base metal will, in many cases, provide good wetting results. Removal of the particular contaminant is often not a simple task, and each case must be treated on its own. Identification of the contaminant is of prime importance, however. Acid pickling may be required to successfully remove many of these contaminants, which adhere tightly to the component surfaces.

CLADDING OF SURFACES WITH FILLER METALS. Base metals clad with filler metals are not uncommon in industry. An example of this is copper preclad with BCuP on one or both sides. Other examples are aluminum alloys clad with Al-Si alloys and stainless steels plated with electroless nickel (NiP). Also, copper clad with BAg-3 is used in so-called sandwich brazing of carbide cutting tools. If precoated components are not commercially available, and if joint designs require that the filler metals flow long distances into a capillary it will often be advantageous to precoat one or both components in a separate tinning operation before assembling to complete the braze.

Tungsten and molybdenum, common materials used in resistance welding electrode tips, are often precoated by the manufacturer to aid in accomplishing the final braze in which they are induction or torch brazed to special copper alloy electrode materials (RWMA Class 1, 2, or 3 materials). The filler metal in the final brazing operation is one of the BAg series, and brazing is accomplished at 1200 to 1500 °F (649 to 816 °C) in air, using fluxes.

Salt Bath Processes for Braze Pretreatment. A proprietary process involving a combination of cleaning, precoating, and carbon removal has been applied to tungsten carbide components. It utilizes a proprietary salt bath treatment prior to brazing of the tips to the tool holders for cutting tools, mining bits, etc. The treatment facilitates wetting of the filler metal on the carbide surfaces. Another proprietary process involves the use of molten sodium hydride or other salt combination baths to preclean components prior to brazing. Its most common use is in the pretreatment of cast irons to remove carbon from the surface to permit brazing with BAg series filler metals. A similar process is also used to remove refractory oxide (islands) from the surface of A-286 turbine components. In this case, however, the components are pretreated in hydrogen at 2100 °F (1149 °C), processed in the salt bath, and then pickled in HF-HNO_3 solution.

Prebraze Inspection of Components. Adequate inspection of the parts in an assembly before brazing can often avoid problems in the brazing operation. Components frequently come from different sources, and checking for proper fit, surface preparation, and cleanliness will prove well worthwhile. A pilot run of a few assemblies will usually pinpoint any trouble spots which can be corrected prior to scheduling for production.

Filler Metal Placement

Face feeding filler metal to the joint area in torch and induction brazing is an accepted procedure. Preplacement of wire rings is also a well established technique. There are many other methods of filler metal placement, some novel and unique, that will answer specific problems in unique parts geometry.

Forms of Wire

SNAP RINGS: Formed on a ring-making machine, these are made possible by pushing a ring over a rod without dropping over the end of the part. A common style would be a 220 deg segment of a circle. The ring opens as it passes the rod diameter and snaps in place (made from cold worked wire) (see Fig. 9.18).

FLAT RINGS: Round or shaped rings flattened by coining, for example, 0.040 in. (1 mm) round wire flattened to 0.020 in. (0.5 mm) thick.

FORMED RINGS: Using a wire-forming machine it is possible to build many special shapes in wire.

WIRE FEED: In this case the same type of wire feed used in feeding welding wire for the gas metal arc (GMA) welding process is modified to deliver a predetermined quantity of brazing filler metal to the joint area for either induction or torch brazing.

CLAD WIRE: Clad or "flushed" wire is usually made by casting filler metal around a copper core. An example is BAg filler metal cast around a core of copper and then drawn into wire. Such a filler is used widely in joining wire screen cloth seams and Fourdrinier wire cloth of the types used in endless belts in the papermaking process. The core holds the melted filler in place around the wires during brazing, preventing the filler metal from falling off the work.

Forms of Sheet: With the exception of the BNi series of filler metals, sheet filler metal is available for brazing. In the case of the BNi series a plastic binder is often employed to make bonded sheet from the powdered metals alloys. The

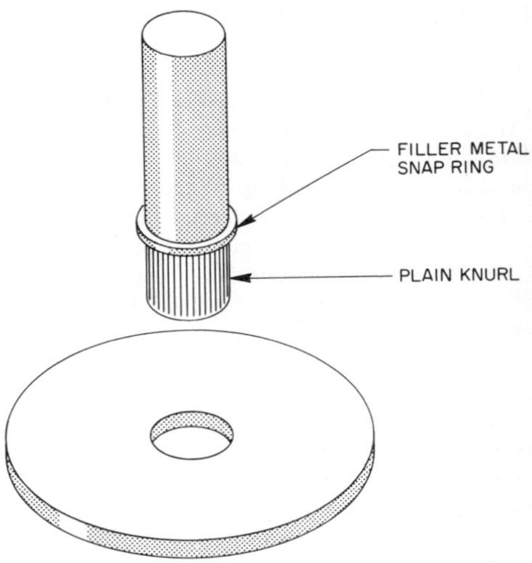

Fig. 9.18—Use of snap ring in brazed assembly. This filler metal ring of BCu is pushed into the upper shaft to a predetermined point. The shaft is then pressed-fitted into the lower member. The brazing filler metal clings tightly to the heavier member and does not melt and flow until both parts are at braze temperature, thus preventing unwanted flow of the filler metal into the surface of the lower member.

binders burn off in atmosphere brazing, and little or no residue results.

CLAD SHEET: Copper clad with BAg-3 is commonly used in the brazing of carbides. The copper core remains in the joint area and acts as a ductile member that plastically deforms during cooling of the brazed joint. This prevents cracking of the carbide at the interface because of large differences in coefficients of thermal expansion. Another popular class of clad material is copper clad with BCuP on one or both sides. This is often used in small electronic assemblies. It is possible to resistance braze two members together using a spot welding machine by reflowing the clad filler metal on each component. BAlSi filler metals are also available, clad on 3003 aluminum and other aluminum alloy sheet. The primary use is in the vacuum furnace brazing of plate fin heat exchangers.

Forms of Powders: Most filler metals are available in powder form. Application to the joint area is made by mixing the powders with suitable binders or flux to make them adhere to the work prior to brazing. Examples are:

ATOMIZED POWDER: BCu, BNi, BCuP, BAg, BAlSi, etc.

FLAKE POWDER: BCu, which is made by ball or rod milling atomized copper powder.

ELECTROLYTIC POWDER: High purity dendritic copper powders used in furnace brazing with dry hydrogen or vacuum. This type of powder will not introduce oxides into the brazing atmospheres.

CAST SHAPES (FROM POWDERS): BNi series filler metals formed into rings by sintering in graphite molds at a temperature below the melting point are widely used on nuclear stainless steel tube assemblies for extremely close tube joints.

FLUX MIXTURES: In special situations it may be possible to allow brazing flux to act as binder for powdered filler metal. One member is dipped in dilute flux, followed by dipping into dry powdered filler metal. A "slurry" of dilute flux and powdered filler metal is one of the oldest known means of preplacement. Examples are the manufacture of filigree decorative silver jewelry, and the joining of hollow handles in tableware in which two half stampings are brazed to form a knife handle.

PASTE MIXTURES: Powdered metals mixed with binders and fluxes become paste brazing filler metals. Special proprietary products are available that can deliver appropriate amounts of these pastes to the assemblies.

OXIDE MIXTURES: Copper oxides are widely

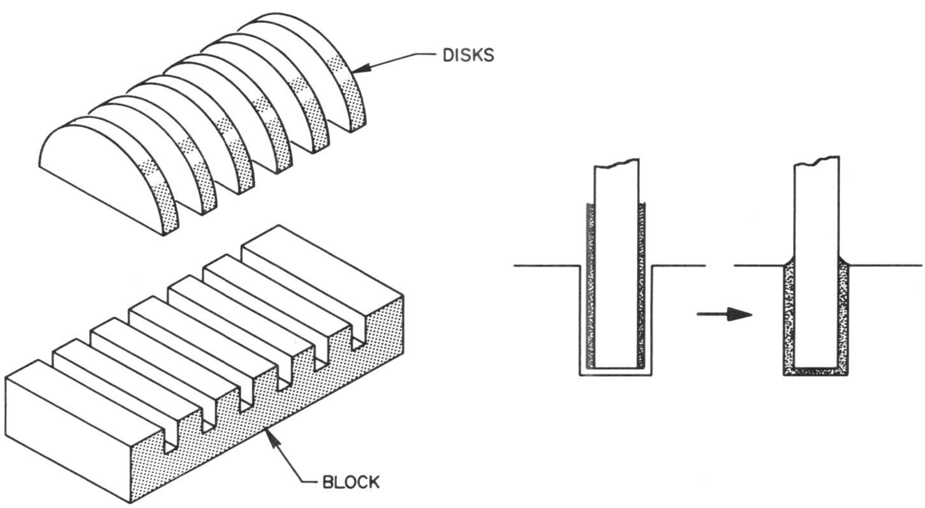

Fig. 9.19—Electroplating one member to supply filler metal. The six plates in this electronic assembly were electroplated with silver prior to assembly into the block by press fitting. All parts are made of copper. Furnace heating in hydrogen or vacuum atmosphere will form the copper-silver eutectic. A low melting brazing filler metal, at points where the parts are in contact, will produce a clean, neat, fluxless brazed joint of high quality and without the need of additional filler metal.

used in furnace brazing of steel assemblies with BCu filler metal. Oxide blends in a suitable binder can be applied to the work by dipping, brushing, roller coating, or extrusion under air pressue. Steel screen (10 mesh) cloth roll coated with such a product will produce a uniform brazed joint at each wire cross joint without filling the holes in the mesh.

SPRAY MIXTURES: Copper-colored lacquer paint in a propellent can is often composed of finely divided flake copper (1-5 μ particle size). Such a mixture sprayed on 100-mesh stainless steel wire cloth will provide ample filler metal and form uniform fillets of approximately 0.001 in. (0.03 mm) radii.

SHOT, CUT WIRE, SHEET: Where a rather large amount of preplaced filler metal is desired, heavy wire, shot, or short strips of sheet may be useful. To keep large pieces from moving or falling away from the joint during brazing, a suitable paste of the same composition as the filler metal is often prepositioned in the joint.

Plating Techniques: Electroplated copper on steel is a common source of brazing filler metal. In many cases only one member of an assembly need be plated. Silver, plated on copper in the joint area, will diffuse into the copper of a mating part and form the lower melting Ag-Cu eutectic melting at 1435 °F (779 °C) (see Fig. 9.19). Electroless nickel plate, which is composed of nickel with 10 to 12% phosphorus, can be used as a source of filler metal for furnace brazing operations on stainless steel, high nickel alloys, and other metals. On mechanical mixtures, a mixture of tin and copper powders (10% tin) suspended in a binder will form bronze filler metal joints on steel after diffusion has formed the desired filler metal composition.

Special Equipment: On large volume production operations, specially designed equipment should be considered as a means of reducing the man-hour labor. Mentioned previously were automatic wire feed to replace hand face fed filler metal, and special automatic applicators to extrude paste-type fillers to the joint area prior to the heating operation. Wire wrap machinery similar to the type used in communication terminal wrapping have been used to wrap filler metal wire at particular points in a brazed assembly.

Automatic assembly of a round filler metal ring is often accomplished by means of vibratory bowl feeders. The filler metal is treated as a detail component in the rapid high speed assembly operation. Such an operation is common in placing a filler metal ring onto the return bends of a tubular heat exchanger prior to brazing.

Cleaning of Filler Metals: Filler metals are a common source of contaminants in brazing. Care should be taken to deliver only clean metal to the braze operation. Oxides formed on filler metal will overburden fluxes and interfere with good wetting. Pickling in acids may be advisable under some circumstances. Because of their large surface area powdered metals must likewise be cleaned on occasion. The silver filler metal powders can be cleaned by light leaching with caustic cyanide solutions. The nickel series may require HF-HNO_3 to remove excessive oxides. They may be also treated below their melting point in a reducing atmosphere or vacuum. In some cases, pastes made of copper oxides may be undesirable since they introduce large quantities of oxygen into the furnace brazing atmosphere, which can inhibit wetting.

CHAPTER 10

POSTBRAZE OPERATIONS

Operations that usually follow brazing include heat treatment and cleaning if required by service surface conditions. Parts that have been brazed without stop-off in a controlled atmosphere, using vacuum or gaseous flux, should not require postbraze cleaning.

HEAT TREATMENT

Various procedures may be used to obtain heat treated brazed assemblies. By selecting a brazing filler metal with a suitable brazing temperature range, the required heat treatment may be accomplished while brazing or while cooling after brazing. It is also practical, when using heat treated parts such as air hardening steels, to braze assemblies after heat treating if the brazing temperature range is kept suitably low by choice of brazing filler metal. Thus, with many combinations of base metals and brazing filler metals, properly heat treated brazed assemblies may be obtained without heat treatment after brazing.

Materials, such as age hardening alloys that have thermal treatment temperatures below the solidus of the brazing filler metal, may be heat treated after brazing. A filler metal should be selected with a solidus higher than the temperature required for the postbraze heat treatment.

Any particular operation will require selection of filler metal and operating sequence to obtain desired results for the particular base metal, joint, and service conditions.

FLUX REMOVAL

If the brazing operation is of a type that the required flux leaves a residue which can cause corrosion, the residue must be completely removed as soon as possible after brazing. Those flux residues which are soluble in hot water may normally be removed by immersion in hot water, with agitation if necessary (see chapter 4). The ease with which flux residues can be removed depends on several variables. Insufficient flux, overheating, and prolonged brazing times all contribute to excessive saturation of flux residues by the metal oxides and make flux removal more difficult. Removal will be facilitated by quenching the brazed parts from an elevated temperature, sufficiently below the solidus to prevent damage to the parts. Caution should be observed in quenching parts, as microcracking may occur in the joint if the cooling rate is too rapid in the higher temperature ranges. Such precautions are especially important in cases where the parts will later be subjected to alternating stresses. In stubborn cases it may be necessary to use a suitable warm acid solution, such as 10% sulfuric acid or 5-10% phosphoric acid. There are also proprietary cleaning compounds which may be advantageous for some applications. Do not use nitric acid where brazing filler metals or base metals containing copper or silver are used. After a chemical dip, parts must be washed free of any chemical and dried.

Fluxes used in the brazing of some materials,

such as aluminum, are not readily soluble in water. These may usually be removed by rinsing in very hot water (above 180 °F, 82 °C) with a subsequent immersion in either an inhibitor alkaline solution or nitric acid, hydrofluoric acid, or a combination of the two acids (see chapter 29). A thorough rinse in water is necessary after the chemical treatment.

OXIDE REMOVAL

The temperatures necessary for brazing are usually high enough to cause oxidation and discoloration of areas which are not protected by flux or a controlled atmosphere. If restoration of these areas to the "before brazing" condition is necessary, it may be accomplished by dipping in a chemical bath and rinse, or by mechanical cleaning methods such as wire brushing or blast cleaning.

REMOVAL OF STOP-OFF MATERIALS

Some brazing applications require the use of "stop-off" materials to prevent the brazing filler metal from flowing onto areas not intended to be wet by the filler metal. When necessary, these materials should be removed after the brazing operation. Most brazing stop-off materials are either of the "parting compound" type or the "surface reaction" type. The first type can be easily removed mechanically by wire brushing, air blowing, or water flushing. The surface reaction type can best be removed by a hot nitric acid-hydrofluoric acid pickle. Strong, hot solutions of sodium hydroxide or ammonium bifluoride can also be used, though these caustic solutions act more slowly than the acids. As previously noted, nitric acid should be avoided where brazing filler metals contain copper or silver, and all chemicals should be thoroughly rinsed off.

CHAPTER 11

INSPECTION

Inspection of the completed assembly or subassembly is the last step in the brazing operation and is essential for assuring satisfactory and uniform quality of the brazed unit. This operation also provides a means for evaluating the adequacy with which the prior steps in the process have been carried out with regard to ultimate integrity of the brazed joint.

The design of the brazement is extremely important to the inspection operation and, where practical, should be such that the completed joints can be examined as easily and adequately as possible. Inspection procedures can in every case be simplified and the number of rejects reduced or eliminated by careful application of the design and production information given in preceding chapters. An intelligent choice of brazing processes, brazing filler metal, joint design, and cleaning methods will also assist in the inspection process.

The inspection method chosen to evaluate the final brazed component should depend on the service requirements. In many cases the inspection methods are specified by the ultimate user or by regulatory codes. In the establishment of these regulatory codes or standards of quality for brazed joints, the same approach should be used as in the setting up of standards for any other phase of manufacturing. These standards should be based, if possible, upon known requirements that have been established by prior service tests or history.

Tests and examinations of brazed joints can be conducted on either test specimens or the finished brazed assembly, and may be either destructive or nondestructive. A brief description of the common tests and inspection procedures used for evaluating brazed joints is presented in the following paragraphs; however, the specific details of these tests are left to the numerous references on inspection and testing techniques (for example: *Welding Inspection* and Section 1, chapter 6 of the 6th Edition of the *Welding Handbook,* both of which are published by the American Welding Society).

Preproduction and workmanship samples are often used for comparison purposes during production. They may be sample specimens made during the development of the brazing procedure to be used or may be samples taken from the actual production. In any case, they show minimum acceptable production quality of the brazed joint. They are used to show the fillet sizes, amount, and extent of allowable voids in a joint, etc., and serve as a ready reference for inspection. Such samples may consist of duplicates of those being brazed in the production run or may consist of a simpler type of specimen, such as an inverted T-joint. The duplicate assemblies tell more about the actual joint to be made, but usually are more expensive to manufacture. The T-joint (see Fig. 11.1) is useful for comparing general flow and erosion characteristics during the brazing cycle. Filler metal is placed at one end of the specimen and flows along the joint during brazing.

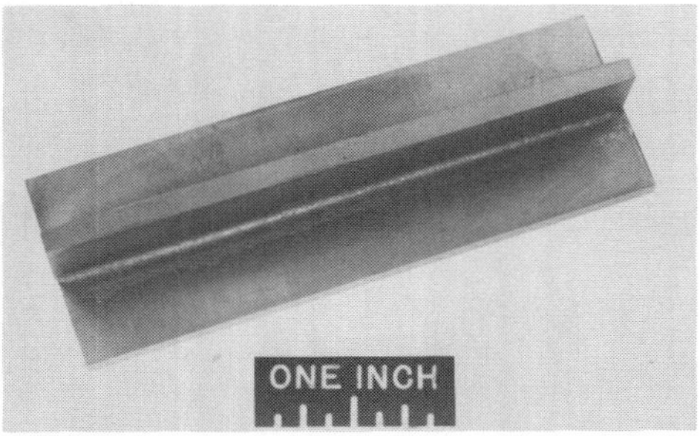

Fig. 11.1—Inverted T-joint for comparing general flow and erosion characteristics during brazing.

INSPECTION METHODS

Nondestructive Inspection

Visual examination (VT) is probably the most widely used of the nondestructive methods of inspection. As with all methods of inspection, however, visual examination will not be effective if the joint cannot be readily examined. Visual inspection is also a convenient preliminary test where other test methods are used.

When brazing filler metal is fed from one side of the joint or preplaced within the joint at or near one side of the joint in such a manner that visual examination of the opposite side of the joint after brazing shows a continuous fillet of filler metal, it can usually be assumed that a sound joint has been attained. If filler metal is applied from the same side of the joint that must be inspected, it is possible, through improper practices, to get a good appearing fillet on this side of the joint without any penetration through the joint. Visual inspection is totally inadequate for detecting such flaws as flux entrapment in the joint and spotty or irregular filler metal flow between the faying surfaces.

Proof testing (RPT) is a method of inspection that subjects the completed joint to loads slightly in excess of the loads to be applied during its subsequent service life. These loads can be applied by hydrostatic methods, by tensile loading, by spin testing, or by numerous other methods. Occasionally it is not possible to assure a serviceable part by any of the other nondestructive methods of inspection, and proof testing then becomes the most satisfactory method.

Leak testing (LT) — Pressure testing is recommended when gas or liquid tightness is required of the brazed joints. Assemblies subjected to low pressure, and not coming under The ASME Pressure Vessel Code requirements, may be tested with air. Assemblies subjected to high pressure and those coming under the pressure vessel code requirements should be tested hydrostatically prior to testing with air. The hydrostatic test pressure should be 1.5 times the maximum gas pressure to which the assembly will be subjected. Test pressures are always greater than service pressures and are usually specified by code or by purchasers' specifications.

Pressure testing with air may be accomplished by one of several methods: (1) closing all openings, pressurizing the assembly, submerging in water, and noting any signs of leakage indicated by rising air bubbles; (2) closing all openings, pressurizing the assembly, brushing a soap solution or commercially available indicator over the joint area, and noting the location where bubbles are formed; or (3) closing all openings, pressurizing the assembly, locking the air under pressure within the assembly by closing the air inlet source, and noting any change in trapped gas pressure over a period of time (corrections for temperature may be necessary).

Pressure testing with helium is frequently preferred because of the relative ease of finding very minute leaks. A mass spectrometer is generally used to detect leakage of helium. In testing for very small leaks, it is important to thoroughly remove all liquid or vapors from the assembly by a suitable drying operation; purging with dry gas while heating the assembly above the boiling point of the liquid is one method.

Pressure testing with Freon is frequently used in the refrigeration industry to determine very small volume leakage. Special Freon detection equipment is available, the simplest being either an air acetylene or air-liquid petroleum-gas halide leak detector. The detector samples the atmosphere from the area being checked, and any leak is indicated by a color change of the detector flame. The more elaborate equipment is of the electronic type and will detect very minute leaks.

Vacuum testing is generally employed in checking assemblies used in refrigeration equipment and small assemblies where it is imperative that the most minute leak be detected. The mass spectrometer is used in this method of testing, as shown in Fig. 11.2, and helium is the sensing medium. The assembly being tested is either placed in a container having a helium atmosphere

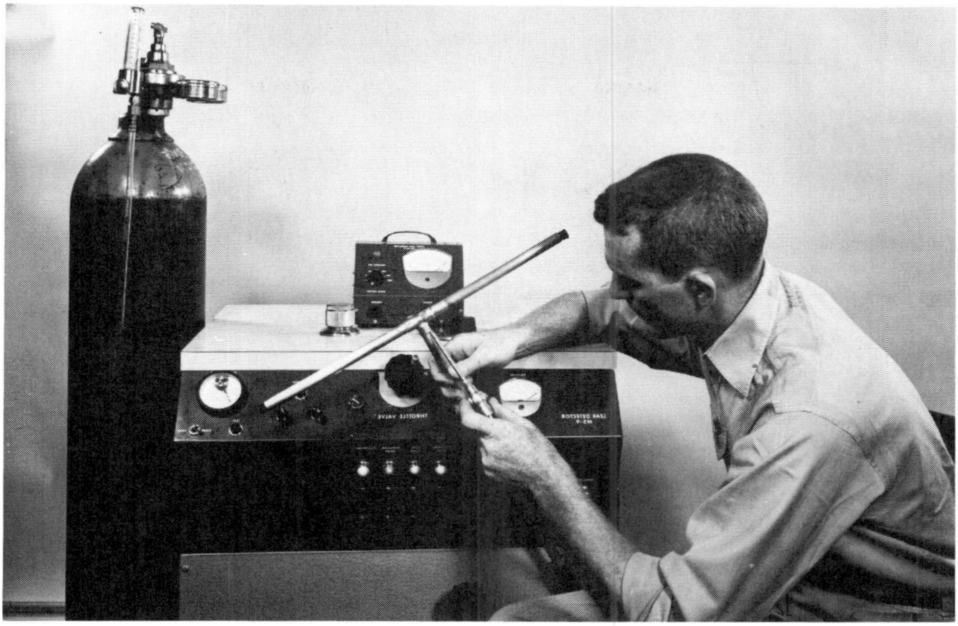

Fig. 11.2—Mass spectrometer leak detection of brazed assembly.

or helium is flushed over the surface. The mass spectrometer is connected to the vacuum side of the assembly, preferably between the auxiliary diffusion pump and the mechanical backing pump. If mass spectrometer helium leak testing is to be used, care should be taken to prevent the use of liquids on the joint prior to the helium leak testing, because of the potential problem of plugging a leak.

Dye- and fluorescent-penetrant inspections (DPT, FPT), although often inappropriate and misleading, are often used for the detection of defects extending to the surface of brazed joints in both magnetic and nonmagnetic materials. Cracks and porosity in the fillet can be detected, but interpretation is sometimes difficult because of small irregularities in the braze fillets. Incomplete flow and partial filleting can also be observed. These types of inspections should not be conducted if subsequent repairs are to be made since the penetrants are often difficult or impossible to completely remove.

Magnetic particle inspection (MT) is not in general used for the inspection of brazed joints since most brazing filler metals are nonmagnetic. However, brazed metal structures, such as honeycomb, can be inspected by this technique. One variation, in which the results can be recorded, utilizes a film containing fine magnetic particles. The film hardens in a few minutes after application and "freezes" the magnetic particles in position. The particle pattern can be observed on the structure being inspected, after which the film can be stripped off and retained as a permanent record.

Radiography (RT) is used extensively for examination of brazed joints. As a general rule, and recognizing that sensitivities of better than 2% are difficult to achieve, sections of uniform thickness and not too thick may be radiographed if an exposure can be made straight through the joint. Possible differences in X-ray absorption characteristics of base metals and filler metals must also be considered in interpreting the radiographs. Special techniques are required to reliably inspect joints of varying thickness. Views that include two sides of a joint, such as one taken through a pipe fitting, must be interpreted with particular care. Both film and fluoroscopic X-ray techniques are used to evaluate brazed honeycomb for brazing filler metal flow and filleting.

The electrical resistance method of evaluating brazed joints depends on the variations in voltage

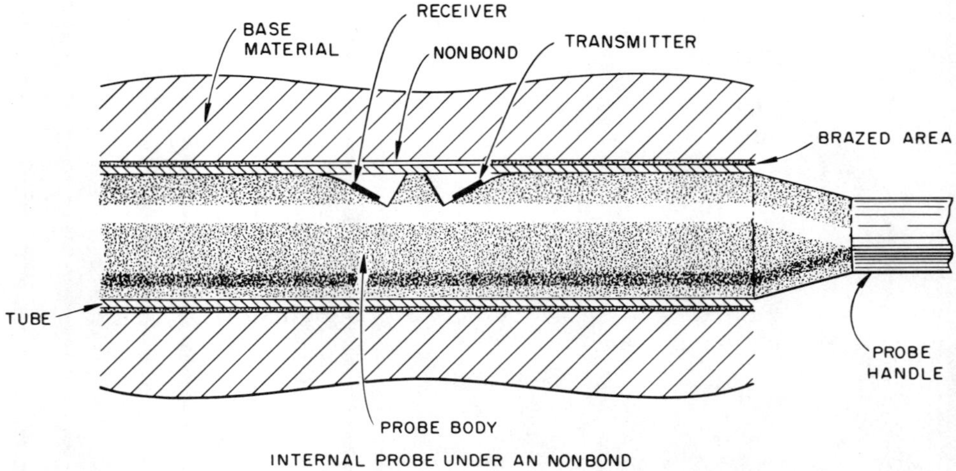

Fig. 11.3—Two-crystal ultrasonic probe for detection of nonbond in brazed tube-to-tubesheet joint.

drop across a constant cross section of metal. Discontinuities create an increased voltage drop, which can be measured by means of a Kelvin Double Bridge.

Ultrasonic inspection (UT) is one of the new inspection methods to be used on brazed joints. The applicability of this method of inspection depends largely on the design of the joint and the configuration of the adjacent areas of the brazed assembly. Ultrasonic techniques have been developed and used for the evaluation of brazed honeycomb. The principal method uses a single transducer and monitors the wave form of the pulse reflected from the surface of the facing sheet. The reflection is different over a brazed cell wall when compared to a point between cell walls or over an unbrazed cell wall. Records in the form of C-scan maps allow ready observation of the variation over a scanned area. Similar techniques are useful for monitoring the quality of braze of a thin sheet bonded to a thick substrate. The ultrasonic frequency should be chosen to promote a "ringing" or modified resonance of the reflection from the outer surface of the material.

Ultrasonic systems were developed to detect lack of bonding in silver brazed pipe joints for submarine construction. Good measurements of percent of bond (on an area basis) can be made by comparing the relative amplitudes of sound reflected from the joint wall that transmitted through the joint. Ultrasonic techniques have been used to evaluate the quality of brazing of a tube inside a thick base material (e.g., a heat exchanger tube back brazed in a tube sheet). Figure 11.3 is a sketch of a two-crystal probe designed to perform such an inspection. When introducing ultrasound into a thin member, such as the wall of the tube, critical tuning of the ultrasonic frequency and angle of incidence of the sound are necessary to achieve the desired results. Figure 11.4 shows a prototype ultrasonic scanning device for inspecting tube-to-tube sheet brazes in place on a heat exchanger mock-up.

Thermal heat transfer inspection is used as an inspection procedure in certain cases. For example, brazed aircraft propeller blades are photographed while still hot a few seconds after leaving the furnace. The covering skins appear to be bright red in areas where they are brazed to the reinforcing rib but are much darker red, or black, in areas where improper brazing has ocurred. Another method of inspecting brazed honeycomb panels utilizes powdered or liquid materials with low melting points to indicate the differences in heat transfer characteristics when the honeycomb assembly is placed under infrared heat lamps. Temperature variations cause the liquid to be repelled from warm areas and to accumulate in cool spots. The core partitions act as heat sinks

Fig. 11.4—Mechanical device for ultrasonic scanning of brazed tube-to-tubesheet joint.

Fig. 11.5—Frost test can determine areas of nonbonding in some types of brazed joints.

causing the fluid to flow to each area of good braze. A simple frost test used on a brazed tube-to-block assembly as shown in Fig. 11.5 can occasionally be used to detect areas of nonbraze when a thin sheet is joined to a thicker base metal. If the assembly is chilled until a film of frost forms, a breeze of warm air will preferentially melt the unbonded areas. Similar but more sophisticated techniques use thermal sensitive phosphors, liquid crystals, and other temperature-sensitive materials. Infrared-sensitive electronic imaging devices with television readout are now commercially available to monitor temperature differences (less than 1 °C) produced by heat transfer differences that would be characteristic of variations in braze quality.

Destructive Inspection

Destructive testing methods are used for random selection or lot testing of brazed joints. In lot testing, a small specified percentage of all production is tested to destruction. The results of these tests are assumed to apply to the entire production, and the joints in the various lots or batches are accepted or rejected accordingly. When used as a check on some nondestructive method of inspection, such as visual examination, a production joint may be selected at regular intervals and tested to destruction so that rigid control of brazing procedures is maintained.

Peel tests are frequently used for evaluating lap joints. One member is held rigid, as in a vise, while the other is peeled away from the joint, as shown in Fig. 11.6. This test may be used as a means of production quality control to determine the general quality of the bond and the presence of voids and flux inclusions in the joint. The permissible number, size, and distribution of these defects will depend upon the service conditions of the joint and may be limited by applicable codes.

Tension or shear tests are usually used to determine the strength of a joint. Such tests are more often used in the laboratory to determine basic strengths of filler metals and the adequacy of proposed joint designs, but they can be used for verifying the relative strengths of the joint and base metal parts. Tension and shear tests have been used to advantage in determining joint strengths for service below and above room temperature. These tests are used more widely

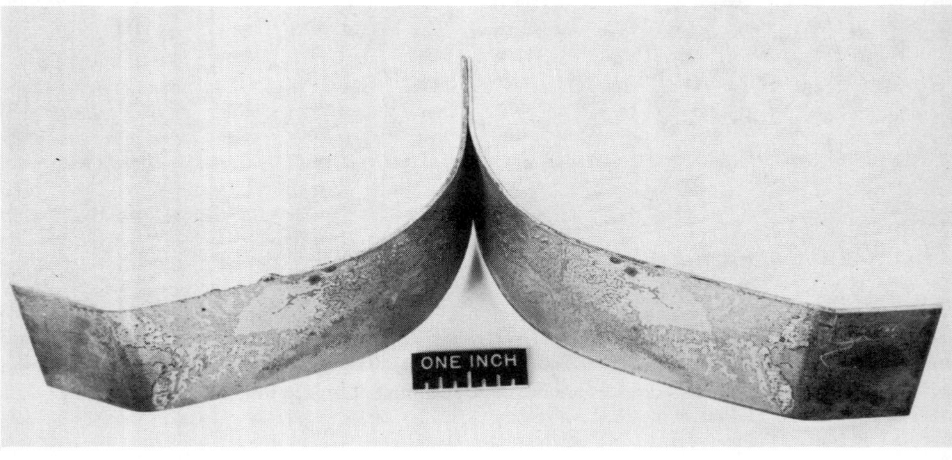

Fig. 11.6—Peel test on brazed specimen. The areas where good bonding had occurred are evident.

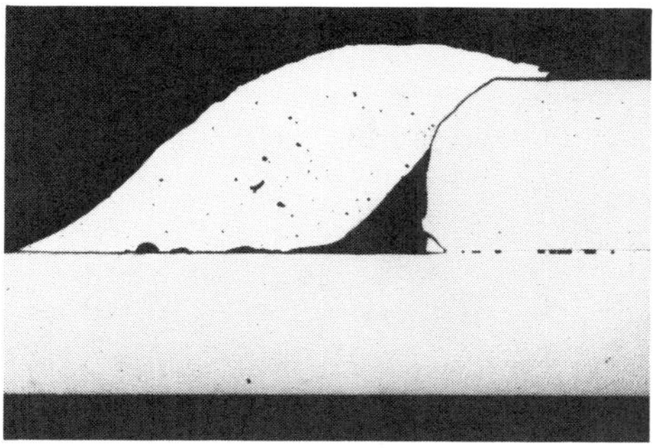

Fig. 11.7—Void in a brazed joint.

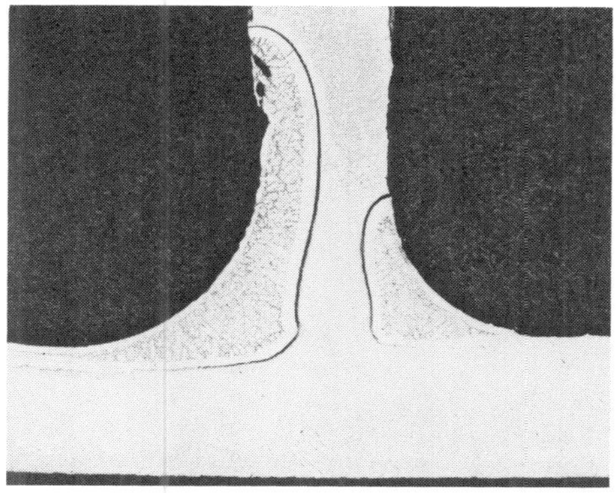

Fig. 11.8—Excessive base metal erosion in a brazed joint.

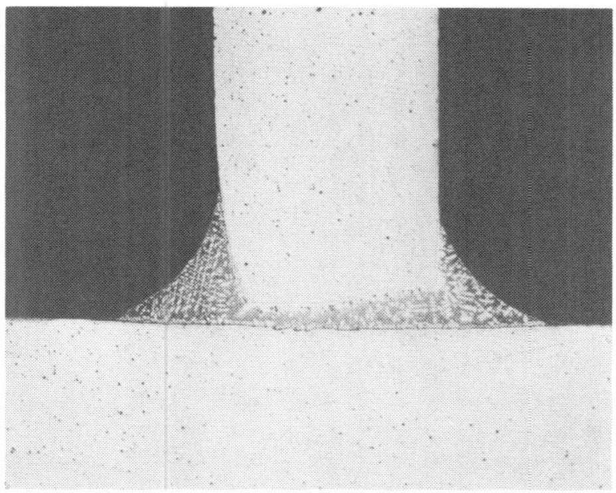

Fig. 11.9—Sound brazed joint.

for development than for production quality control, although they have been used for testing selected samples.

Fatigue tests are used to a limited extent and in most cases are a test of the base metal as well as the brazed joint. As a general rule, fatigue tests require a long time to complete and for this reason are very seldom used for quality control.

Impact tests, like fatigue tests, are generally limited to laboratory work in determining basic properties of brazed joints. As a general rule, the normal notch-type specimens do not appear to be best for brazed joints. Special types of joints may be required to obtain accurate results below and above room temperature.

Torsion tests on brazed joints are used occasionally in production quality control, especially where studs, screws, or tubular members are brazed to heavier sections.

Metallographic examination is frequently used to determine the general quality of brazed joints and to detect such flaws as porosity, poor flow, excessive base metal erosion or improper fit-up. Such an examination may provide a better understanding of the causes of these defects than the other inspection techniques.

TYPICAL DEFECTS AND THEIR CAUSES

Defects found in brazed joints are an indication that some part of the brazing procedure is out of control or that improper technique has been used. Several typical defects have been included in this chapter so that they may be recognized when they occur. The possible cause is given in each case.

Figure 11.7 is a section of brazed copper lap joint in which a large void in the fillet is evident. A defective joint was detected by subjecting the assembly to pressurized air and submerging it. The probable causes of the void were underheating or improper fluxing procedures or both. This is borne out by the irregular flow of the filler metal into the joint.

Figure 11.8 shows a section through a brazed joint in which severe erosion of the base metal occurred. This erosion resulted from overheating of the joint during brazing. Such erosion may not be serious where thick sections are to be joined, but cannot be permitted where relatively thin sections are used. A joint with essentially no erosion is shown in Fig. 11.9.

Figure 11.10a shows a brazed copper assembly. Figure 11.10b shows a section through that assembly. Figure 11.10c is a magnification 25× of the section located by the arrow in Fig. 11.10b.

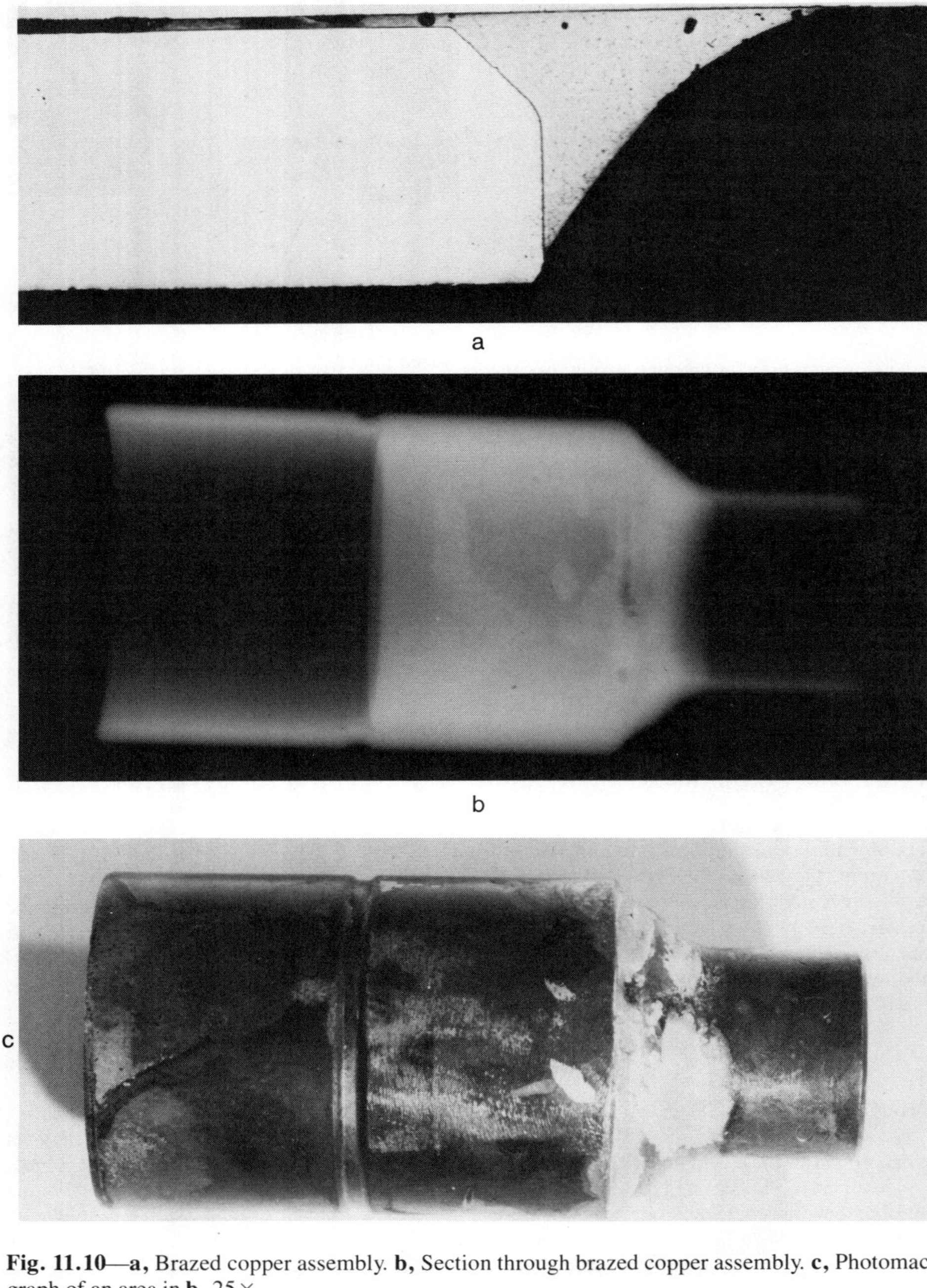

Fig. 11.10—a, Brazed copper assembly. **b,** Section through brazed copper assembly. **c,** Photomacrograph of an area in **b.** 25×.

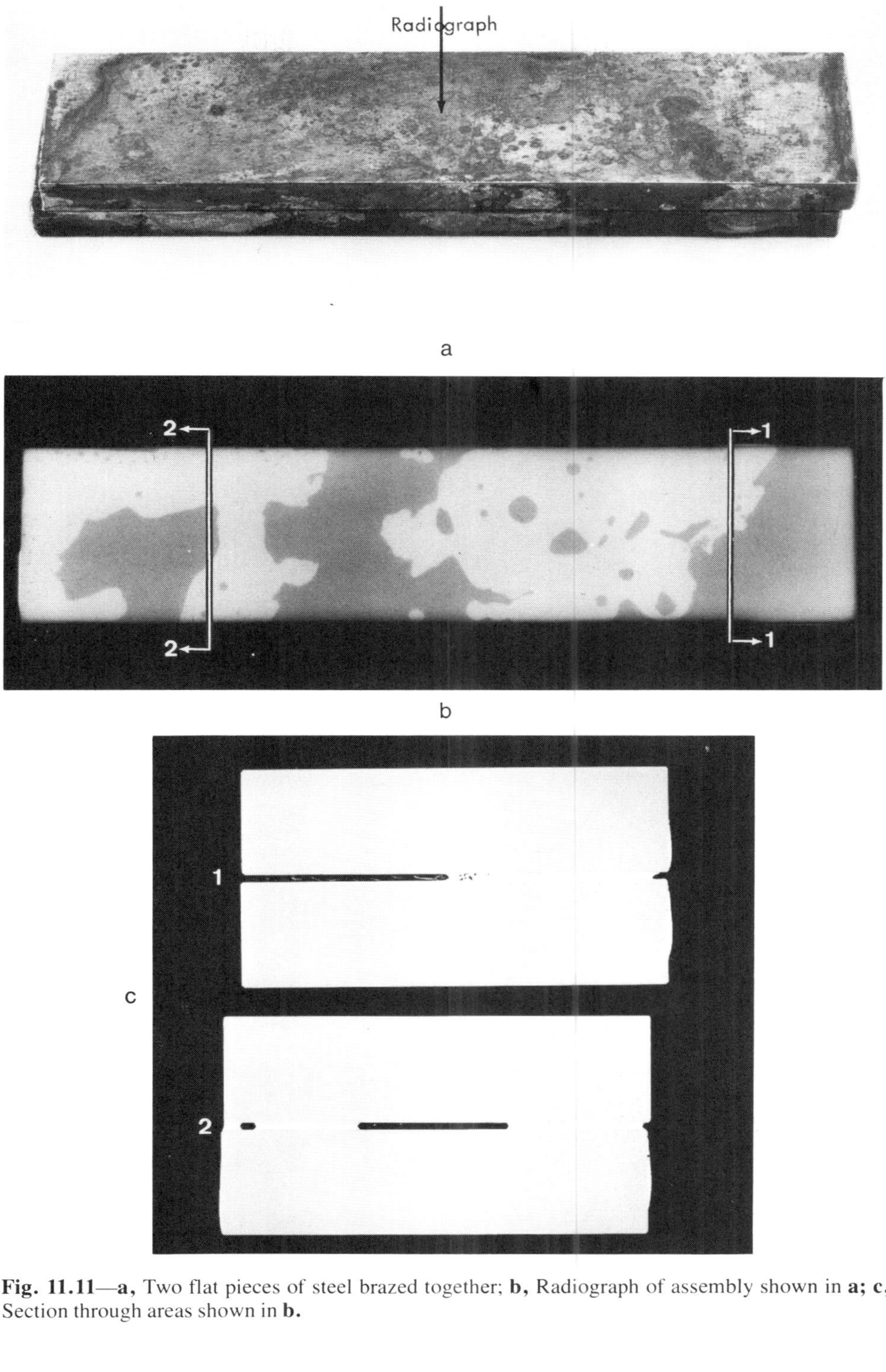

Fig. 11.11—a, Two flat pieces of steel brazed together; **b,** Radiograph of assembly shown in **a; c,** Section through areas shown in **b.**

The voids and incomplete penetration of the filler metal through the joint are caused by insufficient heating. Such an internal defect is detectable by a torsion or other mechanical test. If the joint contains leakage producing voids, these are sometimes revealed by a pressure test.

Figure 11.11 shows a brazed joint that has been radiographed and sectioned. The assembly shown in Fig. 11.11a is of two flat pieces of low carbon steel brazed with BAg filler metal. The radiograph was taken through the joint as shown. In Fig. 11.11b the dark areas indicate voids. The dotted lines indicate the location of the sections in Fig. 11.11c. Note the voids.

CHAPTER 12

ALUMINUM AND ALUMINUM ALLOYS

Aluminum and aluminum alloys are brazed using practices similar to those used for brazing other metals except that different fluxes, different filler metals, and generally lower temperatures are used. The aluminum base metals that can be brazed are listed in Table 12.1. These include a number of non-heat treatable and heat treatable wrought alloys and brazeable cast alloys. The non-heat treatable base metals include the high purity aluminum alloys and those with low content of alloying ingredients which achieve their best mechanical properties through cold working. Alloy 3003, the most common brazeable alloy, is among this group. The heat treatable alloys generally contain magnesium and silicon and are described as the magnesium-silicide type or 6xxx series. These alloys, because of their higher total alloy content, begin to melt at lower temperatures than higher purity aluminum alloys and consequently are generally brazed at lower temperatures than the non-heat treatable alloys. The heat treatable alloys, which include the commonly used 6061 alloy, are thermally treated during fabrication or after brazing to attain their high mechanical properties.

Not all aluminum alloys can be brazed. The high strength wrought aluminum alloys and certain casting alloys contain high amounts of alloying ingredients. These alloys melt at temperatures lower than commercially available filler metals. (See chapter 3 for a listing of filler metals for brazing aluminum.) Some experimental lower melting filler metals have been produced which successfully brazed the "not readily brazeable" alloys 2024, 2219, and 7075. Low melting filler alloys contain zinc, copper, or other metals with or without silicon alloyed with aluminum. Although some of these combinations have been used successfully, none has been developed with adequate corrosion resistance to be recommended for commercial use.

Aluminum can be brazed by most of the standard practices. Most aluminum brazing is done by the torch, dip, or furnace processes. Furnace brazing may be done in air or controlled atmosphere (including vacuum). Other methods, including induction brazing, radiant lamp brazing, and resistance brazing, have been used for specific applications, usually involving the more readily brazeable aluminum alloys. Regardless of the process used, temperature must be closely controlled for successful brazing. With dip or furnace brazing, automatic proportioning control temperature devices are available to maintain the flux bath within ±5 °F (3 °C) and the furnace atmosphere within ±10 °F (6 °C) of the desired brazing temperature. Properly designed dip pots and brazing furnaces will provide temperature distribution throughout the dip bath or furnace chamber within these same limits. In manual torch, induction, or resistance brazing operator skill and judgment maintain required temperature range for brazing based upon flux color and melting and flow of the filler metal.

FLUX APPLICATION

Brazing flux must be used for every brazing process except fluxless inert gas or fluxless vacuum

Table 12.1—Nominal composition and melting range of common brazeable aluminum alloys

Commercial designation	Aluminum assoc. No.	Brazeability rating†	Nominal composition* Copper	Silicon	Manganese	Magnesium	Zinc	Chromium	Approximate melting range °F	°C
EC	EC	A	99.45% min. Al		—	—	—	—	1195-1215	646-657
1100	1100	A	99% min. Al		—	—	—	—	1190-1215	643-657
3003	3003	A	—	—	1.2	—	—	—	1190-1210	643-654
3004	3004	B	—	—	1.2	1.0	—	—	1165-1205	629-651
3005	3005	A	0.3	0.6	1.2	0.4	0.25	0.1	1180-1215	638-657
5005	5005	B	—	—	—	0.8	—	—	1170-1210	632-654
5050	5050	B	—	—	—	1.2	—	—	1090-1200	588-649
5052	5052	C	—	—	—	2.5	—	—	1100-1200	593-649
6151	6151	C	—	1.0	—	0.6	—	0.25	1190-1200	643-649
6951	6951	A	0.25	0.35	—	0.65	—	—	1140-1210	615-654
6053	6053	A	—	0.7	—	1.3	—	—	1105-1205	596-651
6061	6061	A	0.25	0.6	—	1.0	—	0.25	1100-1205	593-651
6063	6063	A	—	0.4	—	0.7	—	—	1140-1205	615-651
7005	7005	B	0.1	0.35	0.45	1.4	4.5	0.13	1125-1195	607-646
7072	7072	A	—	—	—	—	1.0	—	1125-1195	607-646
Cast 43	Cast 443.0	A	—	5.0	—	—	—	—	1065-1170	629-632
Cast 356	Cast 356.0	C	—	7.0	—	0.3	—	—	1035-1135	557-613
Cast 406	Cast 406	A	99% min. Al		—	—	—	—	1190-1215	643-657
Cast A612	Cast A712.0‡	B	—	—	—	0.7	6.5	—	1105-1195	596-646
Cast C612	Cast C712.0§	A	—	—	—	0.35	6.5	—	1120-1190	604-643

*Percent of alloying elements; aluminum and normal impurities constitute remainder.
†Brazeability ratings: A = Alloys readily brazed by all commercial methods and procedures
B = Alloys that can be brazed by all techniques with a little care.
C = Alloys that require special care to braze.

‡Sand castings.
§Mold castings.

brazing processes. Brazing fluxes are mixtures of inorganic chloride and fluoride salts supplied in powder form. Brazing flux is applied dry or is mixed with distilled or deionized water or alcohol for either brushing, spraying, or dipping. Dry flux is sprinkled on the joint area of the assembly being brazed, or else the end of the heated filler rod is inserted into the flux powder. Generally, brazing flux is used in slurry form. A mixture of 3 parts flux and 1 part water is satisfactory for brushing. A thinner mixture is used for dipping or spraying, and the amount of water needed is established by trial to suit the assembly, furnace conditions, or spray gun used. See chapter 4 for more information on fluxes.

Care should be exercised in handling aluminum brazing flux. Flux mixtures are strong chemical agents and can be injurious to the body and corrosive to equipment. Flux spraying should be done in an exhaust or water curtain spray booth or with equipment that can control overspray. After flux spraying, the lines and spray equipment must be rinsed to retard corrosion or clogging of the gun. See chapter 29 for further information on safety and health protection in handling brazing fluxes.

FILLER METAL FORMS

Filler metal for brazing aluminum alloys is supplied in the form of powder, paste, wire, or thin gage shim stock which is either face fed or preplaced at the joint area. Round wire of various diameters is used in the form of rings or in straight lengths as required for preplacement. Preforms are made from flattened wire or from shim stock, and the geometry is adapted to the shape of the parts to be brazed.

A convenient method of supplying filler metal is to use brazing sheet that consists of a core of aluminum alloy and a coating of lower melting filler metal. The coatings are aluminum-silicon alloys (BAlSi series) and may be applied to one or both sides of the sheet. The coating is roll bonded to the core during fabrication. Thus, brazing sheet is a product that can be formed by drawing, bending, or other metalworking processes without removing the coating. The formed parts can be assembled, fluxed, and brazed without placing additional filler metal in the joint. Brazing sheet is frequently used as one member of an assembly with the mating piece made from an unclad brazeable alloy. The filler metal on the brazing sheet flows by capillary attraction and gravity to fill the joint at contact sites.

Brazing filler metal clad tubing made by integrally bonding the filler metal to the tube during fabrication is also produced. Brazing filler metal clad thin-walled tubing made by continuous seam welding brazing sheet is also commercially available. Such products as wire, rod, bar, extruded shapes, rolled shapes, forgings, or castings are not available as brazing filler metal clad items.

MATERIAL PREPARATION

Precleaning requirements vary widely depending on the condition of the metal surface, the thickness of the material, the alloys to be brazed, and finally the liquid or gas tightness required in the finished joint. Light oil or heavy forming lubricant will interfere with the flow of brazing filler metal. Solvent-type cleaning operations to remove surface lubricants have been quite satisfactory. Solvent cleaning is mandatory for fluxless brazing and is desirable for flux brazing, particularly where a large surface area is exposed during brazing. The oil fumes otherwise generated during heating of the assembly deter brazing and may be an explosion hazard.

In many applications, chemical cleaning is not required, particularly on brake-formed sheet parts. However, parts formed in spinning operations usually have a severely worked surface containing embedded oxide fragmented by the spinning tool which prevents uniform wetting by the molten flux and filler metal. The same is true when model parts are produced in small quantities by hand hammering or similar methods. In these cases, conventional caustic etch or nitric-hydrofluoric acid etch is desirable. Etchant-type cleaning may be done with a caustic or acid type of cleaner. Two cleaning procedures and cleaner compositions are shown following. Many proprietary cleaning solutions are equally satisfactory.

Caustic Cleaner

- Degrease: solvent or vapor
- Dip: 5% wt. sodium hydroxide — 140 °F (60 °C) — up to 60 seconds
- Rinse: cold water
- Dip: 50% vol. nitric acid — cold — 10 seconds
- Rinse: hot or cold water
- Dry

Acid Cleaner

- Degrease: solvent or vapor
- Dip: 10% vol. nitric acid plus 0.25% vol. hydrofluoride acid — cold — up to 5 min
- Rinse: hot or cold water
- Dry

Solvent or vapor cleaning is usually adequate for non-heat treatable aluminum alloys. For some heat treatable alloys, however, chemical cleaning is necessary to remove the thicker oxide film. Solvent vapor degreasing is desirable when brazing thin metal. Brazing of fins to tubes or flat surfaces for heat exchanger applications usually involves thin gage brazing sheet. The amount of filler metal available on sheets of such thickness is nominally 10% wt., and consistent quality joints are obtained only if the surface is clean. A light etch can also be used, but care must be exercised so that excessive braze cladding is not removed.

Chemical cleaning of the filler metal before preplacement is often desirable. Filler metal made from wire, or shims made from flattened wire, will flow more uniformly if the oxide coating formed in fabricating the wire is removed by etching.

JOINT TYPES AND FIXTURES

Lap joints rather than butt joints are preferred for brazing aluminum and aluminum alloys (see chapter 6). Press or tightly fitting joints must be avoided in assembling aluminum parts to facilitate filler metal flow and minimize flux entrapment. Figure 12.1 illustrates recommended joint designs.

Joint clearance is governed by the fact that capillary action is an important factor in drawing

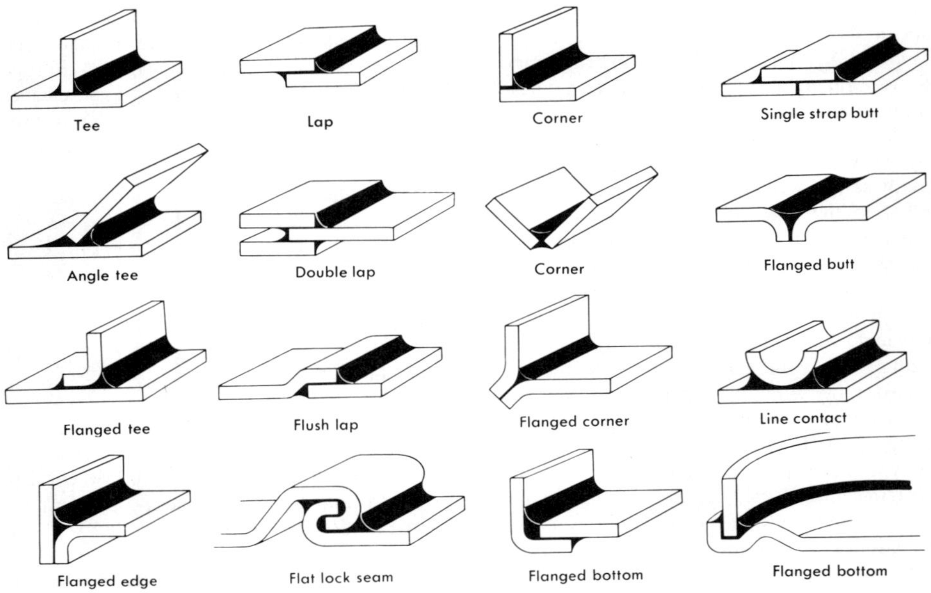

Fig. 12.1—Recommended brazed joint designs for aluminum alloys.

filler material into place. Consequently, the joint fit-up must be carefully controlled to provide good flow. As a rule of thumb, clearances of 0.002 to 0.004 in. (0.05 to 0.10 mm) are suitable for laps less than 0.25 in. (6.4 mm) when dip brazing. For torch, furnace, and mechanized torch or induction brazing with a 0.25 in. (6.4 mm) lap or less, 0.004 in. to 0.010 in. (0.10 to 0.25 mm) clearances can be used. Clearances up to 0.025 in. (0.6 mm) are used for longer laps because the filler metal changes composition by dissolving base metal and becomes sluggish as it flows through long lap joints. The correct clearance for any given joint is best determined by trial.

The joint design should permit easy assembly of the parts prior to brazing. Closed assemblies should be designed to provide for escape of the gases during brazing. It is preferable to design the parts to be self-jigging, as by lock seaming, or to be held in alignment by rivets or projections that remain on the part after brazing (see Fig. 12.2).

Fixtures and jigs must be designed carefully so that the difference in thermal expansion does not force the parts out of line. Pure nickel, Inconel, stainless and special alloy steels, and low carbon steels are commonly used for jigs and fixtures. Low carbon steel will have a short fixture life, but coating with aluminum by dipping or spraying followed by thermal treating to diffuse the coating will increase the useful life.

PERFORMANCE OF JOINTS

In all cases brazing temperatures exceed the temperature at which annealing occurs in the base material. It follows that furnace or flux bath brazed parts made from the non-heat treatable aluminum alloys have mechanical properties corresponding to the annealed temper of these alloys. The heat treatable base metals are also annealed by brazing, but their strength can be substantially increased either by a heat treating operation applied after the parts are brazed or by quenching the part from the brazing temperature. The latter procedure is not always feasible, and the results depend on the geometry of the part. The heat treatable base metals can be quenched in an air blast, a water spray, or a tank of hot or cold water. A slight delay after removal from the brazing temperature is necessary to permit the filler metal to solidify before the part is quenched; otherwise, the dimensional changes that occur on

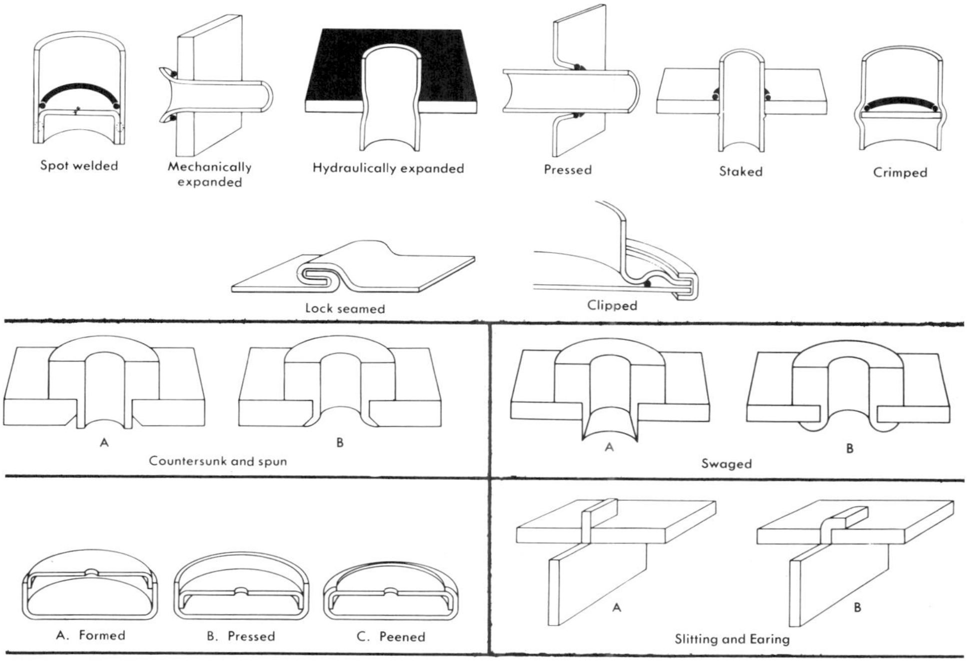

Fig. 12.2—Typical self-jigging joints for aluminum alloys.

quenching will open the joints. An air quench is the slowest quench of those just mentioned.

On some complicated parts, cracking of the joints may occur even though the filler metal has solidified. Complicated assemblies should be allowed to cool and be heat treated in a separate operation. Since proper heat treating times and temperatures depend on the aluminum alloy involved, the aluminum supplier should be consulted for particulars.

No useful purpose is served by using a heat treatable aluminum alloy for brazing unless the part is heat treated after brazing. The non-heat treatable materials are more economical in first cost and are often equal in strength to the annealed heat treatable alloys.

The corrosion performance of brazed aluminum assemblies depends highly upon the adequacy of flux removal. If the cleaning operations after brazing do not remove the flux residues, the combination of flux with moisture can cause corrosion attack of the metal. Such attack may not be enough to damage the part mechanically, but it usually affects the appearance of the parts.

Resistance to corrosion of brazeable aluminum alloys is not lowered by brazing — resistance to corrosion of brazed joints is of the same order as that of welded aluminum joints, and parts exposed to rural, marine, or industrial atmospheres will not show a greater rate of corrosive attack of the brazed joints. Resistance to corrosion of the joints to specific chemicals cannot be described in general terms, but where aluminum welds have been satisfactory, brazed joints will perform similarly. In certain cases, exposure tests under specific conditions are required for performance proof.

FURNACE BRAZING

Equipment for furnace brazing aluminum alloys is much the same as that used for heat treating these materials. Such equipment is designed to operate at temperatures up to 1200 °F (649 °C) and may be heated electrically or by oil or gas. Circulation of the furnace atmosphere to achieve uniformity of temperature increases production rate and improves the flow of the flux and filler metal. Circulation of the furnace atmosphere through radiators and other heat exchangers is necessary to ensure complete brazing at the center of the part. The usual procedure is to adjust the furnace time from 30 seconds to 2 minutes longer than is required to bring the assembly to brazing temperature. Too long a time at temperature may result in excessive silicon diffusion, undesirable flow of filler metal, and attack from flux.

Most furnace brazing of aluminum is done in air atmosphere, but some benefits can be gained by brazing in controlled atmosphere; by controlling moisture in the furnace at low levels, less flux is needed to achieve brazing. New types of brazing sheet designed for fluxless brazing require nitrogen or other inert atmospheres. Aluminum alloys are sometimes brazed to ferrous materials; in such cases, a nonoxidizing atmosphere is used to protect the ferrous material.

Care to prevent contact of the molten brazing flux with metallic furnace parts is desirable because the flux will corrode such parts. This is usually done by providing shallow pans of low carbon steel to catch any flux that may drop. These pans can be replaced at a much lower cost than can the furnace lining or the hearth parts. In some cases brazing aluminum in furnaces that are used periodically for brazing steel with copper filler metal has resulted in rapid attack of the heating elements and furnace lining. The apparent cause was the accelerated attack by the flux, which occurs at the higher brazing temperature used for the steel brazing operations after the furnace has been used for brazing aluminum.

Both continuous belt-type furnaces and batch furnaces have been used successfully. In some furnace designs, a fast rate of temperature rise is obtained by providing a high temperature zone at the entrance of the furnace in which the part is brought up to within 50 to 100 °F (10 to 38 °C) of the brazing temperature and then moved to the brazing zone which is controlled to ±10 °F (6 °C).

FURNACE BRAZING IN VACUUM

Within the past several years, successful brazing of aluminum alloys has been accomplished without flux. Furnaces operating in 10^{-5} torr (0.0015 Pa) range have been used, and the success has stemmed from the use of magnesium as vapor, which acts as a getter of oxide on the aluminum brazing surface. Other work has shown that magnesium in the base aluminum alloy, for example, 6061 aluminum alloy, can produce satisfactory braze joints. New aluminum brazing sheets with magnesium alloyed in the aluminum-silicon claddings BAlSi-6, BAlSi-7, and BAlSi-8 have been developed as well as new braze filler metals. Furnace brazing in vacuum is currently being used for production of automotive and aircraft heat exchangers.

DIP BRAZING

Dip brazing has been used widely in the manufacture of wave guides and heat exchanger assemblies. This process consists of dipping a preheated assembly with preplaced brazing filler metal into molten brazing flux. Thus fluxing and raising the part to brazing temperature are accomplished simultaneously.

Several pot designs and methods of heating the flux are commercially available. Dip pots can be cylindrical but are more commonly rectangular in shape, with either over-the-wall nickel, stainless steel, or carbon electrodes or through-the-wall carbon electrodes. Alternating current flowing between the electrodes through the liquid flux provides heating; thermal stirring of the bath achieves temperature uniformity. Resistance heating of the bath is used almost entirely for production applications although externally fired, gas-heated pots have been occasionally used. High alumina content brick linings have been the only materials with sufficient resistance to provide long life. Metal pots made from pure nickel have exhibited longer resistance than other metals, but not more than six months service can be expected due to intergranular corrosion (of the nickel pot). Repair operations on cracked or leaky pots have not been successful because of the difficulty in determining where the corrosion ends when the corroded area is chipped out (before the repair is welded).

Parts to be dip brazed are preheated prior to immersion in the molten brazing flux. Preheating temperatures range from 900 to 1065 °F (482 to 574 °C), depending on the size and complexity of the assembly. Preheating is done to dry the part (eliminating possible rapid steam generation) and prevent solidification of the flux on the assembly which occurs when a cold part is immersed. Solidification of the flux insulates the part and closes small openings so that the molten flux cannot reach the inside.

After brazing, parts should be suspended over the salt pot a sufficient length of time to allow all possible flux to drain back into the pot. This operation not only reduces flux consumption but also reduces the amount of flux that must be removed from the brazed unit.

Dip brazing is an excellent method for consistently producing high quality brazed joints in complicated assemblies such as heat exchangers and wave guides. Because the supply of flux is virtually inexhaustible, incomplete joints sometimes caused by lack of flux coverage in furnace brazing are nearly nonexistent.

Molten flux must be dehydrated before use to remove chemically combined moisture. Dehydration is accomplished by immersing coils of aluminum sheet in the flux. As long as moisture is present, hydrogen is evolved and bubbles up to the surface. Often it will ignite on the surface of the bath showing the characteristic small orange flames. Several hours of dehydration treatment are usually necessary to remove moisture from a freshly melted bath, depending upon size.

TORCH BRAZING

Torch brazing is generally used for small parts, short runs, and the attachment of fittings to previously welded, furnace-brazed, or dip-brazed aluminum assemblies. Mechanized torch brazing has found application in joining tube to header joints, return bends on heat exchangers, and other similar joint configurations.

All types of air-fuel gas or oxyfuel gas torches can be used. Brazing flux is mixed with water or alcohol and applied to both the work and filler metal by brushing, dipping, or sprinkling. The torch should be adjusted so the flame is slightly reducing. Heat is applied locally to the area to be joined until the flux and filler metal melt and wet the surfaces of the base metal.

DISSIMILAR METALS

Aluminum can be brazed to certain other metals as noted in Table 12.2. Inconel, nickel and beryllium can be brazed directly to aluminum. Steel, titanium, and Kovar must be precoated with another metal to improve alloy flow and protect

Table 12.2—Brazing aluminum to other metals

Metal or alloy joined to aluminum	Brazeability*	Remarks
Ferrous alloys	A	Best results obtained by electroplating or aluminum-coating the ferrous part
Nickel, Inconel	A	Can be brazed directly or precoated with aluminum
Titanium, Kovar	A	Aluminum coat the parts before brazing.
Beryllium	B	Can be wetted directly by aluminum brazing alloys
Monel	B	Can be wetted directly but may tend to be brittle
Copper, brass	C	Difficult, requires special techniques to avoid brittleness; can use a steel transition section
Magnesium	D	Too brittle to be of use

*Brazeability ratings

A - generally brazeable by all commercial brazing procedures and methods.

B - brazeable with special techniques or on specific applications that justify preliminary trials or testing to develop brazing procedure and performance of brazed joints.

C - limited brazeability.

D - not recommended for brazing.

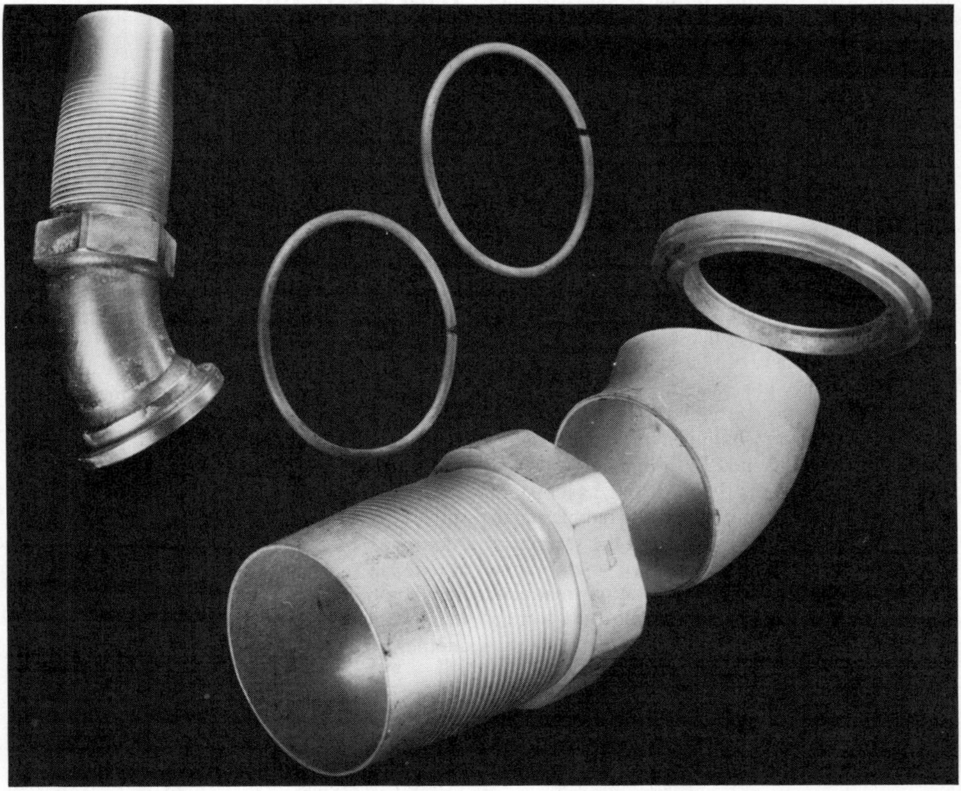

Fig. 12.3—Brazed hydraulic fitting.

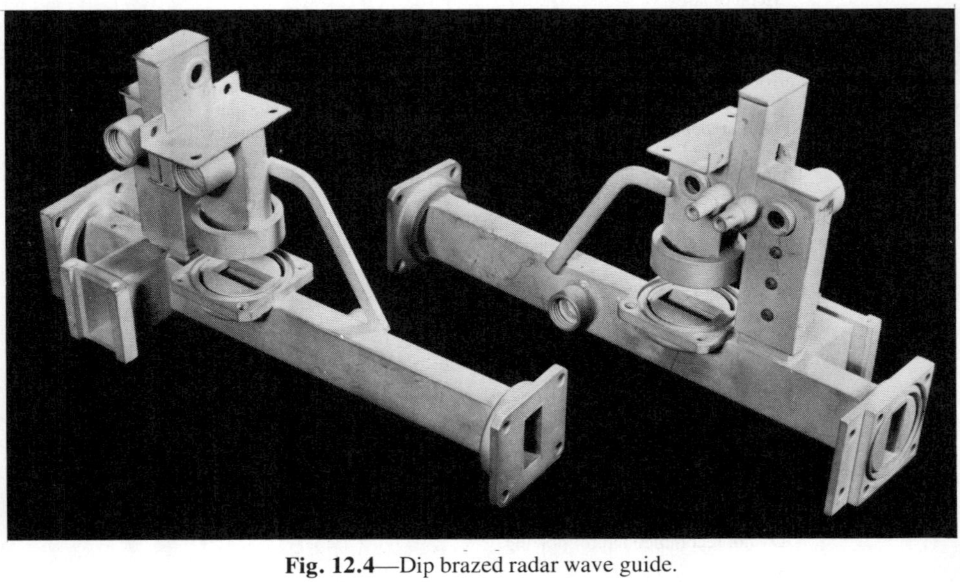

Fig. 12.4—Dip brazed radar wave guide.

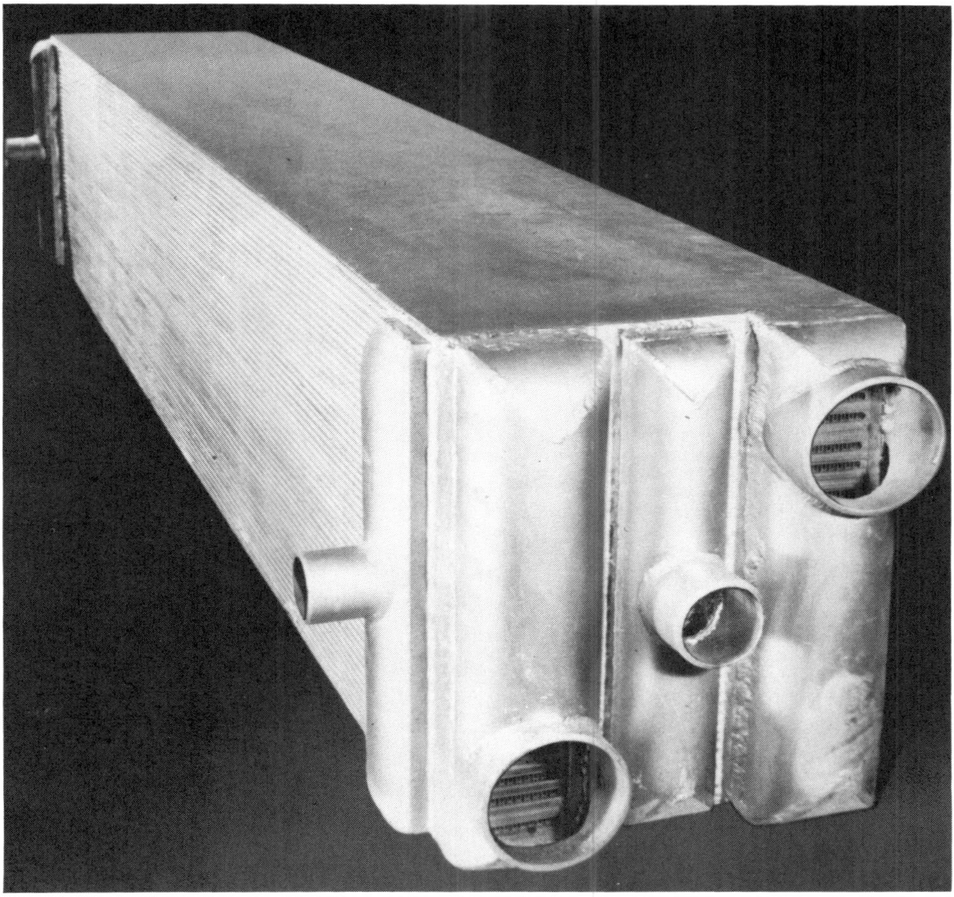

Fig. 12.5—Dip brazed heat exchanger.

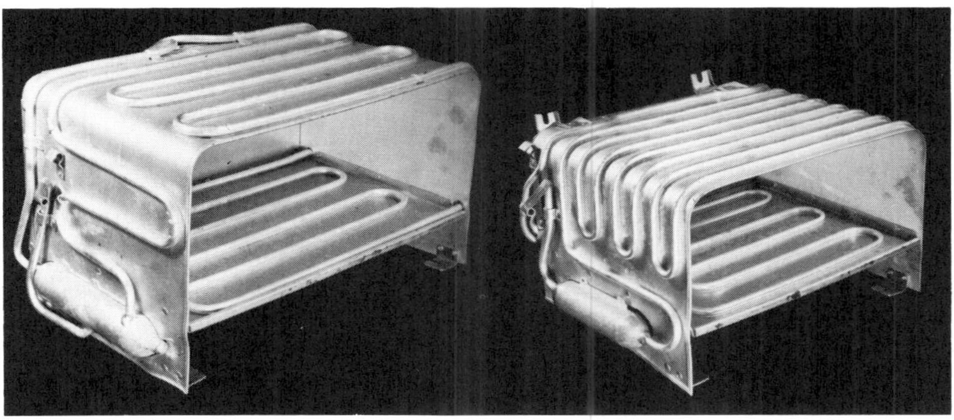

Fig. 12.6—Furnace brazed refrigerator evaporators.

Fig. 12.7—Vacuum brazed heat exchanger.

the surfaces from oxidation. Copper and brass cannot be brazed directly to aluminum.

Steel surfaces must be protected from oxidation during heating. Bare steel parts can be brazed to aluminum in vacuum or very low dew point inert atmosphere. Steel parts can be electroplated with nickel or hot-dip coated with aluminum for torch, furnace, or dip brazing applications. Aluminum-coated steel parts are readily brazed to aluminum using conventional filler metals and commercial flux. During brazing, aluminum and steel will alloy together and form a brittle intermetallic layer at the interface. Consequently, for quality assemblies, preheat and brazing times and temperatures must be kept to a minimum.

Aluminum cannot be brazed to magnesium. Because of their high mutual solubility, an extremely brittle aluminum-magnesium phase forms at the interface.

Bimetal joints require special treatments to insure adequate corrosion performance. Because of differences in electrochemical potential, bimetallic joints could corrode rapidly in aggressive environments. Corrosion performance can be enhanced by painting, encapsulation, or coating with a moisture impervious material.

POSTBRAZE CLEANING

After brazing, the corrosive flux residue must be removed. Flux residues form a hard, brittle layer and can best be removed by dissolution in hot water and by chemical cleaning. Mechanical cleaning, such as wire brushing or grinding, is not adequate; it breaks up the crust of flux into fine particles that become embedded in the surface, and on subsequent exposure to moisture may cause attack.

Several procedures are suitable for the removal of brazing flux. Immersion in hot water before the part has entirely cooled is effective in removing the major portion of the flux. For torch brazed joints particularly, the application of hot running water combined with brushing with a fiber brush is good practice. A five minute immersion in a 10% volume nitric-0.25% volume hydrofluoric acid bath or a two to five minute dip in a 1.5% volume hydrofluoric acid bath applied after a hot water dip will also remove the brazing flux. An acid dip of this type is always required if the joints are inaccessible for brushing. Good results can be obtained with a number of commercial cleaners. All these methods require clean water rinse to remove the cleaning agent.

CHAPTER 13

MAGNESIUM AND MAGNESIUM ALLOYS

Brazing techniques similar to those used on aluminum are used on magnesium alloys. Furnace, torch, and dip brazing can be employed although the latter process is the most widely used.

BASE METALS

The composition and physical properties of brazeable magnesium alloys are given in Table 13.1. Furnace and torch brazing experience is limited to M1A alloy whereas dip brazing has been applied to AZ10A, AZ31B, K1A, M1A, ZE10A, and ZK21A alloys. Because of their low solidus temperatures, other magnesium alloys cannot be brazed with present brazing filler metals and techniques.

FILLER METALS

The filler metals used in magnesium brazing are magnesium base alloys, as shown in Table 13.1. Brazing filler metals BMg-1 and BMg-2 are suitable for the torch, furnace, or dip brazing processes. Because of its lower melting range, the BMg-2a brazing filler metal is usually preferred in most brazing applications.

JOINT DESIGN

Joints should be designed to take advantage of capillary attraction and to allow the flux to be displaced by the brazing filler metal as it flows into the joint. Because of the corrosive nature of the flux, care should be taken in the design of the joints to minimize flux entrapment. Both lap and butt joints can be used on magnesium. Joint clearances of 0.004-0.010 in. (0.10-0.25 mm) at the brazing temperature are satisfactory; however, it is best to design the joint for the least clearance that will permit good flow of filler metal in order to take full advantage of capillary attraction. In dip brazing, the joint should be designed with slots or recessed grooves to accommodate the filler metal and prevent its being washed into the flux bath (see chapter 6).

COMBUSTIBILITY OF MAGNESIUM

The ease of ignition of magnesium depends to a large extent upon the size and shape of the material, as well as on the size or intensity of the source of ignition. In the form of ribbon, shavings, or chips with thin, featherlike edges a spark or the flame of a match may be sufficient to start the metal burning. Heavier sections, such as fabricated assemblies, ingots, or castings, are difficult to ignite because heat is conducted rapidly away from the source of ignition. If the entire piece of metal is raised to its ignition temperature (approximately the solidus temperature of the alloy), self-sustained burning will occur.

The probability of combustion occurring during the torch brazing of magnesium is extremely remote because only the joint area and not the entire structure is heated, and the fluxes used

Table 13.1—Composition and physical properties of brazeable magnesium alloys and filler metals

AWS A5.8 classification	ASTM alloy designation	Avail. forms	Nominal composition, % (bal. Mg) Al	Zn	Mn*	Zr	Rare earth	Be	Specific gravity	Density lb/in³	Density kg/m³	Solidus F	Solidus C	Liquidus F	Liquidus C	Brazing Range F	Brazing Range C	Suitable filler BMg-1	Suitable filler BMg-2a
Base metal																			
—	AZ10A	E	1.2	0.4	0.20	—	—	—	1.75	0.063	1760	1170	632	1190	643	1080-1140	582-616	X	X
—	AZ31B	E,S	3.0	1.0	0.20	—	—	—	1.77	0.064	1790	1050	566	1160	627	1080-1100	582-593		X
—	K1A	C	—	—	—	0.70	—	—	1.74	0.063	1760	1200	649	1202	650	1080-1140	582-616	X	X
—	M1A	E,S	—	—	1.2	—	—	—	1.76	0.063	1760	1198	648	1202	650	1080-1140	582-616	X	X
—	ZE10A	S	—	1.2	—	—	0.17	—	1.76	0.063	1760	1100	593	1195	646	1080-1100	582-593		X
—	ZK21A	E	—	2.3	—	0.60	—	—	1.79	0.064	1790	1159	626	1187	642	1080-1140	582-616	X	X
Filler metal																			
BMg-1	AZ92A	†	9.0	2.0	0.1	—	—	0.0005	1.83	0.066	1850	830	443	1110	599	1120-1140	604-616	—	—
BMg-2a	AZ125A	†	12.0	0.5	—	—	—	0.0005	1.89	0.068	1900	770	410	1050	566	1080-1130	582-610	—	—

E = Extruded shapes and structural sections
S = Sheet and plate
C = Castings
* = Minimum
† = See Table 3.2

prevent burning in the heated joint area. The dip brazing process presents no combustion problems with magnesium because the entire operation is performed in a flux bath. During the 850-900 °F (454-482 °C) preheat of the assembly prior to dip brazing, however, care should be exercised to assure that proper temperature controls are used and that preheat time is the minimum needed to bring the assembly to temperature.

Furnace brazing offers the greatest potential for magnesium combustion because the entire structure is heated to the brazing temperature. However, since M1A alloy with its high solidus temperature (1198 °F; 648 °C) is the only alloy suitable for furnace brazing, the risk of combustion is minimal provided furnace temperature is controlled within the prescribed limits and brazing time is the minimum necessary to achieve filler metal flow.

BRAZING PROCESSES

Furnace Brazing

When brazing AZ31B alloy, electric or gas heating equipment with automatic temperature controls capable of holding the furnace temperature within ±10 °F (6 °C) of the brazing temperature should be used to minimize incipient melting of the base metal and to reduce the potential of a magnesium fire. No special furnace atmosphere is required; however, atmospheres produced by the products of combustion of gas-fired furnaces or SO_2, which is used in heat treating magnesium alloys, reduce filler metal flow and should be avoided. Parts to be brazed should be assembled with the filler metal preplaced in or around the joints. Joint clearances of 0.004-0.010 in. (0.10-0.25 mm) should be used to obtain maximum capillary attraction of the filler. Best results are obtained if dry powdered flux is sprinkled along the joint. Flux pastes made with water or alcohol are unsatisfactory because they retard the flow of filler metal. Flux pastes made with benzol, toluene, or chlorbenzol may be used, but they are not smooth and thus more difficult to apply. Flux pastes should be dried by heating at 350-400 °F (177-204 °C) for 5 to 15 minutes in drying ovens or circulating air furnaces. Flame drying is not desirable since a heavy soot deposit will result.

Brazing time will depend somewhat on the thickness of the materials used and the amount of fixturing necessary to position the parts. The brazing time should be the minimum necessary to obtain complete filler metal flow so as to avoid excessive diffusion of the filler metal and minimize the potential for combustion of the magnesium. Normally, one or two minutes at the brazing temperature is sufficient to make the braze. After brazing, the parts should be allowed to air-cool away from drafts to minimize distortion.

Torch Brazing

Torch brazing is accomplished using a neutral oxyfuel or air-fuel gas flame. Because of the close proximity of the initial melting point of the base metal (solidus temperature), and the flow point of the brazing filler metal (liquidus temperature), manual torch brazing with BMg-1 filler metal is difficult and requires considerable operator skill. For this reason, BMg-2a filler metal is generally preferred for torch brazing. The filler metal should be placed on the joint and fluxed before heating. Face feeding of the filler metal is difficult because of the close proximity of the solidus temperature of the base metal and the flow temperature of the filler metal. Flux pastes made with alcohol give better results and should be used in preference to pastes made with water. Heat is applied to the part until the flux melts and is continued until the brazing filler metal melts, wets the surface, and flows into the joint by capillary attraction. Overheating of the base metal should be avoided because rapid diffusion and drop-thru of the metal will result. Natural gas is well suited for torch brazing because its relatively low flame temperature reduces the danger of overheating.

Dip Brazing

Dip brazing is accomplished by immersing the assembly in a molten brazing flux held at the desired brazing temperature. The flux serves the dual function of both heating and fluxing. When brazing AZ31B alloy, temperature control should be accurate to within ±10 °F (6 °C) of the desired brazing temperature to minimize incipient melting.

Joint clearance in dip brazing should be from 0.004-0.010 in. (0.10-0.25 mm). After preplacing the filler metal, the parts to be brazed are assembled in the fixture (preferably of stainless steel to resist the corrosive action of the flux) and preheated in a furnace to 850-900 °F (455-482 °C). Preheating is necessary to drive off any moisture on the assembly as well as to minimize thermal shock and flux entrapment. Immersion time in the flux bath is relatively short (approximately 30-45 s for 0.063 in. [1.6 mm] base metal) because of the rapid heating of the parts by this method. Large assemblies with more metal mass and/or fixturing may require from 1 to 3 min. immersion. Because of the large volume of flux and uniform heating, more consistent results are achieved with dip brazing than with other brazing processes.

PRECLEANING AND SURFACE PREPARATION

Parts to be brazed should be thoroughly clean and free from burrs. Oil, dirt, and grease should be removed in hot alkaline cleaning baths or by vapor or solvent degreasing. Such surface films as chromate conversion coatings or oxides must be removed by mechanical or chemical means immediately prior to brazing. In mechanical cleaning, abrading with aluminum oxide cloth or steel wool has proved very satisfactory. Chemical cleaning should consist of a 5 to 10 min. dip in hot alkaline cleaner followed by a 2 min. dip in ferric-nitrate bright pickle solution (see Table 13.2).

FLUXES

AWS Brazing Fluxes, Type No. 2 (see Table 4.1), are used to clean the surface of the magnesium, which permits capillary flow. Because of the corrosive nature of these fluxes, complete removal is of utmost importance if good corrosion resistance is to be obtained in brazed joints.

ASSEMBLY

Methods that can be used for prebraze assembling of magnesium components are staking, use of self-positioning and spring-loaded fixtures, or tack welding. Mild steel or stainless steel fixtures are generally used in magnesium brazing. In dip brazing, fixtures should be small and simply designed to minimize heat losses in the flux bath and to allow complete drainage of the flux from the fixture. Tack welding may be desirable to eliminate massive or complex fixtures in short production runs. If tack welding is used, M1A welding rod is recommended because its melting point is above the brazing temperature. The BMg filler metals are usually hot formed to fit the joint by heating to 500-600 °F (260-316 °C). The filler

Table 13.2—Chemical treatment solutions

Treatment	Composition	Method of application
Ferritic nitrate bright pickle	Chromic acid 1.5 lb (0.68 kg) Ferritic nitrate, 5.33 oz (0.15 kg) Potassium fluoride, 0.5 oz (0.014 kg) Water to make one gallon Temperature 60-100 F (16-38 C)	0.25 to 3 min immersion followed by cold and hot water rinses and air dry
Chrome pickle	Sodium dichromate, 1.5 lb (0.68 kg) Conc. nitric acid, 1.5 pt (0.71 l) Water to make one gallon Temperature 70-90 F (21-32 C)	1 to 2 min immersion, hold in air 5 s, followed by cold and hot water rinses and air or forced dry 250 F max (121 C max)
Postbraze cleaner	Sodium dichromate, 0.5 lb (0.23 kg) Water to make one gallon Temperature 180-212 F (82-100 C)	2 h immersion in boiling bath followed by cold and hot water rinses and air dry

metals can be formed over a heated steel mandrel to the desired contour.

POSTBRAZE CLEANING

Regardless of the brazing process used, complete removal of all traces of flux is of utmost importance. Brazed parts should be rinsed thoroughly in flowing hot water to remove the flux from the surface of the part. A stiff-bristled brush can be used to scrub the surface and speed up the cleaning process. The parts should then be given a 1 to 2 min immersion in chrome pickle, followed by a 2 hour boil in postbraze flux remover cleaner (see Table 13.2).

CORROSION RESISTANCE

The corrosion resistance of brazed joints depends primarily on the thoroughness of flux removal and the adequacy of joint design to prevent flux entrapment. Since the brazing filler metal is a magnesium base alloy, the problem of galvanic corrosion is minimized.

INSPECTION

After cleaning, visual inspection of the assembly may reveal incomplete filler metal flow, which might be attributed to inadequate part cleanli-

Table 13.3—Typical mechanical properties of brazeable magnesium alloys

ASTM alloy designation	Form	Temper	% E	TYS ksi	TYS MPa	UTS ksi	UTS MPa	CYS* ksi	CYS* MPa
AZ10A	E	F	10	21	145	35	241	10	69
AZ31B, C	E	F	14	28	193	38	262	15	103
AZ31B	S	H24	15	32	221	42	290	24	165
AZ31B	S	O	21	22	152	37	255	16	110
K1A	C	F	14	8	55	23	159	—	—
M1A	E	F	9	20	138	34	234	11	76
M1A	S	H24	9	27	186	37	255	—	—
M1A	S	O	15	16	110	32	221	—	—
ZE10A	S	H24	12	26	179	37	255	25	172
ZE10A	S	O	23	20	138	33	228	16	110
ZK21A	E	F	10	33	228	42	290	—	—

E = Extruded shapes and structural forms
S = Rolled sheet
C = Sand casting
F = As fabricated
O = Annealed condition
H24 = 0.5 Hard, work hardened condition
*Compressive yield strength

Table 13.4—Tensile properties of magnesium sheet alloys after heating in brazing flux

Heating Time, min.	Temp, F	Temp, C	AZ31B-0 UTS ksi	AZ31B-0 UTS MPa	AZ31B-0 TYS ksi	AZ31B-0 TYS MPa	AZ31B-0 %E	ZE10A-0 UTS ksi	ZE10A-0 UTS MPa	ZE10A-0 TYS ksi	ZE10A-0 TYS MPa	ZE10A-0 %E
None	None		36.2	250	20.7	143	23.5	32.9	227	20.2	139	22.5
0.5	1075	579	34.4	237	17.0	117	19.0	27.3	188	10.8	74	16.5
1	1075	579	33.9	234	16.5	114	17.0	27.3	188	10.6	73	14.5
2	1075	579	34.0	234	16.9	117	17.5	26.6	183	10.1	70	12.0
3	1075	579	33.4	230	15.8	109	15.5	26.2	181	9.9	68	10.0
5	1075	579	32.9	227	16.0	110	13.0	25.2	174	9.0	62	10.5
0.5	1100	593	33.8	233	16.5	114	17.0	27.6	190	11.5	79	15.0
1	1100	593	33.2	228	15.9	110	16.0	27.0	186	11.1	77	14.0
2	1100	593	32.6	225	15.4	106	14.5	26.7	184	11.1	77	13.0
3	1100	593	32.5	224	14.6	101	15.0	26.8	185	11.0	76	12.0
5	1100	593	31.5	217	14.4	99	14.0	25.7	177	10.4	72	10.5
0.5	1125	607	32.5	224	16.1	111	14.0	27.5	190	10.6	73	16.0
1	1125	607	31.4	217	12.1	83	12.0	27.2	188	11.2	77	15.5
2	1125	607	29.2	201	10.8	74	9.0	26.5	183	11.0	76	16.5

ness, insufficient flux, or too low a temperature. In addition, the nondestructive tests, e.g., radiography, fluorescent and dye penetrant; pressure tests; or electrical resistance, outlined in chapter 11, can be used. Flux inclusions can also be detected by exposing the joint to a high humidity atmosphere (95%) for several days and inspecting for indications of flux residue.

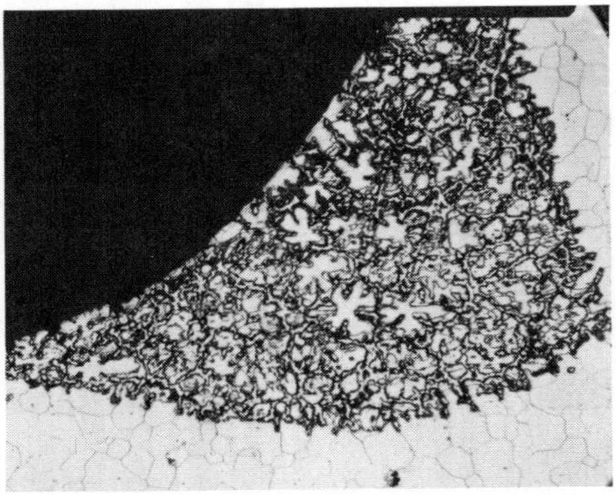

Fig. 13.1—Brazed joint in AZ31B base metal using BMg-2a, 65× (reduced by 20% on reproduction).

The adequacy of the test brazements can be determined by etching a cross section of the joint and examining it under a microscope. Although optimum detail of the brazed joint microstructure is obtained when conventional magnesium metallographic etchants are used, sufficient joint detail is revealed by etching the cross section in a 5 to 10% solution of acetic acid. A typical joint with BMg-2a filler metal is shown in Fig. 13.1.

STRENGTH OF BRAZEABLE MAGNESIUM ALLOYS

Typical mechanical properties of all brazeable magnesium alloys are shown in Table 13.3. The temperatures involved in brazing will reduce the properties of work hardened (tempered) magnesium sheet alloys to the 0 temper (annealed) level. Torch brazing will reduce properties only in those areas heated for brazing; furnace and dip brazing will reduce properties of the entire structure. Properties of the cast or extruded alloys or of the annealed temper sheet alloys are not greatly affected by the heat of brazing because their metallurgical microstructure is already recrystallized. Table 13.4 illustrates the minor re-

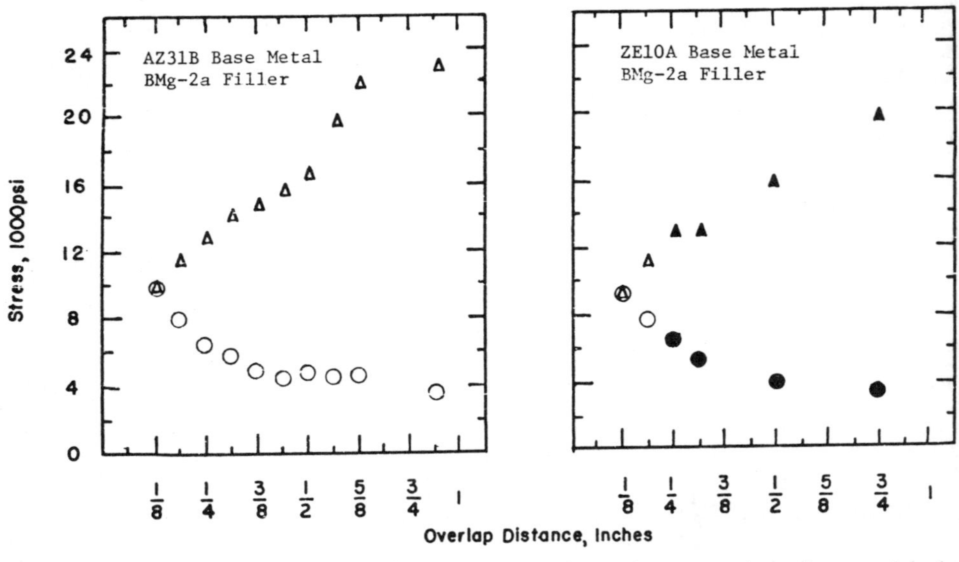

Fig. 13.2—Average unit shear stress in the filler metal (circles) and average unit tensile strength in the base metal (triangles) at failure as a function of overlap distance. Open symbols represent failure in the filler metal; filled symbols represent failure in the base metal. Tested per AWS C3.2.

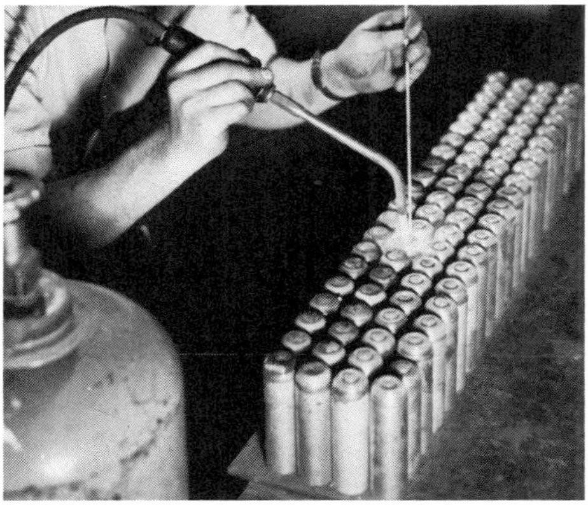

Fig. 13.3—Torch brazing magnesium hydraulic lift float.

Fig. 13.4—Longitudinal section of a completed float.

duction in strength effected on annealed sheet alloys by exposure to various brazing temperatures.

STRENGTH OF BRAZED JOINTS

The strength of brazed joints in AZ31B and ZE10A sheet alloys brazed with BMg-2a filler metal and tested per AWS C3.2, Standard Method for Evaluating the Strength of Brazed Joints, is shown in Fig. 13.2. On base metal 0.125 in. (3.2 mm) thick a 0.25 in. (6.4 mm) overlap is necessary to ensure base metal failure in ZE10A brazements; on AZ31B alloy base metal failure does not occur even at overlaps as great as ⅞ in. (22.2 mm). The difference exists because ZE10A alloy suffers a greater loss of tensile properties on exposure to brazing temperature than does AZ31B alloy (see Table 13.4).

TYPICAL APPLICATIONS

Figure 13.3 shows the torch brazing of M1A magnesium alloy hydraulic lift floats. Filler metal BMg-1 is face fed into the annular groove and upon melting is distributed through the joint by capillary attraction. The flame is directed against the plug so that tube walls are heated largely by conduction. Figure 13.4 shows a longitudinal section of a completed float.

Figure 13.5 shows a ZW1 (British magnesium alloy nearly equivalent to ASTM ZK21A alloy) battery container (0.040 to 0.625 in. [1.0 to 15.9 mm] sheet and plate) dip brazed with BMg-2a filler. Photos (a) and (b) show details of the magnesium assembly as well as details of the stainless steel brazing fixture. Note especially the spring loading of the front and back supports to accommodate differences in thermal expansion between magnesium and stainless steel. Photo (c) shows the battery container after brazing and flux removal.

Figure 13.6 shows an M1A magnesium alloy microwave antenna dip brazed with BMg-2a filler. Photo (a) shows details of the assembly without the feeder and end caps in position. Photo (b) shows the assembly in the stainless steel brazing fixture; an expendable section of magnesium wave guide is used in three locations along the length of the assembly to assist in positioning the parts during brazing. Photo (c) shows the antenna after brazing and flux removal.

a

b

c

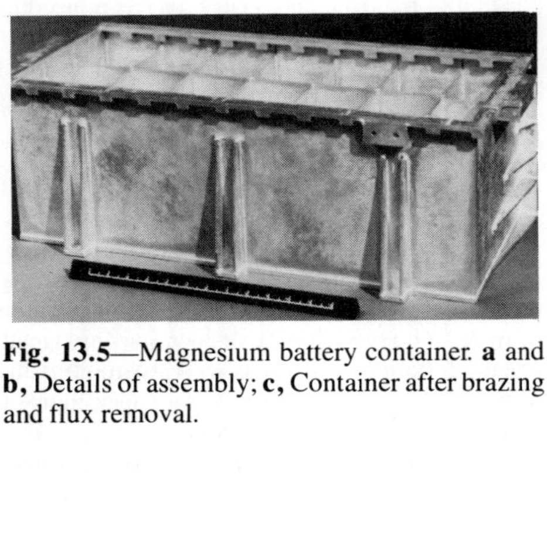

Fig. 13.5—Magnesium battery container. **a** and **b,** Details of assembly; **c,** Container after brazing and flux removal.

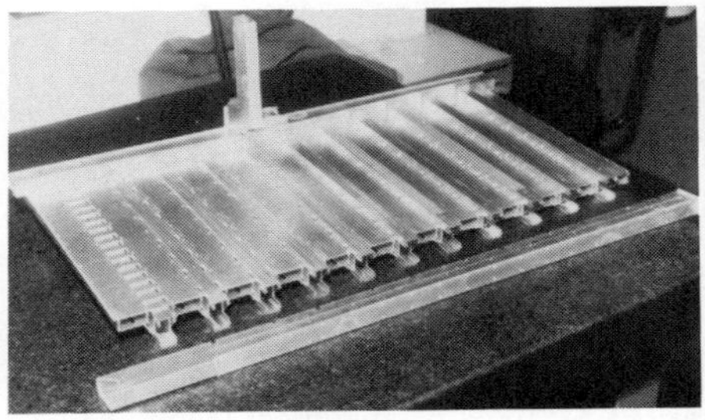

a

b

c

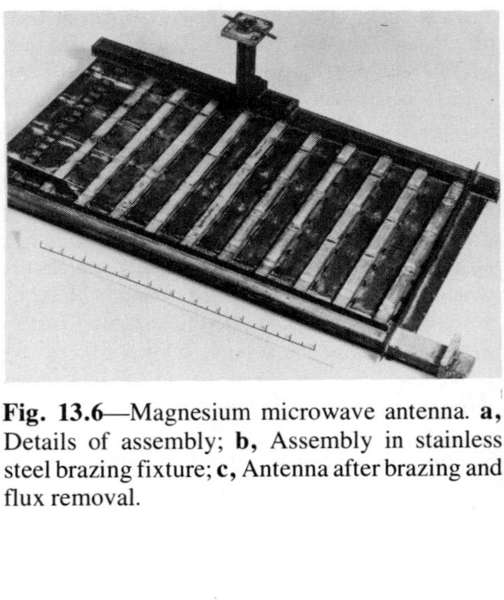

Fig. 13.6—Magnesium microwave antenna. **a,** Details of assembly; **b,** Assembly in stainless steel brazing fixture; **c,** Antenna after brazing and flux removal.

CHAPTER 14

COPPER AND COPPER ALLOYS

Copper and its alloys are selected for diverse applications owing to the wide range of properties that can be obtained from this family of metals. These alloys normally have excellent corrosion resistance, high electrical and thermal conductivities, and formability. Brazing is often used in fabrication and can be performed readily with a few precautions.

The compositions of wrought and cast alloys are shown in Tables 14.1 and 14.2. In comparison to steels, copper and its alloys have higher thermal conductivities and higher thermal expansion. This is an especially important consideration when brazing dissimilar metals.

As with all metals, cracking, distortion, and softening may occur if proper precautions are not taken when copper and copper base alloys are brazed; however, these problems should not discourage the use of these alloys in brazed assemblies. A great volume of copper and copper base alloys is successfully brazed daily in industry. A working knowledge of the factors leading to cracking, distortion, and softening difficulties can do much to eliminate them or help predict the properties of the resulting brazement.

Softening of the base metal frequently occurs during brazing because many copper base alloys derive their properties as a result of heat treatment at relatively low temperatures or cold work, or both. Heat treated alloys, in general, soften less than cold worked alloys for a given brazing brazing temperature and time at temperature. The degree of softening increases with higher temperatures and longer times of exposure to elevated temperatures. Softening of areas remote from the braze can be minimized by cooling the assembly, except for the area to be brazed, by immersion in water, packing with wet rags or wet asbestos, or otherwise providing a heat sink to keep the overall temperature of the part as low as possible. In all cases, brazing at the lowest temperature possible with a low melting brazing filler metal will minimize softening.

Residual stresses arising from cold working, casting, or machining operations can cause cracking during brazing. Heating and cooling can induce additional stresses as a result of uneven expansion or contraction. If the total stresses induced by all of these factors are sufficient, cracking at elevated temperatures can result. The importance of uniform and controlled heating cannot be overemphasized, especially when dealing with some of the brasses, cold worked phosphor bronzes, and cold worked silicon bronzes.

BASE METALS

Oxygen-Bearing Coppers

This group includes the fire-refined and electrolytic tough pitch grades of copper and silver-bearing copper (10-20 oz./ton [300-600 g/tonne]). Fire-refined copper is produced from the nonelectrolytic processing of scrap copper or from the treatment of ores. It contains very small percentages of impurities as well as from 0.02 to 0.05% oxygen in the form of a copper-cuprous oxide eutectic. The copper-cuprous oxide eutectic is scattered as globules throughout wrought metal and appears as an interdentritic structure in

Table 14.1—Wrought coppers and copper alloys

Number series	Description	Composition ranges	Representative electrical conductivities, % IACS
101 to 107	Oxygen-free	99.95% Cu or better	> 100
109 to 142	Tough-pitch and deoxidized	Contain oxygen or deoxidizers	80 to 100
145 to 147	Free-machining	Small additions of S, Te, etc.	> 90
150 to 194	High copper alloys	Neighborhood of 1 or 2% additions of Cd, Be, Cr, Co, Fe, Ni, Zn, and/or Sn	20 to 85
205 to 240	Red brasses	Up to 20% Zn	35 to 60
250 to 298	Yellow brasses	From 25 to 50% Zn	25 to 35
310 to 385	Leaded brasses	From 10 to 45% Zn and up to 4.5% Pb	25 to 45
405 to 485	Tin brasses	To 5.5% Sn, to 48% Zn	25 to 30
502 to 529	Copper-tin alloys (phosphor bronzes)	From 1 to 11% Sn	10-50
532 to 546	Leaded phosphor bronzes	1 to 4% Pb, about 5% Sn, some with additions of Zn	10-20
606 to 642	Aluminum bronzes	From 2.6 to 13% A1, to 5% Fe, some with additions of Si or Ni	10-20
647 to 661	Silicon bronzes	From 1 to 3.5% Si, some with Mn, Si, or Sn	7-12
665 to 697	Alloy brasses	Zinc-containing alloys with additions of Ni, Sn, Mn, Al, and Si	20-25
701 to 720	Copper nickels	From 2 to 40% Ni, additions of Fe, Be, Mn, or Cr	4-10
732 to 798	Nickel silvers	From about 43 to 73% Cu, from 7 to 23% Ni, some with Pb or Mn, balance Zn	5-10

NOTE: For specific compositions and properties see Standards Handbook No. 2, Copper Development Association, N.Y.

cast products. It has no serious effects on mechanical properties or electrical conductivity; however, it does make the copper susceptible to embrittlement when heated in a hydrogen atmosphere. Hydrogen rapidly diffuses into the metal and reacts with oxides to form steam, which expands and creates porosity at the grain boundaries of the copper. Carbon monoxide, which may be present in oxyacetylene flames or reducing environments, can contribute to weakness if moisture is also present, since carbon monoxide may reduce water vapor to hydrogen, which will then diffuse into the metal. A final point to remember on embrittlement is that when oxygen-bearing coppers are heated above 1690 °F (921 °C) for prolonged periods, as during some brazing operations, copper oxide concentrates in the grain boundaries and causes major reduction in strength and ductility.

Oxygen-bearing coppers are of medium strength and low hardness; if the copper oxide is uniformly distributed, these coppers are tough, ductile, and highly malleable. Their electrical and thermal conductivities are the highest of all copper alloys except for the oxygen-free copper grades. They may be softened at temperatures of 450 to 1500 °F (232 to 816 °C), depending on the time and temperature of exposure. Annealing in a hydrogen-containing reducing atmosphere should be avoided because of the potential for the production of voids or fissures, which weaken the material.

Deoxidized and Oxygen-Free Coppers

This group includes phosphorus-deoxidized copper and oxygen-free copper. Phosphorus-deoxidized copper is usually copper from which oxygen has been removed by a precasting addition of 0.01 to 0.04% phosphorus. In some cases phosphorus is added to oxygen-free copper to provide a greater margin of protection against hydrogen embrittlement. If the phosphorus addition is closely controlled to obtain less than 0.01% residual, the copper is a high conductivity (low phosphorus) deoxidized copper. Phosphorus-deoxidized copper with 0.01 to 0.04% phosphorus has lower electrical and thermal conductivities than the low phosphorus grade.

Oxygen-free copper is copper from which the oxygen has been removed by melting and casting copper cathodes under atmospheres that elimi-

Table 14.2—Cast coppers and copper alloys

Number series	Description	Composition ranges	Representative electrical conductivities, % IACS
801 to 811	Coppers	Minimum of 99.70% Cu and remainder Ag	92-100
813 to 828	High copper alloys	Additions of up to about 2.5% Be, Co, Si, Ni, and/or Cr	20-80
833 to 838	Red brasses	83 to 93% Cu, to 12% Zn with lesser amounts of Sn, Pb	115-40
842 to 848	Semi-red brasses	76 to 80% Cu, 8 to 15% Zn with lesser amounts of Sn, Pb	15-20
852 to 858	Yellow brasses	57 to 72% Cu, balance primarily Zn, 1 to 2% Sn, Pb, Ni, or Al	18-28
861 to 868	High-strength yellow brasses	55 to 67% Cu, additions of Fe, Ni, Mn, Al, balance Zn	7-22
872 to 879	Silicon brasses and silicon bronzes	65 to 90% Cu, about 3 to 5% Si, some with large amounts of Zn	6-15
902 to 945	Tin bronzes	3 to 19% Sn, some with large amounts of Pb, less Zn, Ni	7-15
947, 948	Nickel-tin bronzes	About 5% Sn and 5% Ni, to 2.5% Zn, Alloy 948 has 1% Pb	12
952 to 958	Aluminum bronzes	7 to 11% Al, at least 71% Cu, balance Ni, Fe, Mn, and/or Si	3-13
962 to 966	Copper nickels	10 to 31% Ni, about 1% additions of Fe, Cb, Si, Mn and/or Si	4-11
973 to 978	Nickel silvers	55 to 65% Cu, Pb and Sn additions, 12 to 25% Ni, balance Zn	4-5

NOTE: For specific compositions and properties see Standards Handbook No. 7, Copper Development Association, N.Y.

nate oxygen and prevent the formation of oxides during the casting. Since no deoxidant is introduced into the cast copper, copper produced in this way can absorb some oxygen from the air during long heating at very high temperatures and lose its oxygen-free characteristics, at least near the surface.

Phosphorus-deoxidized and oxygen-free coppers are highly ductile. The absence of the copper-copper oxide eutectic improves the cold working properties of these coppers over those of oxygen-bearing coppers, particularly for deep drawing and spinning operations. These coppers can be annealed between 450 and 1500 °F (232 to 816 °C) in reducing atmospheres, since they will not be embrittled by hydrogen. Corrosion resistance is the same as that of oxygen-bearing coppers.

Special Coppers

This group includes some alloys that offer high electrical conductivity in addition to special properties unavailable in ordinary coppers. Lead, tellurium, selenium, and sulfur-coppers have machinability ratings of about 80, compared to 20 for ordinary coppers (based on a rating of 100 for free-cutting brass). However, ductility and workability are reduced somewhat by the presence of the inclusions that impart free machining characteristics. These free-cutting coppers can be supplied with a matrix of deoxidized or oxygen-free copper to prevent embrittlement or gassing when the special coppers are brazed in hydrogen atmospheres. They are widely used in the manufacture of electrical connectors.

High Coppers

The high copper group includes those with enhanced mechanical properties due to the addition of small amounts of alloying elements. Chromium copper combines a tensile strength of about 75 000 psi (517MPa) with a conductivity of about 80% IACS after age hardening. Zirconium copper develops somewhat lower strength than chromium copper, but it has a conductivity of 90% IACS.

Copper-beryllium* alloys are of two types: one has 1.5 to 2% beryllium, while the other has

*Beryllium has been identified as a toxic element. Brazing beryllium containing alloys requires adequate ventilation and cleanliness. See chapter 29.

about 0.5% beryllium. Cobalt additions are made to these alloys to restrict grain growth during annealing. Alloys with about 0.5% beryllium have higher conductivity but lower strength. Beryllium-containing alloys have a moderate tensile strength in the cold rolled (but not age hardened) condition. When age hardened (by heat treatment at temperatures in the neighborhood of 600 °F, 316 °C), they have the highest tensile strength and hardness of all copper alloys. Both cold worked and age hardened alloys can be softened and made malleable by annealing in an exothermic atmosphere at 1450 to 1500 °F (788 to 816 °C), followed by water quenching.

Copper-Zinc Alloys (Brasses)

Copper-zinc alloys are commercially produced with varying ratios of the two elements to provide different properties. Other elements are occasionally added to enhance particular mechanical or corrosion properties; these special brasses are identified by the added element — for instance, aluminum, which gives aluminum brass. Additions of manganese, tin, iron, silicon, nickel, lead, and aluminum, either singly or collectively, rarely exceed 4%, however, these additions may have important effects on the brazing characteristics of brasses.

In general, brasses can be divided into three classifications: low brasses (20% Zn max.), high brasses (more than 20% Zn), and alloy brasses. On sudden exposure to elevated temperatures, brasses in a stressed condition are susceptible to cracking. To avoid such failure, highly stressed material should be gradually brought up to brazing or heat treating temperatures.

Brasses can be softened after cold work operations by annealing in the temperature range of 700 to 1400 °F (371 to 760 °C). If quenched from the annealing temperatures, the alpha-beta brasses show an increased hardness because of the higher proportion of beta phase than would be found after air or furnace cooling. Stress relieving at 500 to 700 °F (260 to 371 °C) will reduce the possibility of residual welding or forming stresses contributing to failure.

Leaded Brasses

Lead is commonly added to brass in amounts up to 5% to improve machinability. This usually has no effect on the tensile strength, particularly of cold worked materials, but it increases the possibility of cracking in materials that are stressed and subjected to heat. Lead has no effect on corrosion resistance, but it impairs the brazeability.

Copper-Tin Alloys (Phosphor Bronzes)

Alloys of copper and tin are properly termed tin bronzes. However, since 0.03 to 0.4% phosphorus is added during casting as a deoxidizing agent and may be found as a residual in the alloys, the tin bronzes have come to be known commercially as phosphor bronzes. Phosphor bronzes have moderate to very high tensile strength, depending on tin content and degree of cold work.

Commercial alloys are available with from 1.0 to 10% tin. In the completely homogenized condition, they are single-phase alloys with a structure similar to alpha brass. The alloys with a tin content over 5% are difficult to cast without dendritic segregation and the formation of a beta phase. During cooling, this beta phase gives rise to a delta phase, which can be embrittling. Electrical and thermal conductivities are low for the low tin phosphor bronzes, and very low for those with high tin content.

All the phosphor bronzes have good cold working properties and high strength and hardness in the hard rolled tempers. After cold working, these alloys can be rendered soft and malleable by annealing at temperature between 900 and 1400 °F (482 and 760 °C), depending on the properties desired. In a stressed condition, the phosphor bronzes are subject to cracking during brazing, and sudden application of high heat must be avoided.

Some wrought phosphor bronzes contain up to 4.5% lead to improve machinability. Many cast bronzes also contain lead, up to 6%, to improve both casting and machining properties. This lead can precipitate out of these metals at brazing temperatures; it tends to prevent wetting by the filler metals and to alloy with these metals to produce brittle joints. Both tendencies increase with temperature and lead content. These alloys also tend to crack readily when stressed at brazing temperatures. It is often necessary to avoid brazing of components and assemblies having sharp corners that concentrate stresses and those with large changes in cross section that produce excessive thermal stresses with temperature changes.

Copper-Aluminum Alloys (Aluminum Bronzes)

Copper-aluminum alloys (aluminum bronzes) are high copper alloys containing 3 to 13% aluminum, with or without varying amounts of iron, nickel, manganese, and silicon. The aluminum bronze alloys are divided into two types. The first includes those alloys that do not undergo phase changes on heat treatment — the alpha or single-phase alloys. The second type includes those that can be hardened by heat treatment — the duplex or multiphase alloys whose properties depend on composition and heat treatment. Both types of aluminum bronzes have low electrical and thermal conductivities.

The alpha phase aluminum bronzes are characterized by relatively high tensile strength, high toughness, good ductility, and moderate hardness. They have a reasonably wide plastic range and can be worked at temperatures between 1450 and 1650 °F (788 and 899 °C). They are also satisfactory for cold working; however, they do work harden rapidly and they are not generally used for severe drawing or stamping. They can be softened by annealing between 900 and 1400 °F (482 and 760 °C) to achieve specific properties desired.

The duplex alloys have very high tensile strengths. As the aluminum content increases, ductility decreases and hardness increases, making these harder grades very desirable for metal-to-metal wear applications and for service under abrasive conditions. The physical properties of the duplex alloys can be improved by quenching in water from 1550 to 1750 °F (843 to 954 °C), followed by heat treating at 800 to 1200 °F (427 to 649 °C), depending on the specific composition of the alloy.

When brazing these alloys, the formation of refractory aluminum compounds creates difficulty. Brazing of these aluminum-containing metals in dry hydrogen atmospheres is not recommended because the aluminum oxides are not reduced and inhibit the flow of the brazing filler metal.

Copper-Silicon Alloys (Silicon Bronzes)

Copper-silicon alloys generally contain 1.5 to 3.25% silicon and 1.25% or less of zinc, tin, manganese, or iron. The addition of iron (possibly due to the formation of iron silicides) increases tensile strength and hardness. Toughness and shear strength increase with increasing silicon up to 3.5%. Electrical and thermal conductivities, however, decrease as silicon content increases.

High silicon alloys work harden rapidly and high tensile strength and hardness can be obtained in cold worked products. All these alloys can be softened after cold working by annealing between 900 and 1400 °F (482 to 760 °C). The silicon bronzes are subject to cracking during brazing, and their susceptibility increases with increasing silicon content. The alloys are all susceptible to stress cracking by molten filler metal (see chapter 2) and to hot-shortness. For both reasons they should be well fixtured to prevent excessive stressing when heated.

Copper-Nickel Alloys

The copper-nickel alloys are commercially available with nickel content ranging from 5 to 30%; those most commonly used in fabrication contain 10, 20, or 30% nickel. These alloys may also contain minor alloying constituents, such as iron, manganese, or zinc.

The copper-nickel alloys have moderately high tensile strength, good ductility, and are relatively tough, though electrical and thermal conductivities are low. The conductivities decrease as nickel content increases. The copper nickels have good hot working characteristics; however, they are not well suited to hot forging. Cold working properties are excellent, and the degree of malleability of those with low nickel content approaches that of copper. Since they do not work harden rapidly, these alloys can be used in severe drawing, stamping, and spinning operations. They are annealed at temperatures from 1200 to 1600 °F (649 to 871 °C), depending on the properties desired.

Copper-Nickel-Zinc Alloys (Nickel Silvers)

Nickel is added to copper-zinc alloys to make them silvery in appearance for decorative purposes and to increase strength and corrosion resistance; the resulting alloys are called nickel silvers (formerly German silvers). These alloys are of two general types: (1) alloys containing 65% copper plus nickel and consisting of a single phase, alpha; and (2) alloys containing 55 to 60% copper plus nickel and consisting of two phases, alpha and beta. Manganese and magnesium are added to these alloys to reduce free sulfur. In the hard (37% cold rolled) temper, the yield strength of these alloys is high, exceeding 70 ksi (480 MPa). Electrical and thermal conductivities of these alloys are quite low.

The single-phase nickel silvers are satisfactory for deep drawing, stamping, and spinning. Because of their good plasticity over a wide temperature range, two-phase nickel silvers may be hot worked by any of the commercial processes. The two-phase alloys are difficult to cold work and are not commonly so processed, though they can be softened by annealing at from 900 to 1300 °F (482 to 704 °C), depending upon the properties desired.

FILLER METALS

Brazing is an effective means of joining copper alloys. Silver base filler metals with brazing temperatures from 1145 to 1600 °F (618 to 871 °C) and copper-phosphorus filler metals, which require 1300 to 1500 °F (704 to 816 °C), are most commonly used. All the RBCuZn, BCuP, BAu, and BAg filler metals are suitable for use with the copper base metals. Any one of them may be used on almost any copper or copper alloy provided its liquidus temperature is sufficiently lower than the melting range of the base metal to make brazing practicable. Filler metal BCu may be used to braze the copper-nickel alloys but has too high a liquidus for use on the other copper base metals.

The RBCuZn filler metals may be used to braze the coppers and the copper-nickel, copper-silicon, and copper-tin alloys. Their liquidus temperatures are too high for brazing the brasses and nickel silvers. They are not useful for the aluminum bronzes because the required brazing temperatures destroy the effectiveness of the available flux types required for these base metals.

The BCuP filler metals are useful for most of the copper base metals, including some of the copper-nickel alloys. They have been used for brazing 90Cu-10Ni, but their usefulness for alloys with nickel content should be established for each application by suitable tests. These filler metals are useful for brazing the high-leaded alloys used in the standard cast-brass pipe fittings if precautions are taken to flux properly and avoid overheating. They are not recommended for brazing beryllium copper due to joint porosity and consequent low strength of the joints.

The BAg filler metals may be used on all copper base metals. The BAu filler metals are used primarily in electronic applications when a low vapor pressure filler metal is necessary.

Corrosion resistance is an important factor in the selection of a filler metal, since quite often a copper base metal is used because of its corrosion resistance. In many cases involving copper-nickel, copper-silicon, and copper-tin alloys, the RBCuZn filler metals will not have adequate corrosion resistance when in contact with these more noble alloys. As another example, it is not advisable to use BCuP filler metals for service above room temperature in sulfurous atmospheres.

Other factors that may influence the selection of filler metal are service temperature and the subsequent brazing operations in step brazing procedures.

JOINT DESIGN

The temperatures encountered in brazing copper base metals are high enough to anneal them; therefore, joint designs must be based only on the annealed strengths of the metals. All the joint designs discussed in chapter 6 are feasible. As shown in Table 6.1 (for the filler metals that would be used), clearances from 0.001 to 0.005 in. (0.03 to 0.13 mm) will develop the maximum available in joint strength and soundness. However, somewhat larger clearances can be used and are tolerable where it is impossible or impracticable to maintain the suggested values. It is possible to compensate for the lower strength developed in wide gap cases if lap joints or variations thereof can be used. In this type of joint, a brazed area at least three times the cross-sectional area of the thinner member will develop the full annealed strength of that member.

Coefficients of expansion for copper base metals encompass a considerable range. They are higher than the coefficients for the nickel base metals and considerably higher than those for carbon steel and cast iron, but they are much less than those of aluminum alloys. The coefficients for stainless steels are similar to those for some of the copper metals but lower than those for others.

PROCESSES AND EQUIPMENT

Copper and copper base alloys can be brazed by a wide variety of processes and techniques. In general, all the processes and equipment described in chapter 1 are applicable, with some precautions:

1. Furnace brazing may not be suitable if base metal and filler metal have similar melting ranges or if they are readily soluble in one another. Prolonged heating may cause excessive reaction and lowering of the base metal melting range to the extent that local melting occurs — for example, the brazing of 90Cu-10Ni with copper (BCu) filler metal. In most cases, where adequate temperature control is obtainable (±20 °F [±11 °C]), all combinations of base metals and filler metals can be brazed without damage provided time at brazing temperature is limited, since dissolution occurs above the solidus of the filler metal when a liquid phase first becomes available.

2. Thermal shock should be avoided when brazing metals, such as phosphor bronze, silicon bronze, and nickel silvers, in the cold worked condition to prevent cracking. Resistance, induction, dip, and oxyacetylene heating processes may cause difficulties if heating is done rapidly.

3. Exposure of oxygen-bearing coppers to hydrogen-containing atmospheres can produce embrittlement of the copper. Torch brazing of large assemblies and furnace brazing in hydrogen-bearing atmospheres should be avoided to preserve the integrity of the assembly. Embrittlement by hydrogen is temperature and time sensitive: higher temperatures and longer times increase the damage.

4. Copper alloys containing lead are susceptible to lead sweating when heated for long times; consequently, torch brazing of large assemblies

Table 14.3—Guide to brazing copper and copper alloys

Material	Commonly used brazing filler metals	AWS brazing atmospheres*	AWS brazing flux No.	Remarks
Coppers	BCuP-2†, BCuP-3†,BCuP-5† RBCuZn, BAg-1a, BAg-1, BAg-2, BAg-5, BAg-6, BAg-18	1 or 2 or 5	3 or 5	Oxygen-bearing coppers should not be brazed in hydrogen-containing atmospheres.
High coppers	BAg-8, BAg-1		3A	
Red brasses	BAg-1a, BAg-1, BAg-2 BCuP-5, BCuP-3, BAg-5, BAg-6, RBCuZn	1 or 2 or 5	3; or for RBCuZn, 5	
Yellow brasses	BCuP-4, BAg-1a, BAg-1, BAg-5, BAg-6 BCuP-5, BCuP-3	3 or 4 or 5	3	Keep brazing cycle short.
Leaded brasses	BAg-1a, BAg-1, BAg-2 BAg-7, BAg-18 BCuP-5	3 or 4 or 5	3	Keep brazing cycle short and stress relieve before brazing.
Tin brasses	BAg-1a, BAg-1, BAg-2 BAg-5, BAg-6 BCuP-5, BCuP-3 (RBCuZn for low tin)	3 or 4 or 5	3	
Phosphor bronzes	BAg-1a, BAg-1, BAg-2 BCuP-5, BCuP-3 BAg-5, BAg-6	1 or 2 or 5	3	Stress relieve before brazing.
Silicon bronzes	BAg-1a, BAg-1, BAg-2	4 or 5	3	Stress relieve before brazing. Abrasive cleaning may be helpful.
Aluminum bronzes	BAg-3, BAg-1a, BAg-1 BAg-2	4 or 5	4	
Copper nickel	BAg-1a, BAg-1, BAg-2 BAg-18, BAg-5 BCuP-5, BCuP-3	1 or 2 or 5	3	Stress relieve before brazing.
Nickel silvers	BAg-1a, BAg-1, BAg-2 BAg-5, BAg-6 BCuP-5, BCuP-3	3 or 4 or 5	3	Stress relieve before brazing and heat uniformly.

*Inert gas, including hydrogen, or vacuum atmospheres are usually acceptable (AWS Type 6 or 9 or 10). See chapter 4.

†Protective atmospheres are not required for brazing copper.

and furnace brazing may prove troublesome because heating times are extended. Excessive sweating of lead from leaded alloys, especially those containing more than about 2.5% lead, can result in defective brazed joints due to embrittlement and incomplete bonding.

5. Furnace brazing of sulfur-bearing coppers may prove troublesome when done in hydrogen or inert gases because sulfur vapor may inhibit wetting of the copper surface.

PRECLEANING AND SURFACE PREPARATION

Consistently high quality in brazed joints cannot be achieved unless the joint surfaces are free of oxides, dirt, and other foreign substances. Standard solvent or alkaline degreasing procedures are suitable for cleaning the copper base metals, and mechanical methods, wire brushing, sanding, etc., may be used to remove oxides. Complete chemical removal of oxides requires proper choice of pickling solution. Typical procedures used for chemical cleaning are as follows:

Aluminum Bronzes

Successive immersions in two solutions needed:

1. Cold 2% hydrofluoric acid and 3% sulfuric acid mixture.
2. A solution of 5% by volume sulfuric acid at 80 to 120 °F (27 to 49 °C).

Repeat until clean.

Chromium Copper and Copper Nickel

Immerse in hot 5% by volume sulfuric acid.

Copper-Silicon Alloys

Immerse in hot 5% by volume sulfuric acid, then in cold 2% by volume hydrofluoric and 5% by volume sulfuric acid mixture.

Brass and Nickel-Silver Alloys

Immerse in cold 5% by volume sulfuric acid.

Coppers

Immerse in cold 5 to 15% by volume sulfuric acid.

Prior to brazing it is often desirable to copperplate alloys containing strong oxide-forming elements to simplify brazing and fluxing requirements. About 0.001 in. (0.03 mm) copper is used on chromium coppers while 0.0005 in. (0.013 mm) copper is plated on beryllium coppers and silicon aluminum bronzes.

FLUXES AND ATMOSPHERES

Fluxes

The AWS flux Types 3A and 3B are suitable for use with BCuP and BAg filler metals in brazing all the copper base metals except the aluminum bronzes. Refractory oxides form easily on the latter, and the more active Type 4 fluxes are needed to cope with them (see Table 14.3). The effectiveness of Type 3A flux may be reduced rapidly at the temperatures needed for RBCuZn filler metals, while it is completely destroyed in brazing with BCu. Type 5 flux may be used with these filler metals except on base metals of aluminum bronze and beryllium copper. More active fluxes are needed for these, and mixtures of Types 4 and 5 may be found satisfactory for the few applications of this kind.

Atmospheres

Combusted fuel gases are economical and useful atmospheres for brazing most copper base alloys. The most obvious exceptions are atmospheres with high hydrogen content, Types 2, 3, and 4, which may not be used on oxygen-bearing coppers because of their embrittling action. Dissociated ammonia and hydrogen are also useful, except on oxygen-bearing coppers. Inert-gas atmospheres, such as argon and helium, as well as nitrogen may be used on all copper base alloys without harmful effects.

A vacuum is a suitable atmosphere for copper and copper alloys when the alloys contain few or no elements having high vapor pressures at the brazing temperature (lead, zinc, etc.). The brazing filler metal should also be restricted to alloys containing few or no elements of high vapor pressure, such as zinc, phosphorus, and cadmium. For high vacuum brazing the vapor pressure temperature curve for copper given in Fig. 2.1 should be consulted to determine maximum permissible temperatures.

ASSEMBLY, PROCEDURE, AND TECHNIQUE

The methods described in chapter 8 may also be used in assembling components of copper base metals for brazing, and the requirements outlined therein should be observed in a design of fixtures. Assembly procedures should start with thorough cleaning of all joint areas. When flux is to be used it should be applied to all joint areas and preferably immediately after the cleaning operations.

Procedures and techniques are determined by the brazing process used, and those suggested by the descriptions in chapter 1 are applicable to all the copper base metals; additional steps may be incorporated as needed for certain applications. For instance, it is often possible to braze heat treatable base metals without sacrificing all the superior mechanical properties available from them by combining the brazing operation with the solution annealing operation. This applies to the brazing of beryllium copper, chromium copper, and zirconium copper. Some of the temper originally present in the work hardened and aged base metals can then be recovered by applying the suitable aging heat treatments again. Of course, this can only be done with brazing filler metals appropriate for the solution annealing temperatures required; it is only practicable when the brazed assembly can be rapidly cooled to maintain the solutioned condition, and requires that the assembly be properly supported during the aging heat treatment to avoid damaging the brazed joints. Information on the required temperatures for solution annealing and aging should be checked before proceeding.

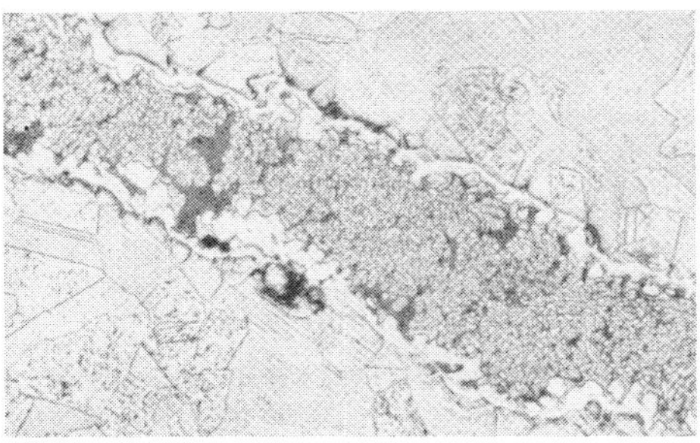

Fig. 14.1 — Copper joint brazed with BCuP-3 filler metal, 120 × (reduced by 20% on reproduction).

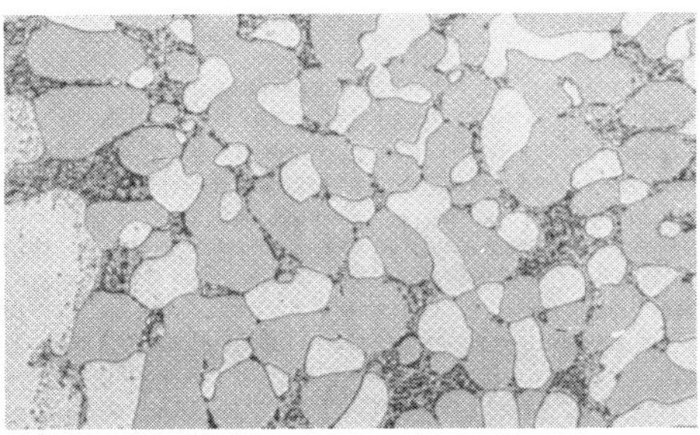

Fig. 14.2—Phases present in copper brazed with BCuP-3 filler metal: copper (light gray); Cu_3P compound (dark gray); and ternary eutectic (fine structure), 1000 × (reduced by 20% on reproduction).

POSTBRAZE OPERATIONS

The only postbraze operations, besides the reheat treatment of age hardenable alloys, are flux removal and cleaning. Flux residues can be a source of corrosion, and removal is usually also desired because of further work to be done on the brazed assemblies, or for the sake of good appearance. These residues are readily dissolved by immersion in hot water. Removal of oxides requires mechanical cleaning, such as wire brushing, or pickling with the appropriate solutions and procedures as outlined under Precleaning and Surface Preparation.

INSPECTION

Most of the inspection procedures described in chapter 11 are feasible for brazed assemblies of copper base metals. Selection depends on the design of the assemblies and the requirements of their applications.

PROCEDURES AND PROPERTIES

Copper

Prior to brazing, the copper oxide surface film is easily removed by abrading. Where large areas are to be brazed, pickling in cold 5 to 15% by volume sulphuric acid is recommended.

Oxygen-free high conductivity copper and deoxidized coppers are readily brazed by furnace or torch methods. Oxygen-bearing (tough-pitch) copper is susceptible to oxide migration and/or hydrogen embrittlement at elevated temperatures; therefore, oxygen-bearing copper should be furnace brazed in an inert atmosphere such as argon, nitrogen, or vacuum, and torch brazed with a neutral or slightly oxidizing flame.

The copper-phosphorus and copper-silver-phosphorus filler metals are self-fluxing on copper; however, flux is beneficial for heavy assemblies where prolonged heating would otherwise cause excessive oxidation. Due to the danger of corrosive attack, joints brazed with filler metal containing phosphorus should not be exposed for long periods at elevated temperatures to sulfurous atmospheres. Microstructures of copper brazed with BCuP-3 are shown in Figs. 14.1 and 14.2.

With the copper-zinc filler metals care should be taken not to overheat the metal, since volatilization of zinc causes voids in the joint. When torch brazing, an oxidizing flame will reduce zinc fuming. The AWS brazing flux type No. 5 should be used. A problem with RBCuZn-type fillers is that their corrosion resistance is inferior to that of copper.

A widely used rule of thumb for design is that in a lap joint an assembly will develop the full strength of annealed copper with a lap three times the thickness of the thinner member. Actually, a deoxidized copper lap joint develops the full strength of the base metal at a lesser overlap, as shown in Fig. 14.3. The knee in the tensile strength-overlap distance relationship at twice the thickness overlap reflects the change in the location of the fracture from the joint to the base metal. At greater than *2T* overlap, short-time tensile fracture generally occurs in the base metal. Even when specimens are deliberately made

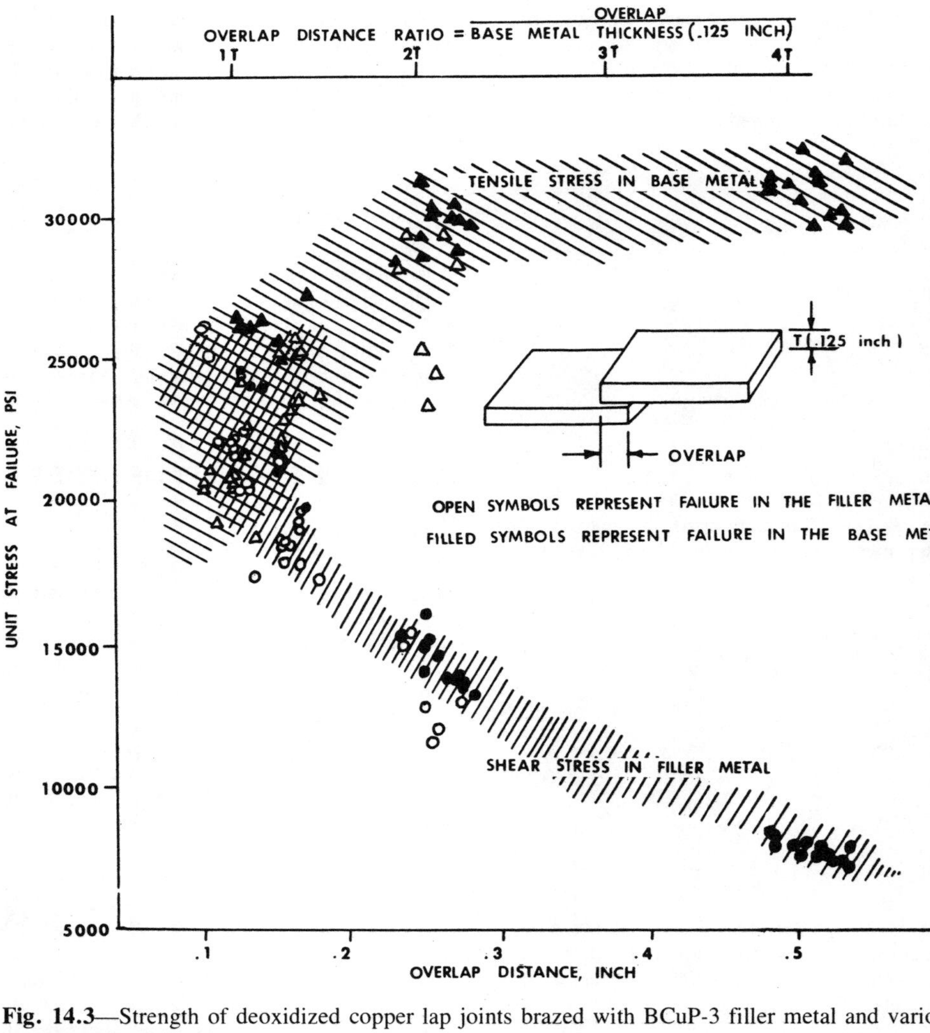

Fig. 14.3—Strength of deoxidized copper lap joints brazed with BCuP-3 filler metal and various overlap ratios (Ref. 1).

to fail in the joint area, the location of failure generally is partly or completely through the base metal in a plane parallel to the joint interface. This is because the room temperature strengths of the BAg and BCuP filler metals are greater than those of annealed base metal. At elevated temperatures, the brazing filler metal decreases in strength more rapidly than the copper, and eventually joint failure occurs through the filler metal. Filler metals for elevated temperature service are recommended in Table 14.4. The properties of joints of electrolytic tough-pitch and phosphorus-deoxidized copper are shown in Table 14.5; creep and fatigue properties for brazed tough-pitch copper are shown in Figs. 14.4 and 14.5.

Special Coppers

The coppers that contain small additions of silver, lead, tellurium, selenium, or sulfur (generally no more than 1%) are readily brazed with the self-fluxing BCuP filler metals, but wetting action is improved when a flux is used and when a sliding motion between components is provided while the filler metal is molten.

High Coppers

The surface must be clean to braze beryllium copper. The oxide scale can be removed by pickling: immersion in 20% by volume sulfuric acid solution at 160-180 °F (71-82 °C); water rinse; quick dip (less than 30 s.) in cold 30% by volume nitric acid solution followed by immediate and thorough rinsing. Chromium copper can be pickled by immersing in hot 5% by volume sulfuric acid, then in a cold mixture of 2-5 oz./gal (15 to 37 g/l) sodium bichromate with 3-5% by volume sulfuric acid. Copper plating to a thickness of 0.001 in. (0.03 mm) prior to brazing facilitates the operation.

The high strength beryllium-copper alloys (2% beryllium) can be brazed by two methods. The more common procedure involves simultaneous furnace brazing and solution treating at 1450 °F (788 °C), lowering of the temperature to 1400 °F (760 °C) to solidify the braze filler metal, a fast quench in water, and aging at 600 to 650 °F (316 to 343 °C). The silver-copper eutectic filler metal BAg-8 is generally used with AWS type 3A flux.

A second method applicable to sections that can be heated rapidly (less than 1 min. is desirable) permits brazing in the solution annealed condition, below the solution annealing temperature, and then aging without reannealing. Sufficiently fast heating rates can be attained with either torch or resistance brazing using BAg-1 filler metal and AWS Type 3A flux.

The high conductivity beryllium-copper alloys (0.5% beryllium), which are annealed at 1700 °F (927 °C) and precipitation-hardened between 850 and 900 °F (454 to 482 °C), can be silver brazed with BAg-1 and BAg-1a filler metals in the age-hardened condition. However, a loss in hardness of about 10 to 15 points on the Rockwell B scale will result.

Chromium copper and zirconium copper are age-hardened at about 900 °F (482 °C), following

Table 14.4—Suggested maximum operating temperatures for brazing filler metals

Filler metal classification	Service temperature: Continuous service F	C	Short time service F	C
BCuP	300	149	400	204
BAg	400	204	600	316
BAu	800	427	1000	538
RBCuZn	400	204	600	316

Table 14.5—Tensile strength of single-lap brazed joints in copper

Brazing filler metal	Tensile strength: Tough pitch copper, Room temperature	Tough pitch copper, −321 F (−196 C)	Deoxidized copper, Room temperature	Deoxidized copper, +400 F (+204 C)
Type BAg, ksi	18-20	24-30	17-19	10-11
MPa	124-138	165-207	117-131	69-76
Type BCuP, ksi	18-20	17-22	18-20	10-11
MPa	124-138	117-152	124-138	69-76

NOTE: Specimens were made from 0.25 in. (6.4 mm) thick sheet; joints had an overlap of 0.15 in. (3.8 mm) and no fillet (Ref. 2).

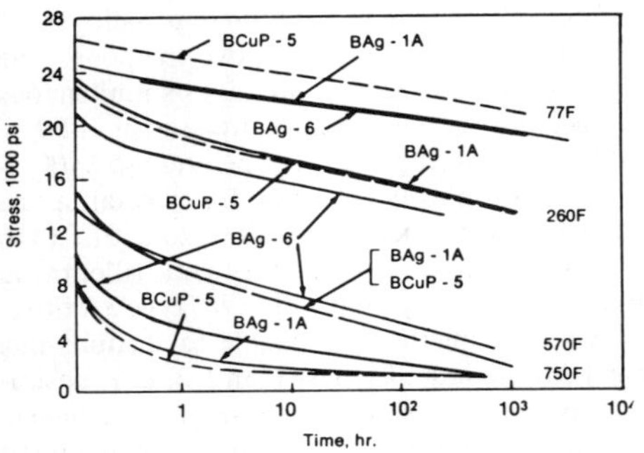

Fig. 14.4 — Stress-rupture strength of copper brazed with several filler metals using modified plug and ring creep specimens. At 77 and 260 °F (25 and 127 °C) failures were largely in copper (Ref. 3).

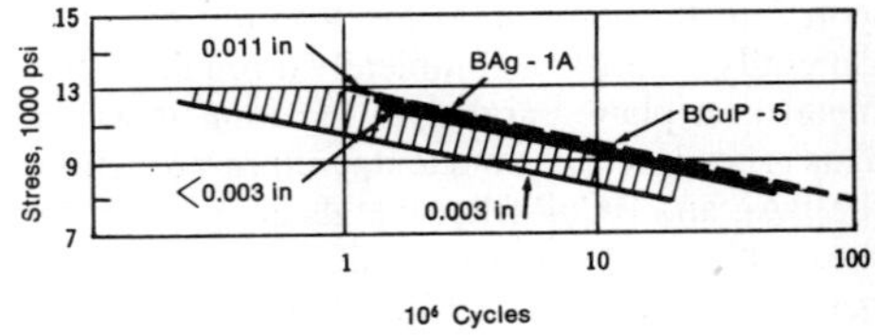

Fig. 14.5 — Comparison of torsional fatigue strengths of two brazing filler metals. Joint clearances from 0.003 to 0.011 in. (0.076 to 0.279 mm) give little difference in fatigue strength of joint (Ref. 3).

solution annealing at 1650 to 1850 °F (899 to 1010 °C). Brazing with silver-bearing filler metal and fluoride-containing flux should be performed after solution annealing and cold working, but before age hardening; however, the heat treated properties obtained in this way are reduced from normal values. The properties of chromium and zirconium coppers after brazing at lower than the suggested solution heat treating temperatures are given in Tables 14.6 and 14.7. Brazing at the solution annealing temperature followed by rapid cooling can give near optimum mechanical properties for chromium coppers; zirconium coppers require an intermediate cold working step, which is not practical for most brazed structures.

Cadmium copper is brazed in the same manner as deoxidized copper, only with added proper safety precautions owing to the toxicity of cadmium. For joining lengths of cadmium copper rod or wire prior to drawing to small sizes, resistance brazing with silver filler metals is recommended.

Copper-Zinc Alloys (Brasses)

All the brasses can be brazed with the BAg and BCuP filler metals, and the higher melting point (low zinc) brasses can also be be brazed with RBCuZn filler metal. The AWS Type 3 flux should be used with BAg and BCuP, and AWS Type 5 flux with RBCuZn filler metal. Even in a protective atmosphere, flux should be used to promote good wetting by the brazing filler metal. Chapter 4 lists the protective atmospheres commonly used in furnace brazing.

When heated above 750 °F (399 °C), brass tends to lose zinc by vaporization. This loss can be reduced by fluxing the parts during furnace brazing and by using an oxidizing flame during torch brazing. Brasses are also subject to cracking and should therefore be carefully and uniformly heated. Sharp corners that concentrate stresses and large changes in cross section that produce thermal strains with temperature changes should be avoided. It is good practice to bring parts to temperature gradually.

Brasses containing aluminum or silicon may require treatment similar to aluminum or silicon bronzes. Lead added to brass to improve machinability may alloy with the brazing filler metal and cause brittleness. The major brazing difficulties occur when the lead content is over 2 to 3%. To maintain good flow and wetting during brazing, leaded brasses require complete flux coverage to prevent the formation of lead oxide or dross. Rapid heating of high leaded brasses may cause cracking. Stress relief annealing before brazing, plus slower uniform heating and cooling, will minimize this cracking tendency.

Copper-Tin Alloys (Phosphor Bronzes)

In a stressed condition these alloys are subject to cracking. It is good practice to stress relieve or anneal the parts before brazing, to support them in a stress-free condition during brazing, and to avoid thermal shock by using slow heating cycles. When the tin content is high or when there are appreciable lead additives, adequate flux protection during brazing is necessary. All the phosphor bronzes can be brazed with the BAg and BCuP filler metals, and the low tin varieties can be brazed with the RBCuZn filler metals.

The phosphor bronzes are sometimes in the form of powder metal compacts; before brazing these require pretreating by painting the surface with a water base or oil base colloidal graphite suspension, baking at a low temperature, cleaning, and degreasing. This procedure seals the pores so that brazing can be performed.

Copper-Aluminum Alloys (Aluminum Bronzes)

Aluminum bronzes can be brazed with silver-bearing brazing filler metals and AWS Type 4 flux. However, the formation of refractory aluminum oxide at brazing temperature in alloys containing more than 8% aluminum presents difficulties. This problem can be avoided by electroplating at least 0.0005-in. (0.013-mm) copper on surfaces to be brazed. In furnace brazing unplated surfaces, flux should be used with a protective atmosphere.

Copper-Silicon Alloys (Silicon Bronzes)

Copper-silicon alloys should be cleaned and flux coated or copperplated before brazing to prevent the formation of refractory silicon oxide. Mechanical cleaning is recommended, or for light oxide the material can be pickled as noted earlier. Silver-bearing brazing filler metals and AWS Type 3 flux are generally used.

Silicon bronzes are subject to intergranular penetration of the filler metal and to hot-shortness under stress. They should be stress relieved before brazing and brazed below 1400 °F (760 °C).

Copper-Nickel Alloys

Copper-nickel alloys are generally brazed with BAg filler metals. Before selecting alloys of the BCuP filler metal class for joining copper nic-

Table 14.6—Mechanical properties of 0.83% chromium copper after simulated brazing cycles

	Tensile strength				0.2% Yield strength				Elongation in 1 in., %		Vickers hardness	
	As cooled		Aged*		As cooled		Aged*					
Brazing cycle	ksi	MPa	ksi	MPa	ksi	MPa	ksi	MPa	As cooled	Aged*	As cooled	Aged*
1110-1190F (599-643C)												
1 min.	68.1	469	66.1	456	61.8	426	56.2	387	17	17	139	132
4 min.	56.4	389	56.7	391	42.5	293	43.9	303	19	19	122	113
16 min.	44.8	309	44.6	308	23.1	159	21.7	150	34	45	86	85
1200-1290F (649-699C)												
1 min.	60.3	416	58.7	405	47.0	324	47.0	334	23	21	125	116
4 min.	44.6	308	45.2	312	22.4	154	22.8	157	—	31	94	85
16 min.	42.3	292	42.3	292	17.9	123	17.9	123	35	38	85	78
1290-1380F (699-u49C)												
1 min.	43.0	246	43.9	303	21.3	147	22.0	152	31	36	86	86
4 min.	42.8	295	42.1	290	17.7	122	17.5	121	37	39	76	79
16 min.	41.4	285	41.2	284	15.9	110	15.0	103	41	39	77	74
1380-1470F (749-799C)												
1 min.	42.8	295	43.5	300	18.8	130	18.1	125	36	39	75	78
4 min.	41.2	284	42.1	290	15.2	105	15.4	106	42	42	72	74
16 min.	41.2	284	41.2	284	13.4	92	13.4	92	46	45	69	70
1470-1560F (799-849C)												
1 min.	41.2	284	45.0	310	14.6	101	20.6	142	42	38	70	88
4 min.	40.8	281	47.5	328	12.5	86	25.8	178	41	35	67	97
16 min.	40.5	279	46.4	320	11.4	79	21.9	151	47	36	64	91
NONE (solution heat treated and aged*)		76.6 ksi		528 MPa		70.3 ksi		485 MPa	16		152	

*Aged at 840 °F (449 °C) for 8 hours.

Table 14.7—Mechanical properties of 0.16% zirconium copper after simulated brazing cycles

	Tensile strength				0.2% Yield strength				Elongation in 1 in., %		Vickers hardness	
	As cooled		Aged*		As cooled		Aged*					
Brazing cycle	ksi	MPa	ksi	MPa	ksi	MPa	ksi	MPa	As cooled	Aged*	As cooled	Aged*
1110-1190 F (599-643 C)												
1 min.	56.4	389	56.2	387	50.8	350	48.6	335	17	17	123	126
4 min.	40.3	278	39.4	272	19.4	134	18.8	130	33	42	74	74
16 min.	39.0	269	38.1	263	15.4	106	14.1	97	42	46	72	67
1200-1290F (649-699C)												
1 min.	41.0	283	44.1	304	21.5	148	26.0	179	—	39	74	76
4 min.	38.5	265	37.4	258	15.0	105	13.0	90	44	46	68	66
16 min.	39.9	275	37.2	256	13.4	92	12.8	88	49	46	70	64
1290-1380F (699-749C)												
1 min.	38.5	265	37.4	258	14.3	99	13.9	96	46	44	70	69
4 min.	37.6	259	37.4	258	13.6	94	12.8	88	46	47	67	59
16 min.	37.9	261	37.0	255	11.6	80	11.6	80	46	46	63	60
1380-1470F (749-799C)												
1 min.	38.3	264	37.0	255	12.1	83	12.5	86	—	45	68	66
4 min.	37.6	259	36.5	252	10.5	72	9.4	65	44	44	64	61
16 min.	37.4	258	36.7	253	9.6	66	9.8	68	44	49	62	55
1470-1560F (799-849C)												
1 min.	37.2	256	37.0	255	9.2	63	9.2	63	44	46	59	61
4 min.	37.4	258	36.7	253	8.5	59	8.7	60	43	48	60	55
16 min.	38.3	264	37.0	255	8.5	59	8.7	60	45	47	59	51
NONE (heat treated and aged*)	68.3 ksi		471 MPa		63.1 ksi		435 MPa		17		155	

*Aged at 840 F (449 C) for 8 hours.

Table 14.8—Braze filler metal selection chart for joining dissimilar metals.

	Al & Al alloys	Mg & Mg alloys	Cu & Cu alloys	Carbon & low alloy steels	Cast iron	Stainless steel	Ni & Ni alloys	Ti & Ti alloys	Reactive metals	Refractory metals	Tool steels
Cu & Cu alloys	—		BAg, BAu, BCuP, RBCuZn	BAg, BAu, RBCuZn	BAg, BAu, RBCuZn	BAg, BAu	BAg, BAu, RBCuZn	BAg	BAg	BAg	BAg, BAu, RBCuZn BNi

kels, one must thoroughly check performance and structure, because a brittle nickel phosphide may have formed. The base metal must be free of sulfur or lead, which might cause cracking during the brazing cycle. Standard solvent or alkaline degreasing procedures may be used. Oxides can be removed by abrading or by pickling in hot 5% volume sulfuric acid and immediately and thoroughly rinsing. The AWS type 3 flux is suitable for most applications.

Copper-nickel alloys are susceptible to intergranular penetration by molten filler metal in the stressed condition. To prevent cracking, they should be stress relieved before brazing, and stresses should not be introduced during brazing.

Copper-Nickel-Zinc Alloys (Nickel-Silvers)

Nickel silvers can be brazed readily with the same procdures used for brazing brass; however, when RBCuZn filler metals are used, great care is required because of the relatively high brazing temperatures. These alloys are subject to intergranular penetration of the filler metals unless they are stress relieved before brazing. The poor thermal conductivity of these alloys tends to lead to local overheating; thus, slow and uniform heating and the use of sufficient flux to prevent undue oxidation are recommended.

Dissimilar Metals

Dissimilar combinations of the copper alloys can be brazed readily; dissimilar combinations can also be made with steel, austenitic stainless steel, and nickel, using BAg filler metals. Suggested brazing filler metals for dissimilar metal systems are given in Table 14.8. Brazing to aluminum and magnesium alloys is not practical. In cases where base metals have widely dissimilar coefficients of expansion, the design and execution of brazed joints having satisfactory characteristics may be extremely difficult or impractical.

SAFETY

Some brazing filler metals contain zinc and cadmium, some fluxes contain fluorides, and some base metals contain beryllium, lead, or zinc. When these materials are used, adequate ventilation must be provided to protect personnel — see chapter 29. Do not attempt to braze without consulting *Safety in Welding and Cutting,* American National Standard Z49.1 (latest edition). In addition, *Beryllium and Its Compounds* (latest edition), Hygiene Guide Series, American Industrial Hygiene Association, is a useful data sheet.

REFERENCES

1. Communication of Unpublished Data, Kawecki Berylco Industries, Inc.
2. Munse, W. H., and Crawford, D. C. 1952. Filler-metal strengths in brazed copper joints. *Welding Journal Research Supplement,* 34, (2).
3. McQuillan, W. P., 1946. High-silver copper braze shows good creep strength. *Materials in Design Engineering,* 59, (1): 146.

CHAPTER 15

LOW CARBON AND LOW ALLOY STEELS

The brazing of low carbon and low alloy steels is a highly developed low cost production process. Filler metals can either be manually or automatically applied or preplaced in the joint. The heat treatment considerations of the low alloy steels are factors in determining the specific brazing filler metal and brazing temperatures to be used.

BASE METALS

This chapter discusses low carbon steel (less than 0.30% carbon) and certain low alloy steel base metals. These steels include a variety of fabricated shapes as well as castings and forgings. The low alloy steel category includes the SAE/AISI 23XX nickel steels, the 31XX nickel-chromium steels, the 41XX chromium-molybdenum, the 43XX nickel-chromium-molybdenum, and several other types containing less than 5% total alloy content.

Leaded steels can be torch brazed using filler metals and flux combinations normally recommended for steel. Tests conducted on low carbon-free machining steel containing 0.25-0.35% lead show no detrimental effects after torch brazing with BAg filler metals and AWS type 3 flux. Satisfactory joints were made in these leaded steels using BCu and BNi filler metals. Laboratory tests indicate that surface lead boils off in vacuum and hydrogen furnace brazing.

FILLER METALS

All the BAg classifications can be used for brazing low carbon and low alloy steels. Those silver filler metals containing nickel usually provide better wettability and are preferred for brazing certain low alloy steels wherein joint strength is most important. The BCu group is used mainly for preplacement in controlled atmosphere furnaces. The RBCuZnA and RBCuZnD are often used face fed but can be preplaced for furnace and induction heating. The high solidus temperatures often allow simultaneous brazing and heat treating of certain low alloy steels. The BNi classification of brazing filler metals has been used for joining low carbon and low alloy steels when special joint requirements so dictate. Brazing is usually restricted to controlled atmosphere furnace.

JOINT DESIGN

Joints should be close-fitting and properly designed (see chapter 6). Clearances on the order of 0.002-0.005 in. (0.05-0.13 mm) produce the best mechanical properties with most filler metals when using a mineral flux. Light press fits are preferred for the BCu and other filler metals, where furnace brazing, with a protective atmosphere, is used (0.000-0.002 in. [0.00-0.05 mm] clearance, see Table 6.1).

It is important to consider the coefficient of expansion of the parts being joined to determine the joint clearance at the brazing temperature recommended for the filler metal chosen. Those filler metals with relatively narrow melting ranges are required for close-fitting joints; conversely, wide melting range filler metals provide the necessary bridging characteristics when wide clearances are involved. Furnace dew point, however, can be used to control the fluidity of BCu when brazing joints with larger gaps.

PROCESSES AND EQUIPMENT

Low carbon and low alloy steels can be brazed using virtually all known processes. Torch, furnace, and induction heating techniques are the most common. Filler metals in the form of continuous wire or strip can be automatically applied using electromechanical wire feeders; powder filler metals blended with flux and paste forming ingredients are automatically applied with pressurized dispensing equipment. For torch brazing the equipment would include standard oxyacetylene and similar-type torches. Furnaces can be of the batch or conveyor type, with or without control of the atmosphere; they can be either electric, gas- or oil-fired and should provide accurate temperature control.

PRECLEANING AND SURFACE PREPARATION

For best results, faying surfaces should be mechancially or chemically cleaned to ensure essentially complete absence of oxides or organic matter; the performance of a given joint can be directly affected by the cleanliness of the joint area prior to brazing. Rough joint edges should be avoided, as well as metal slivers or other superficially attached fragments.

FLUXES AND ATMOSPHERES

Flux or a proper atmosphere is generally required when brazing steels. Selection will depend upon filler metal type: AWS brazing flux types 3A, 3B, and 4 are suitable for BAg filler metals; types 4 and 5 are normally used with the RBCuZn types. Fluxes and atmospheres may be used together. The flux can be used in either paste or powder form or combined with the filler metal. In face fed operation, the hand-held filler metal can be coated with the appropriate flux. In atmosphere brazing, the filler metal is preplaced in or near the joint and the assembly charged into the brazing chamber. Maximum temperature settings and dwell time must be controlled to ensure proper melting and thorough penetration of the filler metal into the joint.

PROCEDURE AND TECHNIQUE

In torch brazing a neutral or slightly reducing flame is usually preferred, as the filler metal is face fed into the prefluxed joint. Flux-coated filler metal is often beneficial. As in all brazing processes it is important not to overheat in brazing to prevent undesirable effects on the base metal, the filler metal, or the flux. Time at temperature is an important consideration, especially when the filler metal contains volatile elements, such as zinc and cadmium. Excess heat might also affect the integrity of the braze and reduce the mechanical properties of the joint. Automated torch and burner-type production equipment is available for high production applications. These units usually use brazing fluxes and BAg filler metals.

In production applications, the filler metal (usually BCu) is preplaced in or adjacent to the joint and the preassembled parts are charged into controlled atmosphere batch- or conveyor-type furnaces for brazing. Induction heating for brazing is also advantageous in that maximum temperature rise is restricted to the immediate joint area by selective coil design and the use of an appropriate frequency for the induction heating circuit.

In joining certain hardenable low alloy steels, it is usually desirable to use the lower melting BAg filler metals, which can be applied below the lower transformation temperature of the steel. However, some local annealing may occur. When postheat treatments are required, the higher solidus filler metals are necessary to preclude the possibility of joint impairment. Low alloy steels are sometimes simultaneously brazed and heat treated using filler metals of the RBCuZn and BAg classifications. The solidus temperature of the filler metal chosen must be above the austenitizing temperature recommended for the base metal prior to quenching. In application of this type, the joint is made at normal brazing temperature, then removed from the heat source to permit a drop to the hardening temperature, at which point the assembly is quenched. The procedure is satisfactory only for base metals compatible with rapid cooling.

POSTBRAZE OPERATIONS

When flux has been used all residues must be removed. Quenching the brazement in water while still hot produces sufficient thermal shock to accelerate flux residue removal providing the base metals are compatible with such treatment. Washing and brushing in warm or hot water should be followed by a mild chemical dip. A more involved procedure might be required in the event the brazement is to be painted or plated (see chapter 10). When protective atmospheres are properly used in furnace brazing, induction brazing, etc., no postbraze cleaning operation is required since parts come out bright and clean. For brazements requiring some type of postbraze heat treatment, the effects of the temperature rise and quench on the filler metal must be considered.

INSPECTION

Most inspection is visual only. Observation under low magnification will usually be satisfactory in detecting surface voids or flux inclusions. There are also nondestructive tests for brazed joints (see chapter 11). For certain production brazing applications destructive tests applicable to the end use of prototype brazements are recommended to establish final manufacturing parameters.

TYPICAL APPLICATIONS

A variety of applications involve the use of brazing filler metals on low carbon and low alloy steels in everyday production, these include many components of such vehicles as automobiles, trucks, bicycles, motorcycles, snowmobiles, all-terrain vehicles, and the like. Other common brazements include the production of window and door frames, ducts, tanks, containers of all types, perforated and expanded steel panels, steel partitions, and shelving. A great variety of tubular steel furniture is production brazed; also cutting tools and industrial knives, hydraulic oil tanks, reservoirs, electronic chassis and supports, hand tools, honing appliances, certain instruments, and steel assemblies of all types.

CHAPTER 16

TOOL STEELS

Tool steels are used in the fabrication of cutting tools and dies where high hardness and resistance to wear are primary requisites. Frequently, the cutting or wearing surface may be only a small section of the tool assembly and it is necessary to join the tool steel to the assembly. Brazing is one commonly used method for doing this.

BASE METALS

As a wide variety of steels is used for tools, a system of classification that groups grades of similar properties has been developed by the American Iron and Steel Institute and the Society of Automotive Engineers (Ref. 1). A few typical tool steels from this table are given in Table 16.1.

For discussion of the brazing of tool steels, it is convenient to group them in two broad classifications: carbon steels and high speed tool steels. The carbon tool steels depend primarily on their high carbon content (0.60-1.40%) for their hardness. Except for thin sections, these steels must be drastically quenched during heat treatment to achieve their optimum properties. Alloying elements may be added to the carbon steels to impart special properties, such as reduced distortion on heat treatment, greater wear resistance, and toughness or better high temperature properties. Such steels are referred to as alloy tool steels; they are known by various trade names and grades, and their properties are adequately covered in manufacturers' published information and in various handbooks.

High speed steels, although logically falling in the alloy steel group, are classified separately because their properties depend on relatively high percentages of such alloying elements as tungsten, molybdenum, chromium, and vanadium. Their carbon content is normally much lower than that of the carbon tool steels. The high speed steels are widely used in industry as metal cutting tools. A common type known as 18-4-1 contains 18% tungsten, 4% chromium, and 1% vanadium.

Hard sintered carbides are also used extensively as cutting tools but do not fall under the classification of steels. Brazing of the carbides is discussed in chapter 23.

FILLER METALS

The choice of the filler metal to be used depends on the properties of the tool steel being brazed and the heat treatment required to develop its optimum properties. Practically all brazing filler metals of the BAg, BCu, and RBCuZn classifications are used at various times. The best filler metal to use should be determined for the specific application.

JOINT DESIGN

Lap or sleeve type joints are generally used in brazing tool steels. With BAg or RBCuZn filler metals, it is best to allow for clearances of 0.002-0.005 in. (0.05-0.13 mm) at the brazing temperature. Preplacing the filler metal where it is protected from direct contact with the heat source is recommended. Methods of preplacing

Table 16.1—Types of tool steels

AISI designation	Type	Composition, % (balance iron) Carbon	Manganese	Silicon	Chromium	Vanadium	Tungsten	Molybdenu
W1	Water hardening	0.06-1.40	....	....	....	....	....	....
W2	Water hardening	0.06-1.40	....	....	....	0.25	....	....
S1	Shock resisting	0.50	....	....	1.50	....	2.50	....
S5	Shock resisting	0.50	0.80	2.00	....	....	....	0.40
O1	Oil hardening	0.90	1.00	....	0.50	....	0.50	....
A2	Air hardening medium alloy	1.00	....	....	5.0	....	....	1.00
D2	High carbon-high chromium	1.50	....	....	12.0	....	....	1.00
H12	Chromium hot work	0.35	....	....	5.0	0.40	1.50	1.50
H21	Tungsten hot work	0.35	....	....	3.5	....	9.50	....
T1	Tungsten high speed	0.70	....	....	4.0	1.00	18.0	....
M1	Molybdenum high speed	0.80	....	....	4.0	1.00	1.50	8.50

are described in chapter 6. Where brazing and hardening operations are combined, the joint should be designed so that the brazing filler metal is under compression during quenching, as tensile stress on the joint at temperatures near the melting point of filler metal may crack the joint.

EQUIPMENT

Torch, furnace, or induction heating are the procedures most commonly used in brazing tool steels. Available equipment is frequently the main factor in deciding the heating procedures.

SURFACE PREPARATION

The base metals must be clean, that is, free from oil, oxide, or other foreign material, in order to secure proper flow of the filler metal. A machined or roughened surface is always preferable to a smoothly ground or polished surface, since it is more difficult for the flux and brazing filler metal to wet and flow on an extremely smooth surface (see chapter 7).

FLUXES AND ATMOSPHERES

In general, AWS Brazing Flux Types 3A and 3B (see Table 4.1) are used for brazing tool steels. Some modification of the flux may be required depending upon the particular tool steel used and the brazing temperature required.

Controlled atmospheres, with possible elimination of fluxes, are being used more and more to prevent oxidation during heating and to avoid the necessity of postbraze cleaning. If a controlled atmosphere is used, steps should be taken to guard against carburization or decarburization (see Table 4.2).

TECHNIQUE OF BRAZING

The brazing of carbon tool steel is best done prior to or at the same time as the hardening operation. The hardening temperature for carbon steels is normally in the range of 1400-1500 °F (760-816 °C). If the brazing is done prior to the hardening operation, it is customary to use a filler metal which has a solidus well above this range so that the assembly will have sufficient strength when reheated to the hardening temperature to be handled without failure of the joint. A BCu filler metal is frequently used for this purpose. However, at times the high temperature required for copper brazing adversely affects the structure of the steel. Silver (BAg) and copper-zinc (RBCuZn) filler metals are available that can be used at brazing temperatures of 1700-1800 °F (927-982 °C).

When brazing and hardening are combined, a filler metal having a solidus close to the quenching temperature is generally used. Particular attention must be given to the design and handling in this process as the joint strength is extremely low at the quenching temperature. Any stress during quenching should put the joint under compression rather than tension.

The brazing of the alloy carbon steels depends on knowledge of the particular steel involved. The alloy tool steels cover a wide range of compositions and, therefore, wide differences of behavior on heat treatment and heating for brazing.

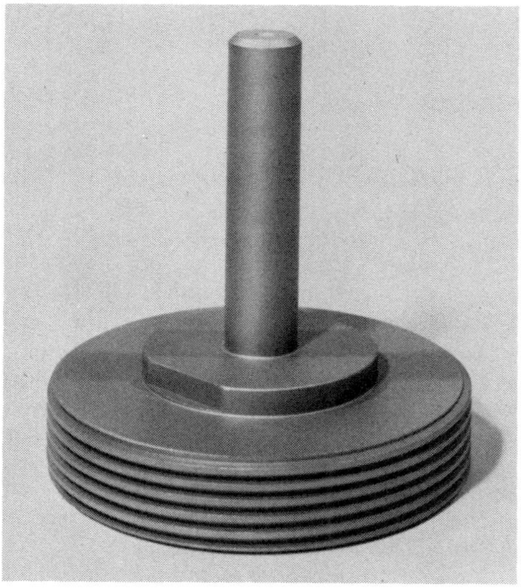

Fig. 16.1—Furnace brazed alloy tool steel.

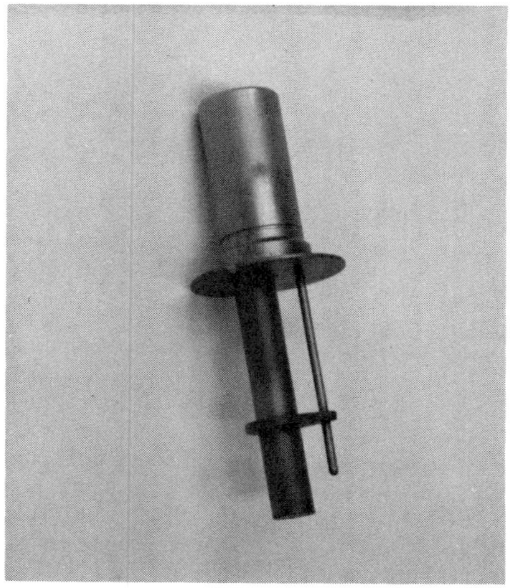

Fig. 16.2—Alloy tool steel assembly.

Fig. 16.3—Photomicrograph of joint in Fig. 16.2.

Fig. 16.4—Torch brazing high alloy tool steel.

The alloy steel in question should be studied carefully to determine its proper heat treating cycle, the kind of quench necessary (water, oil, or air), the best brazing filler metal, and the proper technique for combining the heat treating and the brazing operation so as to achieve maximum properties in service life.

Quenching sets up steep temperature gradients which produce differential expansions and contractions that may rupture a brazed joint. Austenite-martensite transformations in certain steels may result in parts of any assembly being quenched to first contract, then expand, and finally contract again. When the joint can be properly supported during quenching, a filler metal melting below the temperatures at which these changes occur can be used.

The high speed tool steels require hardening treatments at temperatures above the usual BAg brazing temperatures. It is, therefore, common practice to harden the steels prior to brazing and braze during or after the secondary tempering treatment. Tempering is usually done in the range of 1000-1200 °F (538-649 °C). Brazing filler metals, such as BAg-1 or BAg-1a, can be used at temperatures above 1150 °F (621 °C) and, if care is taken to use short brazing cycles, hardened high speed tool steel may be brazed without overtempering.

Broken high speed tools may be repaired by brazing. This frequently results in avoiding delays in production which may occur in obtaining a replacement. Broaches, circular saws, and milling cutters are examples of tools salvaged by brazing.

The fundamentals of brazing outlined in previous chapters are applicable to the tools steels and should be observed. Since each job will require a knowledge of the service requirements, as well as of the mechanical and metallurgical characteristics needed in the particular application, it is not practicable to outline in detail specific procedures here.

TYPICAL APPLICATIONS

Figure 16.1 shows a furnace brazed assembly of 1.5% carbon, 12% chromium tool steel. The joint is at the juncture of the vertical and horizontal components. A copper brazing filler metal was used in a hydrogen atmosphere.

Figure 16.2 shows a furnace brazed assembly of 1% carbon, 5% chromium, and 1% molybdenum tool steel. A BAg filler metal was used in a hydrogen atmosphere. Hardening was accomplished in conjunction with the brazing operation. Figure 16.3 shows a photomicrograph of the brazed joint shown in Fig. 16.2.

Figure 16.4 shows the torch brazing of a high alloy tool steel assembly.

REFERENCE

Metals Handbook, Vol. 1, 8th Ed., (Table I, p. 638), American Society for Metals (1961).

CHAPTER 17

CAST IRON

The several kinds of cast irons include white, gray, malleable, and ductile; typical compositions are given in the appendix. There are many applications in which it is desirable to braze gray, malleable, and ductile irons either to themselves or to dissimilar metals. The white irons are seldom brazed.

It is only in the last few years that the brazing of ordinary gray cast irons using silver alloy filler metal has become commercially practical; the development of a satisfactory surface pretreatment has opened up many design possibilities. Intricate forms can be built up from simple castings or by joining castings to standard wrought forms, such as tubing, rolled shapes, etc. Foundry work may be simplified by making intricate castings in several parts to reduce coring. Gray, ductile, and malleable irons can be joined to most of the commercially important metals and alloys. The most common wrought metals used with these castings are ferrous base products.

In the brazing of ductile and malleable irons certain precautions are imperative, and these are the same for either type. If ductile or malleable irons are heated above 1400 °F (760 °C) the metallurgical structure may be damaged; brazing should thus be done below this temperature. Figure 17.1 shows the relationship of time at temperature to the changes in malleable iron; ductile iron behaves in a similiar manner.

METALLURGICAL CONSIDERATIONS

When a gray, malleable, or ductile cast iron is heated above its critical (transformation) temperature, the structure normally present begins to transform into austenite. Upon cooling, this new structure is converted either to martensite, if the cooling rate is high, or to a fine pearlite structure having a cementite network. In either case, the metallurgical structure becomes extremely unfavorable in the heat affected areas. The critical temperature varies with the composition and rises gradually with increased amounts of silicon.

WETTING

Graphitic carbon present in cast irons will not be wet by brazing filler metals and may prevent a good metallurgical bond. This creates a difficulty with gray irons, but not with malleable and ductile irons. Where wetting difficulties are encountered there are several cleaning methods available. One such method is a proprietary electrochemical surface cleaning method that produces a surface essentially free of graphite, sand, silicon, oxides, etc. The process involves using a catalyzed molten salt bath operating at 850-950 °F (454-510 °C). Direct current is passed through

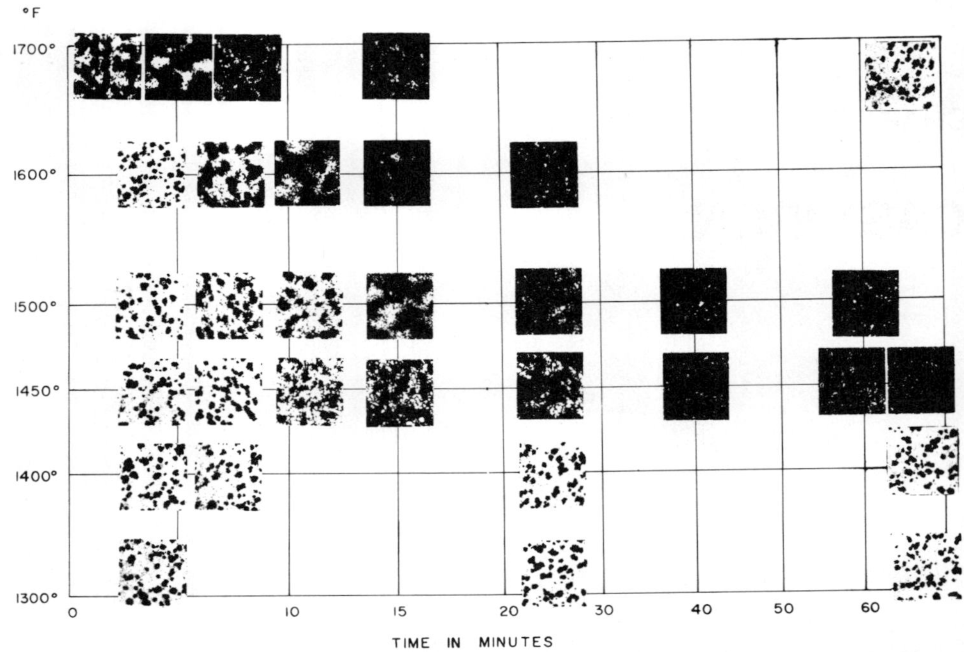

Fig. 17.1—Relationship of time at temperature to the changes in malleable iron. Inset (top right) shows typical original microstructure.

the bath using the work as one electrode and the steel tank as the other; the direction of flow is occasionally reversed to produce reducing, oxidizing, and again reducing effects. A water rinse completes the surface treatment. Other methods of cleaning include searing with an oxidizing flame, grit blasting, and chemical cleaning.

PRECLEANING AND SURFACE PREPARATION

It is usually desirable to machine or file the surface of a casting to secure uniform joint clearance. The casting "skin," which contains sand inclusions, is difficult to braze without proper cleaning. The surfaces to be brazed should be free of dirt, oil, grease, etc. (see chapter 7). If necessary, the exposed graphite is removed following surface cleaning by the electrochemical method or one of the other methods mentioned previously.

FILLER METALS AND FLUXES

With proper surface preparation, any filler metal suitable for use with iron or steel can be used for the cast irons. Those preferred, however, are the lower melting BAg filler metals: those containing nickel, such as BAg-3 and BAg-4, have greater affinity for cast irons and therefore produce higher strength joints. Copper and copper-zinc filler metals can be used, but great care must be exercised because of their higher brazing temperature ranges. Filler metal containing phosphorus (BCuP) are not suitable for joining cast iron due to the formation of brittle iron-phosphide constituents — such joints, if made, would be extremely brittle. The AWS brazing flux types 3A and 3B are used with the recommended BAg filler metals.

BRAZING PROCESSES

Any of the processes described in chapter 1 are

Fig. 17.2—Steel sprocket brazed to cast iron hub.

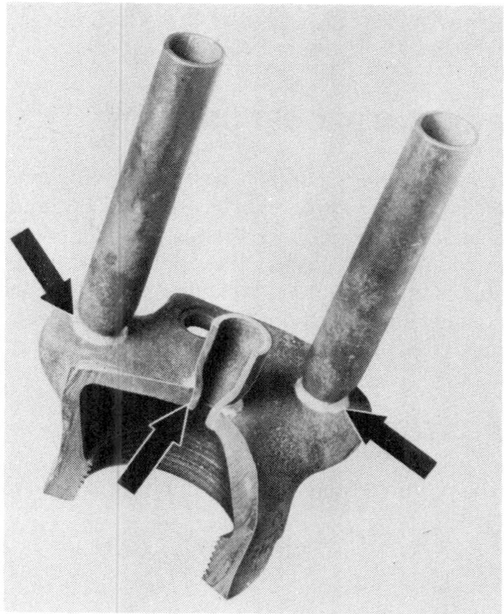

Fig. 17.3—Steel tubes brazed to malleable-iron header.

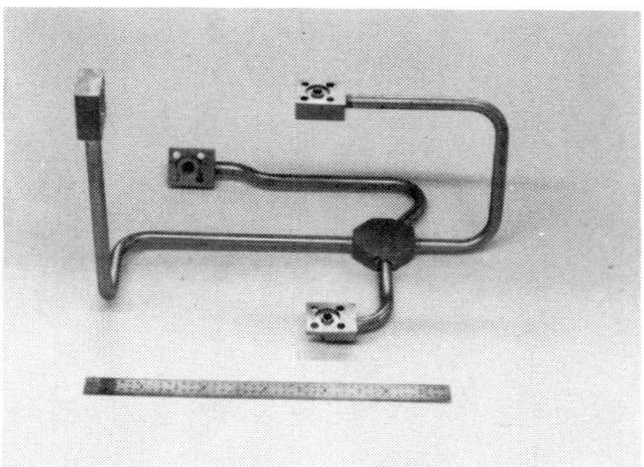

Fig. 17.4—Cast iron fittings brazed to steel tubing.

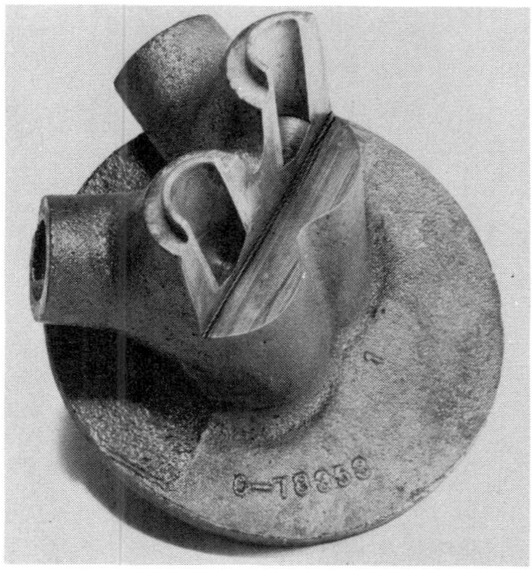

Fig. 17.5—Steel tubing brazed to cast iron base.

applicable to the brazing of cast irons; the particular choice will depend upon the metals being joined, the brazing filler metals used, the design of the joint, and the relative masses of the parts. Those processes that lend themselves to automatic temperature control are most desirable, for overheating should be avoided.

JOINT DESIGN

The general factors associated with joint design have been covered in chapter 6. Joint clearances should be determined for specific applications, considering the thermal expansion coefficients of the metals being joined, the methods of heating, type of filler metal, etc. The recommended joint clearance for optimum results is 0.002-0.005 in. (0.05-0.13 mm), with a maximum spacing of 0.01 in. (0.25 mm). The variation of shear strength with joint clearance follows the same general pattern shown in Fig. 6.4. The proper tolerances are easily maintained in smaller-sized components, but some sacrifice must usually be made as the parts to be joined become larger in size.

BRAZING PROCEDURES

In general, the handling of cast irons during brazing is the same as for other base metals, assuming adequate preparation of the surfaces as outlined previously (see chapter 9). As the gray, malleable, and ductile irons expand rather readily and conduct heat quite poorly, their heating and cooling cycles deserve a great deal of consideration.

POSTBRAZE OPERATIONS

The postbraze operations, including the removal of excess flux, are described in chapter 10. Warm water is usually adequate for washing off the flux normally used with silver base filler metals. Inspection operations may then be made on the cleaned joints.

TYPICAL APPLICATIONS

Figure 17.2 shows a steel sprocket brazed to a cast-iron hub.

Figure 17.3 shows an assembly of steel tubes brazed to a malleable iron header.

Figure 17.4 shows several diesel engine subassemblies made of cast-iron fittings brazed to steel tubing.

Figure 17.5 shows a tripod section made of steel tubing brazed to a cast-iron base.

CHAPTER 18

STAINLESS STEELS

The term "stainless steels" is now used to describe a wide variety of chromium-containing, iron base alloys used primarily in applications demanding heat or corrosion resistance. The brazing of virtually all these alloys is now performed as a routine operation. Despite the increased understanding of stainless steel brazing, the degree of process control required in brazing these alloys is generally greater than in the brazing of carbon steels; this is because of the chemical differences between carbon and stainless steels, and because stainless steel service environments are generally more arduous than the service environments of carbon steel components. Success in the fabrication of stainless steel components by brazing is dependent on a knowledge of the characteristics of the various types of stainless steels and rigid adherence to certain items of process control required by these characteristics.

BASE METALS

Stainless steels may be grouped into four categories: (1) austenitic nonhardenable steels, (2) ferritic nonhardenable steels, (3) martensitic hardenable steels, and (4) precipitation hardening steels. All these alloys are iron base and contain chromium, the basic element that imparts corrosion resistance. The corrosion resistance of the stainless steels varies widely from one alloy to another, and for any given alloy varies widely from one corrosive medium to another. If doubt exists on the proper stainless steel to use in a given environment, standard reference works (Refs. 1, 2) or manufacturers' representatives should be consulted.

Austenitic Nonhardenable Stainless Steels

These steels are alloys of iron, chromium, and nickel (or manganese). The nickel or manganese additions stabilize austenite, the high temperature phase in steel, down to room temperature, and cause these alloys to be nonmagnetic and nonhardenable. This class of stainless steels is the one possessing the highest heat and corrosion resistance. Its members are designated as AISI Types 300 and 200. The most commonly used is Type 302 (of the 18-8 series), an alloy of iron, 18% chromium, and 8% nickel. The Type 200 stainless steels are a relatively recent innovation wherein some of the nickel is replaced with manganese. Type 202, the parallel to Type 302, contains iron, 18% chromium, 5% nickel, and 9% manganese.

The Type 300 stainless steels are widely used for both torch and furnace brazed assemblies. A major precaution in brazing these materials is that the stablized grades, such as Types 302 and 304, are subject to carbide precipitation when held at temperatures in the range of 800-1500 °F (427-876 °C), with a consequent loss of corrosion resistance. The extent of carbide precipitation in these steels is governed by the time that they are held in this sensitizing range; thus, carbide precipitation can be minimized by making the brazing thermal cycle as short as possible. With short time cycles, such as would result from the torch brazing of small parts, these grades can be brazed without serious loss of corrosion resistance.

The deleterious effects of carbide precipitation in stainless steels are virtually eliminated by the addition of the carbide-stabilizing elements titanium and columbium. Type 321 (18-8 + titanium) and Type 347 (18-8 + columbium) stainless steels are stabilized grades that may be brazed with no danger of impairing their corrosion resistance; thus, they are commonly used in aerospace and nuclear reactor components. Several of the unstabilized grades are available in extra-low carbon analyses (for example, Type 304L), which are relatively insensitive to carbide precipitation and can be brazed with essentially the same range of filler metals and heating cycles as the stabilized types. When high melting point filler metals are used, precipitated carbides in stainless steels can be redissolved by heat treatment following brazing. For a more complete discussion of carbide precipitation, see chapter 2.

While the Type 300 stainless steels have been widely used for brazed components, the Type 200 steels are newer and have not seen as much use. Due to their high manganese content the Type 200 steels are more difficult to furnace braze in hydrogen atmospheres than the Type 300 steels, and therefore hydrogen atmospheres of lower dew point are necesary. Manganese forms an oxide that is not easily reduced by dry hydrogen at the furnace brazing temperatures normally used for stainless steels; consequently, wetting problems on incompletely cleaned surfaces occur with the Type 200 steels.

All the chromium-nickel steels are subject to stress corrosion cracking in the presence of molten brazing filler metals. This phenomenon occurs when the base metal is under stress, either residual or resulting from applied loads while the braze is being made. One form of applied load results from thermal gradients during brazing. The filler metal penetrates the base metal along the grain boundaries at the points of stress, producing a greatly weakened base metal — best results, therefore, are obtained on stress relieved material. This stress relieving may occur prior to or during the brazing cycle; if it occurs during the cycle, it must be done below the solidus temperature of the filler metal used. Parts should be assembled and supported in a manner so as to avoid stress during the brazing cycle. For a further discussion of this subject, see chapter 2.

Ferritic, Nonhardenable Stainless Steels

These stainless steels are basically low carbon alloys of iron and chromium, where sufficient chromium has been added to the iron to stabilize ferrite, the low temperature phase in steels, over a wide temperature range. The more common AISI stainless grades in this category are Types 405, 430, and 446. Type 430 is a widely used grade that is particularly subject to a form of interface corrosion when brazed with some BAg filler metals; the corrosion is apparently caused by electrochemical action whereby the bond between the base metal and the filler metal is destroyed. This action has been found to occur in many cases in the presence of tap water. It has been found that the addition of small percentages of nickel to silver base filler metals prevents interface corrosion of brazed joints in most stainless steels. The nickel-containing silver base filler metal is not completely effective, however, with Type 430, even though its use greatly reduces the rate of attack. A special silver base filler metal (63Ag-28.5Cu-6Sn-2.5Ni) has been developed for Type 430.

Martensitic, Hardenable Stainless Steels

These are iron-carbon-chromium alloys of two basic types: the low chromium, low carbon grades (Types 403, 410, and 416) and the high chromium, high carbon grades (Types 440 A, B, and C). These steels are closely related to the ferritic nonhardenable grades, but have their alloy compositions so balanced that they are capable of being hardened by heat treatment. The primary precaution in brazing components made of these alloys is that the brazing thermal cycle must be compatible with the heat treatment required by the alloys. This is usually done by selecting brazing filler metals with brazing temperatures such that austenization of the base metal is performed at the brazing temperature. To secure proper transformation of the austenite to martensite, rather rapid cooling from the brazing temperature is needed.

Precipitation Hardening Stainless Steels

Stainless steels have been developed for applications where high strength plus heat and corrosion resistance are required. These alloys are basically stainless steels to which alloying elements such as aluminum, titanium, copper, and molybdenum have been added. The addition of these elements makes it possible to precipitation harden the alloys by special heat treatments. The basic stainless steels to which the alloying additions have been made are all the chromium-nickel austenitic type; however, the alloys are not all completely austenitic, as the hardening reactions are quite complex, and sometimes involve some martensite formation.

Precipitation-hardening stainless steels have as yet no standard AISI type designations. Some of the designations of these alloys are: 17-7PH, PH14-8Mo, 15-5PH, AM350, 17-4PH, PH15-7Mo, and A-286. As in the case of the martensitic hardenable stainless steels, brazing thermal cycles used in joining these alloys must be compatible with their heat treatments, and as the heat treatments vary rather widely specific brazing procedures are required for each alloy. Suppliers of these alloys and of brazing filler metals should be consulted for specific brazing procedures. The manufacture of honeycomb structural panels for supersonic aircraft and spacecraft is a significant application of brazing these alloys and is described more fully in chapter 25.

FILLER METALS

A wide variety of filler metals is commercially available for the brazing of stainless steel parts. Due to the wide variation in melting point, base metal interface reaction, cost, and resulting service properties, the selection of a filler metal for stainless steel brazing is dependent on the end use of the part. Commercial brazing filler metals are available that have copper, silver, nickel, cobalt, platinum, palladium, manganese, and gold as the base or as addition elements; these may be conveniently grouped according to service temperature.

Filler metals suitable for service at temperatures up to about 800 °F (427 °C) are the BCu types. For temperatures up to 700 °F (371 °C) the BAg types find extensive use. Filler metals suitable for general use on stainless steels from 800 to 1000 °F (427 to 538 °C) are the copper-manganese-nickel type. Above 1000 °F, the BNi filler metals or certain types containing the precious metals gold or palladium are used. Filler metals for service above 800 °F are generally furnace brazed in high purity reducing or inert atmospheres or in a vacuum. Of the copper base filler metals only those in the BCu classification are recommended for brazing stainless steels. They are generally used in high purity reducing atmospheres of low moisture content. The RBCuZn and BCuP filler metals are not recommended for use on stainless steels.

Of the BAg filler metals, BAg-1 and BAg-1a are general purpose filler metals with good brazing characteristics; BAg-2 is also a general purpose filler metal, requiring a somewhat higher brazing temperature. It is particularly useful when tolerance cannot be as closely controlled as is necessary for BAg-1 and BAg-1a. The BAg-2 filler metal should not be used where the heating time is excessive due to its tendency to liquate. Where improved corrosion resistance is needed, BAg-3 and BAg-4 are recommended over silver base filler metals not containing nickel. General purpose metals for higher brazing temperatures are BAg-5 and BAg-6 and they are often used in applications where cadmium is prohibited. The BAg-7 is useful for fabricating food handling equipment due to its white color and lack of cadmium. The BAg-8, the silver-copper eutectic, is often used for vacuum and atmosphere brazing where freedom from volatile cadmium and zinc is required. This filler metal with 0.2-0.5% lithium (BAg-8a) shows increased wettability on stainless steels. The BAg-19 is basically sterling silver with a lithium addition to promote self-fluxing behavior. It has been extensively used for the brazing of lightweight stainless steel honeycomb panels for high speed aircraft. For additional information on honeycomb panel brazing, see chapter 25.

Nickel base filler metals are primarily used where extreme heat and corrosion resistance are required. They are commonly used in the manufacture of jet and rocket engines, chemical processing equipment, and nuclear reactor components. These filler metals are normally supplied as powders; however, they can be obtained as sintered or cast rods, preforms, plastic bonded sheet, plastic bonded wire, and tape. Some filler metals are available in foil as well. The BNi filler metals are commonly used on stainless steels for oxidation resistance to temperatures up to 1800-2000 °F (982-1093 °C). Filler metals BNi-1, BNi-2, BNi-3, and BNi-4 tend to erode thin sheet metal because of their interaction with many base metals. Time at brazing temperature should, therefore, be minimized when these filler metals are used. The boron-free filler metals BNi-5, BNi-6, and BNi-7 are suitable for use in nuclear reactor components where boron cannot be tolerated because of its absorption of neutrons and reduction of reactor efficiency. Oxidation resistance is good up to 2000 °F (1093 °C). The BNi-5 has the highest melting point of all the nickel filler metals.

Filler metals based on gold, platinum, and palladium, and their alloys, such as gold-nickel, gold-nickel-chromium, copper-platinum, silver-palladium-manganese, and palladium-nickel-chromium, have been found useful for brazing heat and corrosion resistant components. The advance of brazing technology in such industrial fields as jet and rocket propulsion and nuclear energy will continue the introduction of new and unique filler metals. These have not, as yet, become sufficiently standardized to be firmly classified. Suppliers of brazing filler metals should be consulted where certain specific requirements must be met.

PROCESSES AND EQUIPMENT

Stainless steels can be brazed with any of the various brazing processes. A large volume of controlled atmosphere brazing is being performed on stainless steels, and the success of this type of brazing can be attributed to the development of reliable atmosphere and vacuum furnaces. The primary equipment requirements are that the furnaces must have good temperature control (plus or minus 10 °F [6 °C] at brazing temperature is desirable), and be capable of fast heating and cooling rates. All gases used in atmosphere furnaces must be of high purity. Commercial vacuum brazing equipment operates at pressures varying from 10^{-5} to 10^{-1} torr (0.0015 to 13.5 Pa). The vacuum requirements will depend upon the materials brazed, the filler metals being used, the area of brazing interfaces, the degree to which gases are expelled by the base metals during the brazing cycle, and the leak rate of the facility. Partial pressures employing dry atmosphere are used in specific conditions.

PRECLEANING AND SURFACE PREPARATION

Stainless steels require more stringent precleaning than do carbon steels, as the tenacious oxide films that impart corrosion resistance to stainless steels are more difficult to remove by fluxes or reducing atmospheres than the oxide films that form on carbon steels.

Precleaning of stainless steels for brazing should include a degreasing operation to remove any grease or oil films. The joint surfaces to be brazed should also be cleaned mechanically or with an acid pickling solution. However, wire brushing should be avoided, especially with a carbon steel wire brush. Care should be taken after cleaning to prevent dirt, oil, or fingerprints from soiling the cleaned surfaces. The best practice is to braze parts immediately after cleaning. When this is not possible, it is desirable to enclose the cleaned part in sealed polyethylene bags to exclude moisture and other contaminents until the part can be brazed.

FLUXES AND ATMOSPHERES

Assemblies of many stainless steels can be furnace brazed in atmospheres of dry hydrogen, argon, helium, and dissociated ammonia without the aid of flux. Though fluxes must be relied upon in production for certain components, some fluxes may require postbraze cleaning, and their use increases the probability of weak joints due to entrapment of residue.

The chromium oxides that impart corrosion resistance to stainless steels are more difficult to remove than the oxides formed on carbon steels. Atmospheres having dew points of −40 °F (−40 °C) or lower are necessary to have water vapor contents below the metal-to-metal oxide equilibrium* for these and other common metal constituents at the brazing temperature. Therefore, the formation of oxide films on heating must be retarded by the high purity of the atmosphere, or else it may be necessary to rely on special high activity fluxes for stainless steels. Chapter 4 should be consulted for the details of these fluxes.

Vacuum atmospheres promote the dissociation of the base metal oxides at brazing temperatures. A vacuum of the proper quality and pressure will protect metals from reoxidation during brazing. The vacuum required depends upon many variables: type of equipment, heating rate, base metal and brazing filler metal outgassing of the part and fixtures, and furnace leak rate. Manufacturers of vacuum equipment, base metals, and filler metals should be consulted for specific recommendations.

A precaution to be observed in the use of dissociated ammonia atmospheres is that at some brazing temperatures certain stainless steels can be inadvertently nitrided. Nitriding produces a hard surface that can be beneficial or detrimental, depending on the service requirements of the components. Nitriding can be detected by an

*See Fig. 4.1.

increase in surface hardness or by metallographic examination.

Oxides of aluminum and titanium cannot be reduced at ordinary brazing temperatures in atmosphere furnaces. If these elements are present in small amounts, satisfactory brazes can be obtained by the use of high purity gas atmospheres and vaporizing flux. When these elements are present in quantities exceeding 1 or 2% the metal surface can be cleaned and nickel-plated in lieu of using fluxes or vacuum atmospheres. Nickel plating will not only prevent the formation of detrimental oxides, it can prove to be an effective means of limiting embrittlement and erosion of base metals. The thickness of the electrolytic nickel plating should be kept in the range of 0.0002-0.002 in. (0.005-0.05 mm), since the joint strength will decrease as the thickness of plating increases and failure may occur in the plating layer.

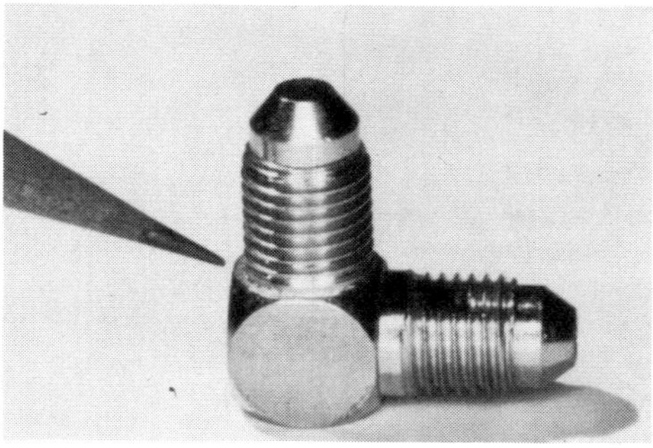

Fig. 18.1—Brazed Type 347 fitting.

POSTBRAZE OPERATIONS

Flux and stop-off residue removal and heat treatment are major postbraze operations that may be necessary in the brazing of stainless steels. No postbraze operations are required on nonhardenable stainless steel parts furnace brazed in reducing or inert atmospheres when no flux is used and where removal of stop-off, if used, is not necessary.

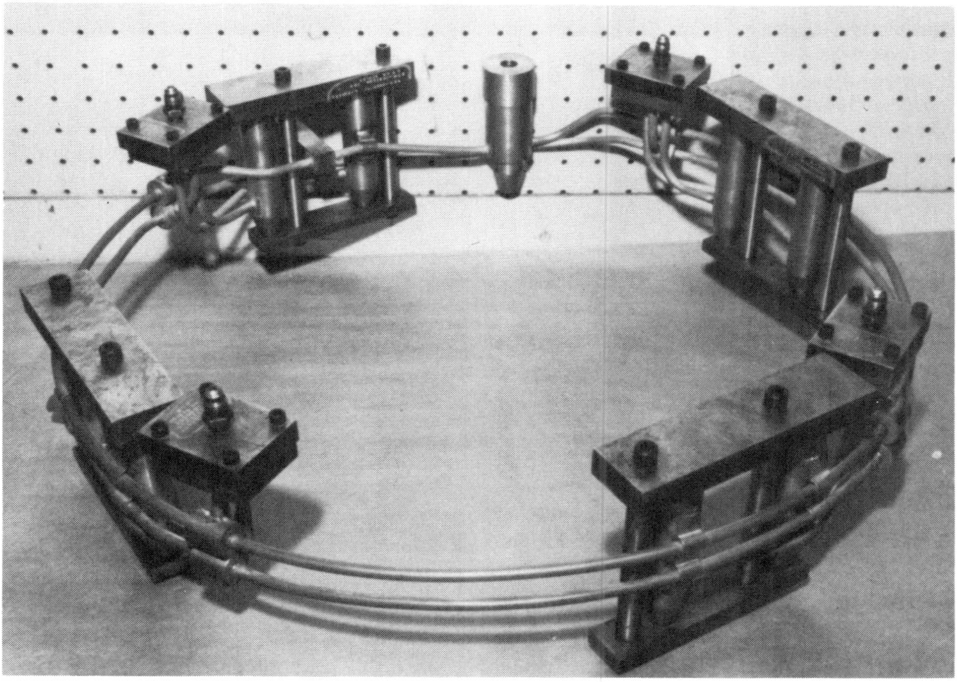

Fig. 18.2—Brazed Type 304 microwave guide assembly.

Fig. 18.3—Brazed Type 347 rocket motor thrust chamber.

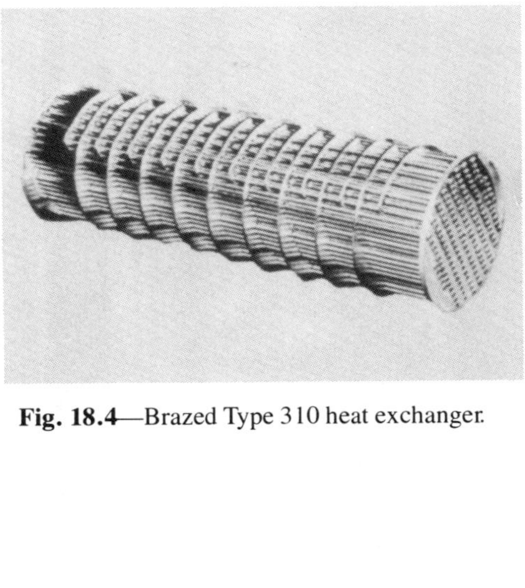

Fig. 18.4—Brazed Type 310 heat exchanger.

Stop-off materials may be removed simultaneously with removal of flux residue. However, if no flux is used, removal of stop-off may be by mechanical or chemical methods. Chapter 10 can be consulted for additional information on post-braze operations.

Depending on the flux and the brazing process used, removal of the flux residue may be accomplished by water rinsing and chemical or mechanical cleaning. If abrasive cleaning is used for flux removal or removal of oxide films in heated areas adjacent to the braze, sand or other nonmetallic grits should be used. Metallic shot other than stainless steel should be avoided as the particles may become embedded in the stainless steel surface and cause rusting or pitting corrosion in service.

Heat treatment after brazing is required on parts made of the martensitic hardenable and the precipitation-hardening stainless steels. These treatments vary so widely that no general rules can be made other than that the recommendations of the suppliers of these materials should be followed.

TYPICAL APPLICATIONS

Figure 18.1 shows a Type 347 stainless steel fitting vacuum brazed with BNi-2.

Figure 18.2 shows a Type 304 stainless steel microwave guide assembly brazed with BAu-4.

Figure 18.3 shows a Type 347 stainless steel rocket motor thrust chamber brazed with a BNi filler metal powder.

Figure 18.4 shows a Type 310 stainless steel corrugated heat exchanger vacuum brazed with a nickel base filler metal. What appears to be a solid bar is actually a tube containing 0.027 in. (0.67 mm) corrugations.

REFERENCES

1. Zappfe, C. A. 1949. *Stainless Steels*. American Society for Metals.
2. American Society for Metals. 1961. *Metals Handbook*. Vol 1, 8th Ed.

CHAPTER 19

NICKEL BASE AND COBALT-CONTAINING ALLOYS

Nickel and the high nickel alloys can be joined readily by most of the commonly used brazing processes. The alloys that are normally considered high nickel are those that contain over 50% nickel. The severe service conditions to which these alloys are generally subjected is one reason that every phase of the brazing procedure requires careful control. The chemical composition, mechanical and physical properties of nickel and high nickel alloys should be considered so that the assembly may be fabricated without encountering problems. Some of the factors that must be considered are the effects of sulfur and low melting point metals and the problem of stress cracking. The important properties of nickel and the high nickel alloys are tabulated in the appendix.

A section on the cobalt-containing alloys is included at the end of this chapter. In general, these alloys can be brazed with the same filler metals and utilize the same procedures and equipment required for the nickel and high nickel alloys. Very little data have been developed specifically for the cobalt-containing alloys.

NICKEL BASE ALLOYS

Base Metals

Nickel and the high nickel alloys may be divided into the following important classes: (1) commercially pure nickel, (2) nickel-copper alloys, (3) nickel-chromium-iron alloys, (4) nickel-chromium-molybdenum alloys, and (5) thorium-dispersed nickel alloys. Alloys in groups (1) and (2) are used primarily for applications where corrosion resistance is important, or where product purity must be maintained. Alloys in groups (3), (4), and (5) also have good corrosion resistance in many media and, in addition, have high strength and oxidation resistance at elevated temperatures. To utilize these properties, it is important to determine the effect of service conditions on the brazed joint.

There are very few, if any, general statements regarding corrosion or oxidation resistance of high nickel alloys which can be valid for all circumstances; hence, the selection of an alloy for a given application should be determined by actual or simulated service tests. Some limitations on service at elevated temperatures are mentioned in succeeding paragraphs dealing with specific brazing filler metals.

Stress Cracking

Many metals are susceptible to stress cracking in certain corrosive environments when the level of tensile stress is high. Nickel and the high nickel alloys are particularly resistant to this phenomenon in the usual aqueous solutions, but one environment that will cause stress cracking is molten brazing filler metal. Hence, nickel alloys must be as free as possible of residual or applied tensile stress during the brazing operation. If residual stresses from prior cold working are present, it is advisable to stress relieve or anneal the assembly before the brazing operation. Depending on the annealing temperature of the material, it may be possible to combine this operation within the brazing cycle. It is important that the brazing procedure be such that tensile stress will not be introduced due to fixtures used or thermal gradients during the brazing operation. See chapter 2 for more details on stress cracking.

Age Hardenable Alloys

The high temperature strength properties of some nickel alloys can be improved by the addition of titanium and aluminum e.g., Rene 80, Rene 120. These improvements are effected by the introduction of a stable, hard phase or phases, such as intermetallic compounds. The principal hardening compounds present in the high nickel alloys are the intermetallics Ni_3Al and Ni_3Ti. The age hardenable nickel and nickel-copper alloys are usually aged at 1100 °F (593 °C) while the nickel-chromium-iron alloys are usually aged at 1300 °F (704 °C).

Due to the inherent stresses present, age hardened nickel alloys are susceptible to stress cracking. These metals should be brazed in the annealed or solution-treated condition with a relatively high melting brazing filler metal (preferably above 1600 °F, 872 °C) that has sufficient strength to withstand handling during the subsequent age hardening treatment.

Filler Metals

The brazing filler metals normally used for ferrous metals are suitable for joining nickel and high nickel alloys. The selection of the brazing filler metal depends on the service conditions of the finished assembly. It is also important to consider any heat treatments that may be required for the base metal, so as to be certain that the brazed joint will withstand the temperatures involved. The brazing filler metals are covered in detail in chapter 3, and Table 3.1 lists the common brazing filler metals that may be used for joining nickel and high nickel alloys to each other and to other metals and alloys.

The BAg brazing filler metals may be used to join nickel and high nickel alloys to themselves and to many other metals and alloys. With proper design and technique, the brazed joints will develop the full strength of the base metal. The low melting filler metals, BAg-1, BAg-1a, and BAg-3, are commonly used, but for many corrosive environments filler metals containing at least 50% silver are preferred. The BAg-7 filler metal is useful where stress cracking might occur.

Most high nickel alloys are capable of being brazed with BCu filler metal using the same equipment as used for steels with only minor changes in the brazing procedures due to the different characteristics of these alloys. The BCu filler metals will alloy more rapidly with nickel alloys than with steel; therefore, copper will not flow far before it has picked up enough nickel to raise its liquidus and reduce its fluidity. To eliminate this problem the filler metal should be placed as close to the joint as possible, and there should be a sufficient reservoir to fill the joint. A slightly rough or lightly etched surface presents better conditions for the capillary flow of copper filler metal. Such surfaces will be wet for relatively great distances, whereas polished surfaces will resist the flow of the copper filler metal. If too much copper is present and the excess flows over a thin member, it may perforate this member.

Phosphorus combines with many metals to form brittle compounds known as phosphides. For this reason BCuP filler metals are not usually used with any iron base or nickel base alloys; however, BNi-6 and BNi-7 filler metals are used for brazing nickel base alloys. The BNi filler metals have been widely used for brazing nickel base and high nickel heat-resistant alloys. Joints with very high oxidation resistance, corrosion resistance, and elevated-temperature strength can be obtained when properly brazed with these filler metals. The BNi filler metals will interact with the base metal at the brazing temperature. This property has the advantage of increasing the joint remelt temperature, joint strength, oxidation, and corrosion resistance. Caution should be used, however, to apply the nickel base filler metals so that erosion and excessive interaction with the base metal do not occur.

Nickel base filler metals containing palladium or platinum additions and gold base, palladium base, and platinum base filler metals have also been used successfully to braze high nickel alloys. These filler metals generally have good wetting and flow characteristics. They have a low interaction rate with most nickel base metals and are used to advantage in many special applications where they produce joints having good ductility, high strength, and good oxidation resistance. Filler metals consisting of base material plus additions of silicon and boron have been used successfully for brazing Rene series alloys.

Processes and Equipment

Most commercial brazing processes may be used on nickel and high nickel alloys. The most common of these are torch, furnace, induction, and resistance brazing; salt bath and metal bath dip brazing have limited application. While the BAg filler metals may be used in torch brazing, the

BCu and BNi filler metals are usually used in controlled atmosphere brazing. See chapter 1 for further details on the various types of furnaces available. In furnace brazing, careful control of the temperature is important. The furnace should be designed to permit a suitable heating and cooling rate and accurate control of the brazing temperature. It is of utmost importance that the furnace be tight, so that the proper atmosphere or vacuum can be maintained. The walls and other internal parts of the furnace should be designed so that outgassing will be kept to a minimum when a vacuum is used.

Precleaning and Surface Preparation

Precleaning and surface preparation are important parts of the brazing procedure, since it is not the primary purpose of a flux or atmosphere to remove oxides, oil, grease, dirt, or other foreign materials. The usual requirements for precleaning and surface preparation, outlined in chapter 7, apply to the nickel alloys. Precleaning the nickel alloys just prior to brazing is particularly important since they are subject to attack by low melting point elements, particularly lead and sulfur, at elevated temperatures. Material which has been embrittled by these elements cannot be salvaged. Because grease, oil, paint, shop dirt, and other foreign material usually contain these harmful elements they must be entirely removed before brazing.

The oxide films formed on nickel alloys are tenacious and wire brushing may not remove them; however, they may be removed by emery cloth or grinding. Uniform oxide removal by pickling cannot be expected unless the high nickel alloy being processed is first thoroughly cleaned of all foreign material. The pickling procedures recommended for these alloys are given in Table 19.1. The cleaning processes for the nickel alloys must be thorough and are the same as for cleaning of other metal products. Soaps can be removed with hot water. Soluble oils, tallow, fats and fatty acid combinations are removed with a hot (180-200 °F, 82-93 °C) 10-20% solution of equal parts of sodium carbonate (soda ash) and trisodium phosphate; sodium hydroxide (caustic soda) may be used instead of sodium carbonate. Parts are immersed in this solution for up to one-half hour and then rinsed in water. Mineral oils and greases are readily soluble in trichloroethylene and other solvents.

Fluxes and Atmospheres

When using BAg filler metals, AWS types 3A and 3B fluxes are suitable for most alloys not containing aluminum. An AWS Type 4 flux may be used with aluminum-containing nickel alloys, while the standard reducing atmospheres listed in chapter 4 may be used to braze nickel and high nickel alloys.

Elements such as aluminum and titanium form refractory oxides which are not reduced in the normal furnace atmosphere. Therefore, age hardenable alloys are more difficult to braze than standard alloys. A procedure is necessary which will either prevent the formation of the oxide or will flux away any oxide formed during heating. Some methods of handling such a situation are as follows:

1. The parts may be copper plated to prevent the formation of harmful oxides. The plating procedure is somewhat special in that oxides must not be present on the surface of the parts under the plating. This can be accomplished by preparing the surface in the normal procedure and then using a nickel chloride strike followed by copper plating in the usual manner. About 0.0003 in. (0.008 mm) of copper is desired. This copper will prevent the formation of oxides and, in addition, will become part of the brazing filler metal.

2. Another method is to nickel plate the parts to prevent the formation of refractory oxides. Nickel plating is generally used in conjunction with BNi-1 filler metal and the age hardenable alloys containing aluminum and titanium. The thickness of the plating required depends on the time and temperature of the brazing cycle. The general range of plating thickness is 0.0002-0.002 in. (0.005-0.05 mm), although at least 0.0005 in. (0.01 mm) is preferred.

3. The use of dry, oxygen-free atmospheres can be used to prevent the formation of most oxides and in most cases will reduce those present. However, if the oxides of titanium or aluminum are involved, it is best to perform the brazing in a vacuum at a pressure of 10^{-3} torr (0.15 Pa) or less. When using inert gases, such as argon or helium, the parts must be free of surface oxides when put in the furnace. In some cases, pure dry hydrogen may be used as the atmosphere. The dew point of the hydrogen to be used is determined by the oxides which can form or must be reduced.

Vacuum brazing in a leak-free, metallic element furnace is widely used for the age hardenable alloys.

Fig. 19.1—Four-step nickel honeycomb seal.

Assembly, Procedures, and Techniques

Silver (BAg) filler metals — In general, when using BAg filler metals the brazing procedures outlined in chapters 1 through 11 apply. The joint clearance should be maintained between 0.002 and 0.005 in. (0.05-0.13 mm) when flux is used; when brazing in a controlled atmosphere, clearances as low as 0.0005 in. (0.01 mm) are feasible. As a rule, wider clearances will result in lower joint strengths.

Copper (BCu) filler metals — When brazing the nickel alloys with BCu filler metals, the steps are similar in many respects to those used on steel. The tolerances for assembly range between a light press fit to 0.002 in. (0.05 mm) maximum clearance. Brazed butt joints will provide strengths approaching those of the annealed base metals if proper clearances are maintained.

Nickel (BNi) filler metals — These filler metals follow the same general requirements as the BAg group. While joint clearances between 0.002 and 0.005 in. (0.05-0.13 mm) are feasible, better strengths are obtained if the joint clearance is less than 0.002 in. (0.005 mm). Filler metals are available for brazing joint clearances to 0.010 in. (0.25 mm) (wide gap alloys). In general, when using BNi filler metals the brazing is performed in a furnace using a suitable atmosphere. These filler metals are commonly applied in the form of a powder mixed with a volatile vehicle, such as an acrylic resin for application by brush, spray, or extrusion. Precautions must be taken in using the vehicle so that no harmful residue will remain. When properly used, the vehicle should completely disappear during the heating cycle. Filler metal is also available in tape and foil forms that permit the application of a controlled quantity of braze filler metal to joints. Adjustments for density must be made when using tape filler metals.

Postbraze Operations

For many applications no postbraze treatment is required. If oxidation has occurred, a pickling operation may be required to clean the assembly.

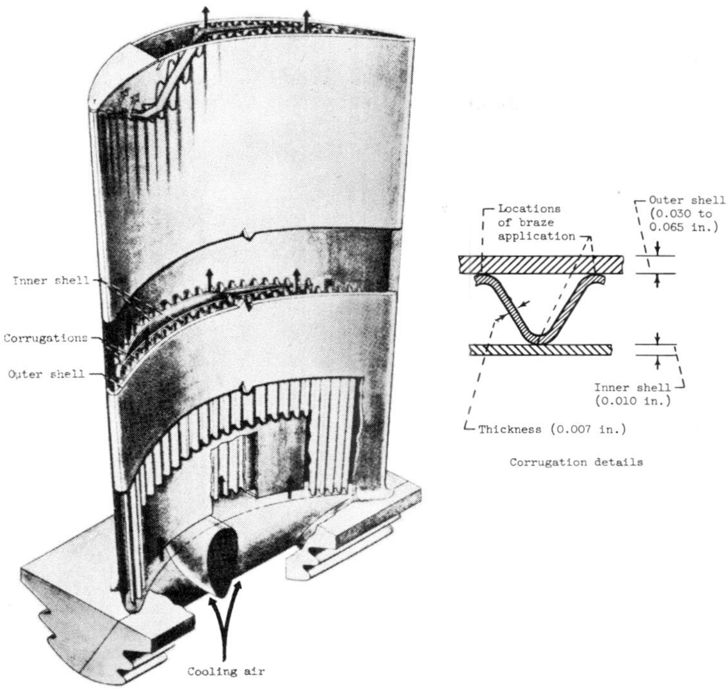

Fig. 19.2—Air-to-air heat exchanger assembly.

If flux has been used, a flux removal operation may be necessary. Flux removal is particularly important on parts that are to operate at elevated temperatures or in corrosive environments.

The age hardenable alloys of nickel may be aged after brazing. It is necessary to select a brazing filler metal that has a melting temperature above the temperature at which the base metal must be heated — see the Age Hardenable Alloys section of this chapter. Most nickel alloy assemblies brazed in a suitable furnace atmosphere require no postbraze cleaning. If excess copper is present from the brazing filler metal it can be removed in a cold bath of 20 parts of ammonia and 1 part hydrogen peroxide or a solution of 15 gal. of water, 1 gallon of sulfuric acid, and 80 lb. of chromic acid thoroughly mixed and used at room temperature. The length of time required in the bath depends on the thickness of copper to be removed.

Inspection

The general techniques outlined in chapter 11 should be followed when inspecting brazed joints in any of the nickel and high nickel alloys. The inspection requirements of components for elevated temperature and corrosion environments will be more rigid because of the service conditions involved. X-ray and ultrasonic inspection techniques have been widely used.

Typical Applications

Figure 19.1 shows a nickel base alloy honeycomb seal.

Figure 19.2 shows a section of an air-to-air heat exchanger fabricated from a 0.015-in. (0.37-mm) thick nickel base alloy using BNi-1 filler metal.

Figure 19.3 shows basic nickel alloy honeycomb sandwich elements.

Figure 19.4 is a nickel-chrominum-iron alloy tubular heat exchanger which was vacuum brazed with BNi-2.

COBALT-CONTAINING ALLOYS

Cobalt, which is similar to nickel in most of its properties, is used primarily as an alloying ele-

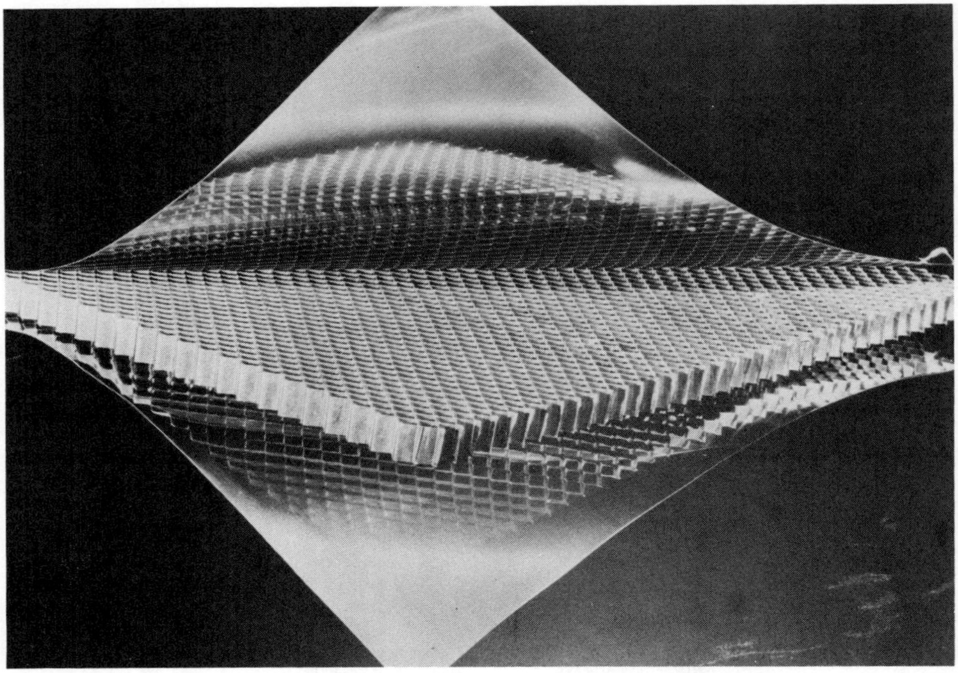

Fig. 19.3—Basic nickel alloy honeycomb sandwich elements.

Fig. 19.4—Vacuum brazed tubular heat exchanger.

ment. In general, these alloys are used for their corrosion and heat resistant properties. Some of the cobalt-containing iron base alloys are used in metal-to-glass seal applications. There are three popular groups of alloys in which cobalt is used: iron base, nickel base, and cobalt base. A few cobalt-containing alloys are listed in the appendix. In general, the same principles as outlined for the nickel and high nickel alloys apply to the cobalt-containing alloys. However, when using the BAg and BCu filler metals some consideration should be given to strength and oxidation resistance as the service temperature is increased. The use of the BNi filler metals is recommended for applications where the service temperatures are high.

The oxide film that forms on these alloys during heat treatment strongly adheres to the base metal. This oxide can be removed by molten caustic baths and acid solutions. It is also important to remove foreign materials before heating to avoid embrittlement of the base metal.

Table 19.1—Pickling recommendations for nickel and high nickel alloys

	Formula to be used for the following types of oxides		
Base metal	Light tarnish	Reduced oxide	Heavy oxide
Nickel	3	4	5* and 3†
Nickel-aluminum alloy (Duranickel)	3	4	5* and 3†
Nickel-copper alloy (Monel)	1 and 2†	4	5 and 6†
Nickel-copper-aluminum alloy (K Monel)	1 and 2†	4	5 and 6†
Nickel-chromium-iron alloy (Inconel)	....	...	7A
Nickel-chromium-iron-titanium alloy (Inconel X)	....	....	7A

Make-up of formulas

	Formula No.						
	1	2	3	4	5	6	7A
Water, gal(l)§	1(4)	1(4)	1(4)	1(4)	1(4)	1(4)	3(12)
Nitric acid (38° Bé), qt(l)	4(4)	4(4)	9(9)	....	....	....	4(4)
Sulfuric acid (66° Bé), pt(l)	....	....	12(3)	¾(⅜)	....	¾(⅜)	....
Hydrochloric acid (20° Bé), qt(l)	....	....	....	....	2(2)	....	....
Hydrofluoric acid (49%), pt(l)	....	....	....	....	....	....	1¼(⅝)
Sodium chloride, lb(kg)	½-¾(¼-⅜)	....	¼(⅛)¶	1(½)	....	....	....
Sodium nitrate (crude), lb(kg)	....	....	....	½(¼)	....	....	....
Sodium dichromate, lb(kg)	....	....	....	....	....	1.1(½)	....
Cupric chloride, lb(kg)	....	....	....	....	¼(⅛)	....	....
Temperature, F	70-100	70-100	70-100	180-190	180	70-100	120-140
Temperature, C	(21-38)	(21-38)	(21-38)	(82-88)	(82)	(21-38)	(49-60)
Time	Not over 5 sec	Not over 5 sec	5-20 sec	30-90 min	20-40 min	5-10 min	15-90 min
Container materials	Crocks, glass, ceramic	Stainless, glass, ceramic	Crocks, glass, ceramic	Crocks, glass, ceramic	Crocks, glass, ceramic, brick	Crocks, glass, ceramic, rubber	Carbon, brick

NOTE: All work should be degreased prior to pickling and should be neutralized as a last operation in a 1-2 vol. % ammonia solution. Fumes from these pickle solutions are strongly irritant. Ventilating hoods should be employed (see chapter 29).

*Will require 1-2 hours.

†Hot water (180 °F [82 °C]) rinse between dips.

§Approximate conversions.

¶Allow to cool before adding salt.

CHAPTER 20

TITANIUM, ZIRCONIUM, AND BERYLLIUM BASE ALLOYS

Titanium, zirconium, and beryllium are reactive metals that have made the transition from laboratory curiosities to engineering materials in the past two decades. A number of commercial titanium alloys exist today, and the metal and its alloys are being increasingly used in aircraft and numerous corrosion resistant applications. Zirconium and beryllium have been used primarily in nuclear applications, although beryllium possesses significant potential as a high efficiency structural material for critical weight aerospace applications.

BASE METALS

The common characteristic of the three metals is their reactivity. Titanium, zirconium, and beryllium combine readily with oxygen at elevated temperature and react to form brittle intermetallic compounds with many other metals. Their tenacious surface oxide film requires that these metals be very carefully cleaned before brazing, and maintained in a clean condition until brazed. The reactivity with other metals imposes rather stringent limitations on brazing filler metals. Still another problem arises from the reactivity of titanium and zirconium with hydrogen, nitrogen, and oxygen. Titanium and zirconium can become embrittled by interstitial absorption of these gases whereas beryllium is only superficially affected.

Three types of titanium alloys are commercially available today. These are classified as all-alpha, all-beta, and alpha-beta alloys, the designations referring to the metallurgical structures present. Pure titanium and a few alloys are classified as all-alpha. These materials are not heat treatable, and brazing operations have very little effect on their properties. Several all-beta base metals are commercially available. When designed for use in the annealed condition, they are virtually unaffected by the brazing thermal cycle. If they are to be heat treated, the brazing temperature may have an important effect on the properties of the beta alloy. The best ductility is obtained if the base metal is brazed at the solution treatment temperature. As the brazing temperature is increased above this temperature, the ductility of the base metal decreases.

Depending on composition, the mechanical properties of alpha-beta titanium alloys may be altered greatly by heat treatment and variations in microstructure. Wrought alpha-beta titanium alloys are generally fabricated to obtain a fine-grained equiaxed duplex microstructure, which results in maximum ductility. In brazing alpha-beta alloys, it is desirable to maintain this microstructure by requiring that the brazing temperature not exceed the beta phase transformation temperature ("beta transus") which may vary from roughly 1650 to 1900 °F (900 to 1038 °C) depending upon specific base metal composition. Alpha-beta alloys may be used in the annealed or solution treated and aged condition. If an annealed structure is desired after brazing, three alternatives are possible.

1. Anneal then braze at or below the annealing temperatures

2. Braze above the annealing temperature and incorporate a step-cooling operation in the brazing cycle to obtain an annealed structure

3. Braze above the annealing temperature and anneal after the brazing operation is completed. Most of the commercial filler metals flow at temperatures above titanium alloy annealing temperatures so that the latter two alternatives are most commonly used.

Zirconium base metals of commercial importance are the pure metal and several alloys (Zircaloy) containing small percentages of tin, iron, chromium, and nickel. These structural alloys were developed for corrosion resistance in nuclear applications, notably pressurized water nuclear power reactors.

Since much of the early development work on beryllium was for nuclear energy applications, high purity was vital. Today, in addition to premium nuclear grades, beryllium is available as massive hot pressed block, large wrought plate, and sheet stock up to 4 × 15 ft (1.2 × 4.6 m). From a modulus/density standpoint, beryllium is about four to six times as efficient as conventional metal alloy structures. It is little wonder that this unique element finds increasing uses in space hardware and certain terrestrial low inertia instruments.

Beryllium presents two problems in fabrication that titanium and zirconium do not. First, the metal has limited ductility at room temperature, which requires that components be handled with care to avoid damage. Second, beryllium and its compounds are toxic when in the form of dust, vapor, and solutions (see chapter 29). Improperly handled, they can cause berylliosis, which may be fatal. Safe exposure limits for beryllium and its compounds have been determined, and the metal may be safely handled by observing those limits, which are normally attainable by installing ventilation and filtering systems to control airborne contamination. Brazing or any other fabrication operation with beryllium should not be attempted without first instituting proper safety standards. Beryllium suppliers should be consulted regarding approved procedures and equipment.

FILLER METALS

Filler metals initially used for brazing of titanium and its alloys were basically silver with alloying additions of lithium, copper, aluminum, or tin. Later developments produced a number of successful commercial fillers including silver-palladium, titanium-nickel, titanium-nickel-copper, and titanium-zirconium-beryllium. Higher strengths and improved resistance to crevice-type corrosion were desirable characteristics. In joining applications requiring a high degree of corrosion resistance, the 48Ti-48Zr-4Be and 43Ti-43Zr-12Ni-2Be type filler metals are outstanding. A silver-palladium-gallium filler metal (Ag-9Pd-9Ga) that flows well at 1650-1675 °F (900-913 °C) is available in wrought forms as sheet and wire. Joints in high pressure aircraft hydraulic systems exhibit a high degree of durability, fatigue strength, oxidation, and corrosion resistance.

An achievement which stems from supersonic transport materials and process technology is titanium honeycomb sandwich brazed with aluminum filler metal. Such aircraft structures up to 23 ft (7 m) in length have been braze fabricated successfully using 3003 brazing foils. The use of aluminum brazing filler metal (3003) provides satisfactory strength up to about 600 °F (316 °C). At 1000-1100 °F (538-593 °C) high strength, corrosion resistant Ti-Zr-Be or Ti-Zr-Ni-Be is recommended.

The development of filler metals for zirconium has been directed toward those that would produce sound, corrosion resistant joints in zirconium alloy tubing exposed to pressurized water at elevated temperatures in nuclear power reactors. Joints brazed in high vacuum with a 4-5% beryllium addition to the zirconium alloy composition have shown excellent resistance to corrosion in 680 °F (360 °C) water at up to 3000 psi (20 700 kPa) pressure. Pure silver, BAg-18, BAg-19, pure aluminum, and BAlSi filler metals have been reported to be satisfactory for brazing beryllium for special applications.

PROCESSES AND EQUIPMENT

As stated in Base Metals, titanium, zirconium, and beryllium are all highly reactive metals. It is thus important to use brazing processes that do not allow the joint surfaces to come in contact with air during heating. Induction brazing and furnace brazing in inert-gas or vacuum atmospheres can be used successfully, but torch brazing of these base metals is difficult and requires special precautions and techniques. Induction brazing of small, symmetric parts is very effective because its speed minimizes reaction between filler metal and base metal. For large, precise assemblies, furnace brazing is favored as uniformity of temperature throughout the heating and cooling cycle can be readily controlled. Special equipment and furnace fixturing is required to maintain reliable production schedules and consistent quality brazements. Whereas titanium and zirconium assemblies are frequently brazed in high vacuum, cold wall furnaces, beryllium assemblies braze better in gettered argon or helium. Beryllium in particular requires great skill in braze processing because stable, stress-free assemblies with precise tolerances are vital conditions for this high modulus material. Critical-use applications like the one illustrated in Fig. 20.6 are of large size and are primarily load-bearing structures.

PRECLEANING AND SURFACE PREPARATION

Thorough surface cleaning prior to brazing is necessary. Heavy scale formed on titanium heated above 1000 °F (538 °C) should be removed mechanically or by molten salt bath cleaning prior to the operations just mentioned. Light oxide film can be removed by pickling. Pickling solutions for titanium should contain 20-45% nitric acid plus 2% hydrofluoric acid. At lower nitric acid concentrations, hydrogen may be absorbed by the base metal. When degreasing is required, nonchlorinated solvents, such as acetone, MEK, or alcohol, should be used. Zirconium parts should be cleaned by procedures similar to those used for titanium.

Beryllium components should be degreased and pickled in 10% hydrofluoric or nitric-hydrofluoric acid mixtures prior to brazing. Ultrasonic rinsing with deionized water is a further requirement. A precoating of the joint faying surfaces by special vacuum metallizing techniques or electroplating prior to brazing has been found very effective. Silver, titanium, copper, aluminum, or BA1Si-4 precoats provide reliability inasmuch as beryllium is exceptionally difficult to wet unless extremely high purity atmospheres are used for brazing.

FLUXES AND ATMOSPHERES

Argon, helium, and vacuum atmospheres are satisfactory for brazing titanium, zirconium, and beryllium. For torch brazing special fluxes must be used. Fluxes for titanium are primarily mixtures of fluorides and chlorides of the alkali metals, sodium, potassium, and lithium. Torch brazing of beryllium should not be attempted because of its toxicity.

Vacuum and inert-gas atmospheres, such as helium and argon, will protect titanium during furnace and induction brazing operations. A vacuum of 10^{-3} torr (0.15 Pa) or better is required to braze titanium, whereas a dew point of −70 °F (−54 °C) or less is necessary to prevent discoloration of the titanium in a brazing temperature range of 1400-1700 °F (760-927 °C). Fabricators of titanium brazements generally favor liquid argon from cryogenic storage vessels because of its inherent purity.

Although scant information is available on brazing zirconium as compared to titanium, the two metals are very similar chemically. In general, the brazing techniques used are applicable to both systems.

Beryllium is most commonly brazed in purified argon although helium and vacuum of less than 10^{-3} torr (0.15 Pa) are also suitable. Two fluxes have been used for furnace or induction brazing with good results: (1) a 60% LiF-40% LiC1 flux and (2) a tin chloride flux. However, no fluxes should be used with vacuum atmospheres. It should again be emphasized that beryllium and beryllium compounds as flux residues are toxic. Only approved installations should consider brazing of beryllium regardless of the methods used.

APPLICATIONS

Brazing of titanium and titanium alloy components requires care and stringent process controls. This high performance material is being widely used for aerospace hardware. Commercial brazed assemblies include jet engine inlet vanes, hydraulic tubing or fittings, and honeycomb sandwich panels. Figure 20.1 is representative of 50 000 sheet metal gas turbine inlet guide vanes fabricated from 6A1-4V titanium base metal. The anti-icing airfoil has a wrap-around skin and corrugated core member, which is brazed in vacuum with Ti-Cu-Ni type of filler metal.

Titanium plumbing systems providing additional weight savings were a logical outgrowth from XB-70 and lunar excursion module brazed stainless steel high pressure fluid line joints and fittings. Application of hinged induction heater tools and protective argon atmosphere permitted rapid, reliable brazes to be made in the production of air force and navy high performance aircraft. One wide-body U.S. jetliner utilizes over 250 brazed joints involving commercially pure

Fig. 20.1—Large fan turbojet engine uses brazed sheet metal vanes.

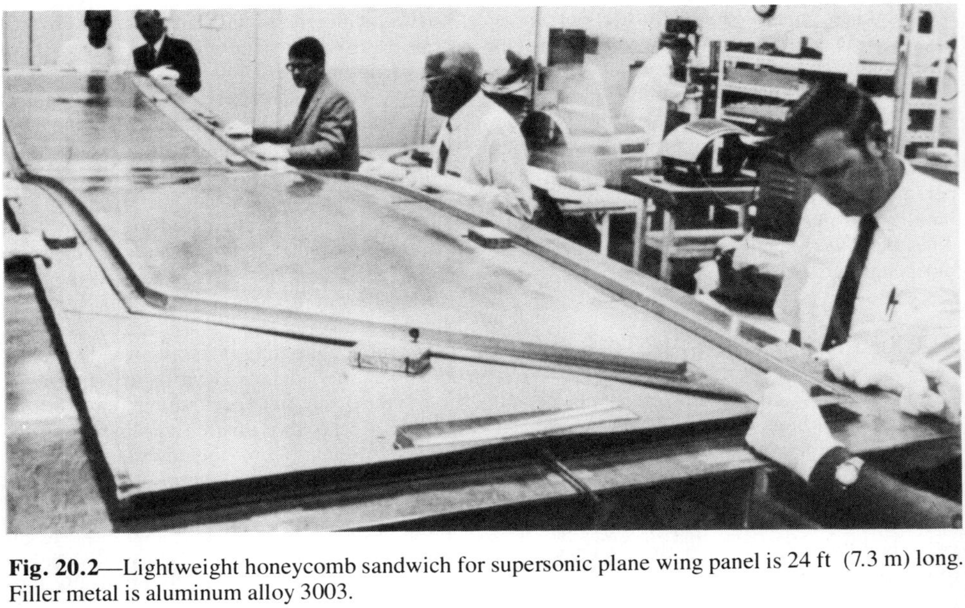

Fig. 20.2—Lightweight honeycomb sandwich for supersonic plane wing panel is 24 ft (7.3 m) long. Filler metal is aluminum alloy 3003.

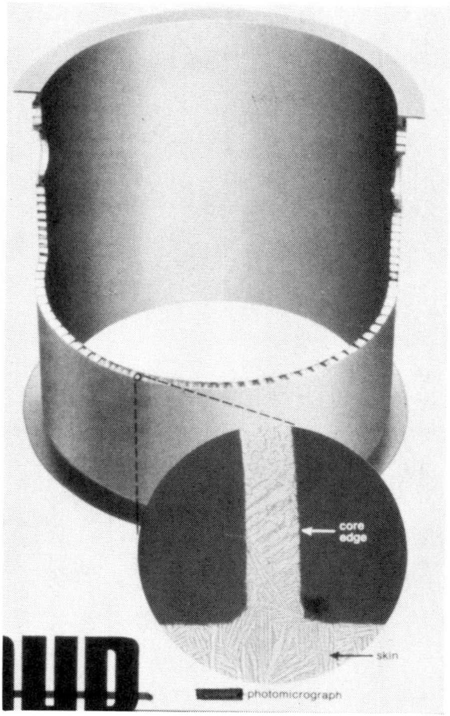

Fig. 20.3—Electroplated eutectic brazing system is used in bonding of titanium alloy honeycomb sandwich cylinder.

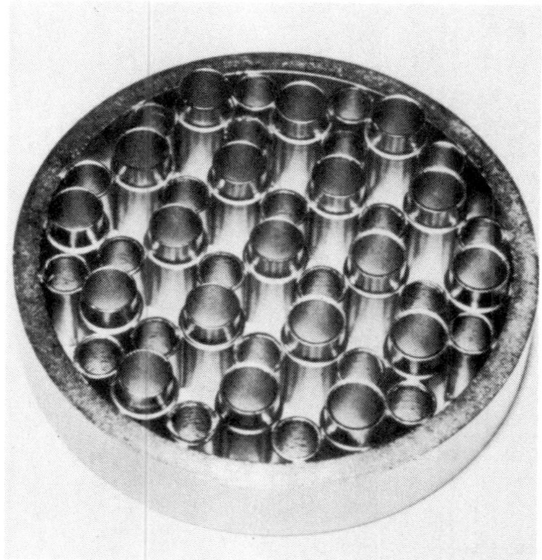

Fig. 20.4—Zircaloy-2 end ring vacuum brazed at 1868 °F (1020 °C) using Zr-5.18Be filler metal (as brazed for CVTR).

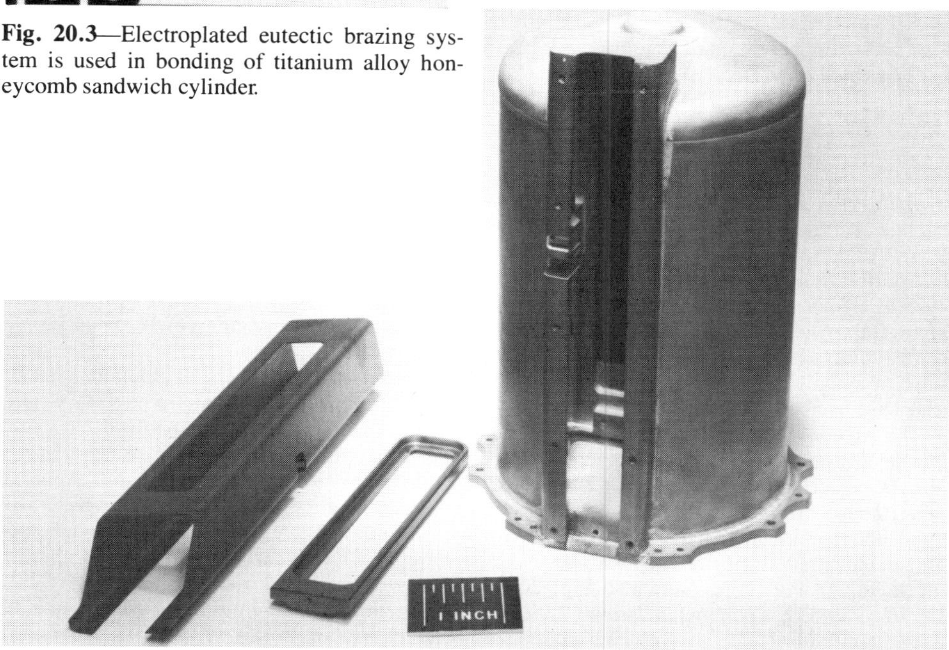

Fig. 20.5—Camera housing brazement for Viking Lander has 0.15 in. (0.33 mm) thick walls. Assembly is joined with aluminum base filler metals.

Fig. 20.6—Ten beryllium parts joined with BA1Si-4 in one furnace brazing operation. The 14 lb (6.4 kg) fabricated ring is main support for NASA's 140 lb (63.7 kg) VISSR satellite instrument.

titanium tubing and Ti-6A1-4V fittings. Sizes include ¼, 5/16, ⅜, and ½ inch; filler metal was 90Ag-10Pd.

Another manufacturer brazes with 48Ti-48Zr-4Be filler metal. A new filler metal which offers improved producibility and metallurgical compatibility consists of 82Ag-9Pd-9Ga. The results of impulse fatigue, burst, and corrosion tests have been outstanding.

The American supersonic transport design relied heavily on high efficiency honeycomb structures. One very promising material for fuselage, wing, and control surfaces was metallurgically bonded titanium sandwich panels. A 24 ft (7.3 m) long demonstration wing test panel is shown during lay-up in Fig. 20.2. (See also Fig. 20.3.) Honeycomb core is commercially pure 0.002 in. thick titanium foil; facings are chem-milled Ti-6A1-4V of 0.010 in. (0.25 mm) thickness. Aluminum alloy 3003 filler metal foil has been placed between all faying surfaces. This compound curvature assembly was successfully brazed using ribbon heated ceramic tooling and argon atmosphere at rarified pressure. The U.S. Air Force F-15 fighter aircraft also has this aluminum brazed titanium alloy honeycomb sandwich system. Engine cowls are in current production.

Applications for brazing of zirconium and zirconium alloys usually relate to nuclear requirements. One noteworthy use involved Zircaloy-2 fuel tube cap assemblies 3 in. (76.2 mm) in diameter by 2 in. (50.8 mm) long. Approximately 500 of these 19-tube clusters were brazed in vacuum. Figure 20.4 shows a typical joint structure. Excellent mechanical and thermal properties were achieved.

Brazing of beryllium is extensively used in the fabrication of complex, high performance hardware. Due to its remarkable modulus-to-weight ratio and the commercial availability of large size wrought sheet, beryllium is used in

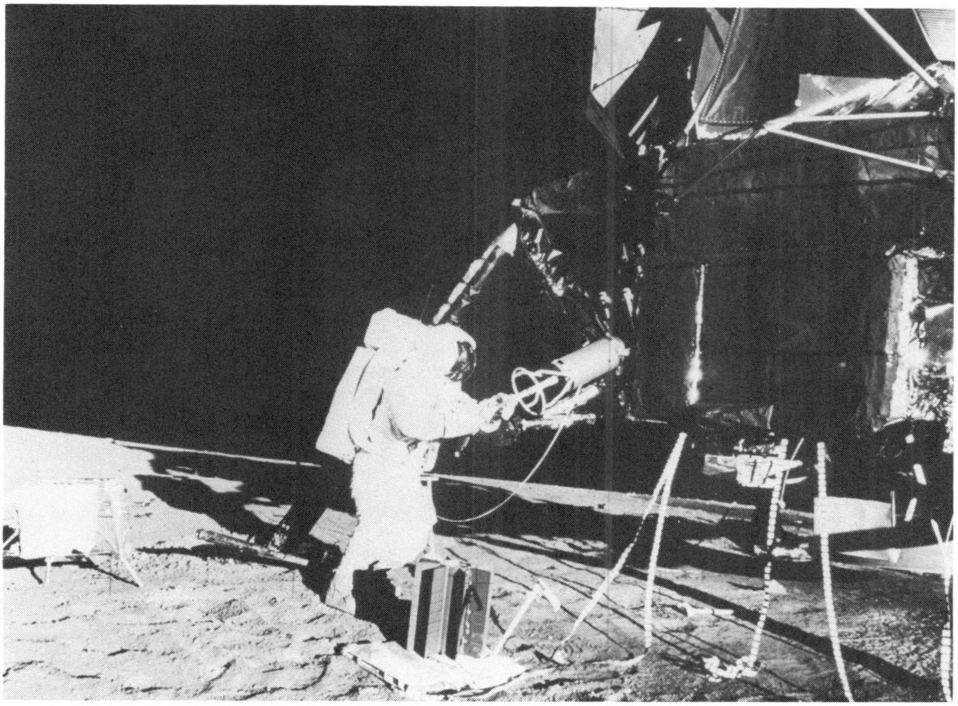

Fig. 20.7—Photomicrograph of BAg-19 joint between stainless steel ring and hot pressed block beryllium in SNAP-27 nuclear power generator casing. Five of these complex brazements are successfully operating at the five Apollo Lunar Science Stations. Beryllium fins are also brazed to the case using BAg-19.

ever-increasing quantities for spacecraft structure and instrument payloads. Several filler metals which afford strong joints with good manufacturability, are in general use.

The electronic camera pressure housing of Fig. 20.5 measures 4 in. (102 mm) in diameter by 8 in. (203 mm) high and weighs but 105 g. A machined hot pressed block beryllium base member is also shown. Kovar sleeves are first brazed to the beryllium using BAg-19; gas tungsten arc welding of the alumina/Kovar electrical lead-through completes the mass spectrometer tight assembly.

An early primary structural brazement in beryllium was the main support ring for VISSR satellites. Figure 20.6 shows the various machined components that are assembled by furnace brazing. A metallographic section of a joint brazed with BA1Si-4 is shown in the inset.

Brazing of beryllium to itself and other metals has developed into a common manufacturing technique. Honeycomb sandwiches, truss structures, built-up I beams, laminated spars, lapped tubes, and radiation windows are important beryllium applications.

Five Apollo Scientific Stations are powered by SNAP-27 nuclear generators. Figure 20.7 shows a critical beryllium housing brazement. Beryllium sheet fins are brazed to a machined beryllium case with BAg-19. A brazed stainless steel transition ring at each end of the case is secured and a mass spectrometer vacuum tight joint is also shown.

CHAPTER 21

TUNGSTEN, MOLYBDENUM, TANTALUM, AND COLUMBIUM

Tungsten, molybdenum, tantalum, and columbium all melt above 4000 °F and are characterized as refractory metals. Because of the demand of the aerospace and nuclear industries for high temperature materials, these refractory metals have graduated from laboratory curiosities to accepted engineering materials. Basic information on these refractory metals is given in the appendix. Alloys of these metals are constantly under development to obtain improved properties and many have become commercially available. In addition, there is extensive use of these elements as pure metals. The technology of brazing the refractory metals is still in the developmental stages with improved techniques and brazing materials being reported continually.

Refractory Metal Resistance Welding Electrode Tips

Often it is desirable to braze refractory metal base tips to the copper base current carrying portion of resistance welding electrodes. These tips are used to minimize electrode deformation where adequate cooling cannot be applied and to achieve heat balance where metals of different thickness are being welded. Electrode sticking is also minimized. These tip materials are included in RWMA (Resistance Welder Manufacturer's Association) Classification Group B.

Tungsten-Copper Tips

These materials are not true alloys but are made by powder metallurgy processes. A normal analysis of one of the materials is 80W-20Cu. The tips have two areas of application.

1. For resistance welding electrodes that are usually brazed to the copper-chromium alloy, which constitutes the main current carrying portion of the electrode. Any of the BAg series of filler metals will be found suitable for this application. The copper-tungsten tips are usually applied by induction or torch heating.

2. For electrical discharge machining (EDM) in which the same or similar composition of tungsten-copper (W-Cu) is used for the finish machining electrode. Here again, only the end of the electrode is made from the tungsten-copper alloy. A base of phosphorized copper or OFHC is used to attach the electrode to the current supply. The large mass and high thermal conductivity of the brazement present problems during joining; in most cases brazing is accomplished in a neutral or reducing atmosphere. The most suitable filler metal is BAg-8; to hold the parts in alignment pins are incorporated in the design, and filler metal is preplaced in the form of shim stock located between the mating parts.

Tungsten and Molybdenum Tips

Resistance welding electrodes that are tipped with pure tungsten or molybdenum are especially useful when welding nonferrous, high conductivity materials. Welding of braided copper wire is a typical use.

Any of the BAg series of brazing filler metals can be used to braze molybdenum or tungsten tips to the copper alloy base. Torch or induction heating is used. For best results it is often desirable to precoat the electrode tip-faying surfaces. The tip maker usually does this by using one of the silver based filler metals.

Precoating of the tungsten surface aids in a uniform flow of the filler metal between the faying surface of the joint. Without this pretreatment poor wetting of the tungsten by the filler metal often results, causing premature joint failure under the high pressures and temperatures in the resistance welding.

TUNGSTEN

Base Metal

Tungsten is one of several materials being used advantageously for components designed to operate at high temperatures. Available in pressed and sintered, arc-cast and wrought forms, it has the highest melting temperature of any of the known metals: 6170 °F (3410 °C). With a density of 0.697 lb/in.^{3} (19 000 kg/m^3) tungsten is a very heavy material that possesses a higher modulus of elasticity and has a coefficient of thermal expansion about equal to that of molybdenum. A combination of high melting point and high temperature strength makes tungsten advantageous for elevated temperature applications. An unusually high ductile-to-brittle transition temperature of about 500-700 °F (260-371 °C) prevents the working of tungsten at room temperature. The recrystallized metal exhibits reduced strength and ductility properties at elevated temperatures. Tungsten can be brazed to itself and to other metals and nonmetals when proper procedures and processes are used.

Because of its extremely high melting point, low vapor pressure, and reduced tendency for material transfer under arcing and sparking conditions, tungsten is utilized in electrical contact applications (e.g., automobile voltage control contacts, make and break type of contacts, relays, and circuit breakers). Tungsten has also been used in electrodes for resistance welding equipment. In recent years, a considerable amount of emphasis has been given to tungsten in the aerospace industry (e.g., nozzle liners, leading edges, heat sinks).

Filler Metals

Filler metals must be chosen for the particular application intended. A wide variety of filler metals and pure metals used to join tungsten are given in Table 21.1. The liquidus temperatures of the brazing filler metal alloys range from 1200 to 3500 °F (649-1927 °C). Low melting filler metals, such as BAg-3, BCuP, and BCu, are used to join tungsten for electrical contact applications, but as the service requirement at extremely elevated temperatures would considerably reduce their usefulness, they are limited in their applications. The use of pure metals, such as tantalum and columbium, is warranted in some cases by the extremely high temperature service of the assembly. Nickel base and manganese base, high temperature filler metals have been used successfully in aerospace applications for joining tungsten. Filler metals based on platinum and boron have been used to 3900 °F (2150 °C) with a diffusion treatment following at 2000 °F. Brazing at relatively low temperatures followed by such diffusion treatments permitted use of joints to around 5500 °F (3038 °C).

Joint Design

The joint clearances required in many instances for joining tungsten are governed by the brazing filler metal being used. The clearances listed in Table 6.1 should be used as guides in the determination of joint dimensions. Very often, the engineering design, base metal, and manufacturing operation may automatically predetermine the conditions of the assembly prior to brazing.

Process and Equipment

Tungsten can be brazed with furnaces (inert-gas or reducing atmosphere and vacuum), torch, resistance, and induction heating. Furnace brazing operations are usually used when the parts are larger than those considered practical for the induction and resistance processes.

Precleaning and Surface Preparation

Thorough cleaning prior to brazing is essential, and both mechanical and chemical cleaning can accomplish this purpose. Cleaning methods found acceptable for tungsten are given in Table 21.2. The most effective cleaning procedure will depend upon the tenacity of the oxide film. In cases where wrought tungsten sheet has been mill cleaned, degreasing is sometimes the only cleaning operation necessary prior to brazing. However, the optimum conditions for preparation of tungsten should be determined for each particular application. In some cases, electroplating of tungsten with nickel and other elements has been satisfactorily used to stop diffusion of elements that form brittle intermetallic compounds with the base metal. A hydrogen atmosphere furnace cleaning operation at 1950 °F (1065 °C) for 15 min has been effective in reducing light oxide films. Tumble polishing in a slurry is commonly used for tungsten contact material.

Table 21.1—Brazing filler metals for refractory metals

Brazing filler metal	Liquidus temperature F	Liquidus temperature C	Brazing filler metal	Liquidus temperature F	Liquidus temperature C
Cb	4380	2416	Co-Cr-Si-Ni	3450	1899
Ta	5425	2996	Co-Cr-W-Ni	2600	1427
Ag	1760	960	Mo-Ru	3450	1899
Cu	1980	1052	Mo-B	3450	1899
Ni	2650	1454	Cu-Mn	1600	871
Pd-Mo	2860	1571	Cb-Ni	2175	1191
Pt-Mo	3225	1774	Pd-Ag-Mo	2400	1316
Ag-Cu-Zn-Cd-Mo	1145-1295	618-702	Pd-Al	2150	1177
Ag-Cu-Zn-Mo	1325-1450	718-788	Pd-Ni	2200	1204
Ag-Cu-Mo	1435	779	Pd-Cu	2200	1204
Ag-Mn	1780	971	Pd-Ag	2400	1316
			Pd-Fe	2400	1316
Ni-Cr-B	1950	1066	Au-Cu	1625	885
Ni-Cr-Fe-Si-C	1950	1066	Au-Ni	1740	949
Ni-Cr-Mo-Mn-Si	2100	1149	Au-Ni-Cr	1900	1038
Ni-Ti	2350	1288		Brazing temperature*	
Ni-Cr-Mo-Fe-W	2380	1304	V-Ta-Cb	3300-3500	1816-1927
Ni-Cu	2460	1349	V-Ta-Ti	3200-3350	1760-1843
Ni-Cr-Fe	2600	1427	Ti-V-Be	2280	1249
Ni-Cr-Si	2050	1121	Zr-Cb-Be	1920	1049
Mn-Ni-Co	1870	1021	Zr-Ti-Be	1830	999

NOTE: Not all the filler metals listed are commercially available.
*Depends on specific composition.

Fluxes and Atmospheres

Tungsten can be brazed in either an inert-gas atmosphere (helium, argon), reducing atmosphere (hydrogen), or in a vacuum. Two precautions should be observed when vacuum brazing: the vapor pressures of the compositional elements of the brazing filler metals should be compatible with the temperatures and pressures involved, and the effect of outgassing on the soundness of the base metals and the deposited filler metal should be evaluated. Conventional low temperature fluxes are used when brazing tungsten for electrical contact applications when silver base and copper base filler metals are used.

Techniques of Brazing

Tungsten parts must be handled with care during assembly operations because of their inherent brittleness. An excess of nickel base filler metals should be avoided since the interaction between tungsten and nickel results in the recrystallizing of the tungsten. Contact with graphite should be avoided to prevent the formation of brittle tungsten carbides. Prestressed tungsten parts, caused by previous forming or welding, should be stress relieved prior to brazing. In most cases, filler metals and brazing cycles should be selected to provide minimum interaction between the filler metal and the tungsten in order to maintain the integrity of the base metal.

MOLYBDENUM

Base Metal

Molybdenum and its alloys are available in the pressed and sintered, arc-cast and wrought forms. The former is used for electrical contacts, while the latter forms find application where high strength at elevated temperatures is required. Molybdenum has a high modulus of elasticity and is about as strong as steel. It is half as dense as tungsten and has good thermal conductivity. Like tungsten, it has excellent high temperature properties; however, poor oxidation resistance requires coating protection at higher temperatures. The presence of minute quantities of oxygen, nitrogen, and carbon lower the ductility of molybdenum, while recrystallized molybdenum has reduced strength and ductility and should be avoided if possible. Some grades of molybdenum are brittle at room temperature, but they can be formed at moderately elevated temperatures (300-500 °F [149-260 °C]). The addition of an

Table 21.2—Cleaning methods for refractory metals

Tungsten		Molybdenum	
Method 1	Immerse in 20% potassium hydroxide solution (boiling)	Method 1	Immerse in a glass cleaning etch composed of 95% H_2SO_4, 4.5% HNO_3, 0.5% HF and Cr_2O_3 (equivalent to 18.8 gm/l)
Method 2	Electrolytic etching in a 20% potassium hydroxide solution	Method 2	a. Immerse in alkaline bath of 10% NaOH, 5% $KMnO_4$ and 85% H_2O by weight operating at 150-180 °F (66-82 °C), 5-10 min.
Method 3	Chemical etching in a 50 vol. % HNO_3 — 50 vol. % HF solution		b. Immerse 5-10 min. in second bath to remove smut formed during first treatment. Bath consists of 15% H_2SO_4 15% HCl, 70% H_2O, and 6-10 wt. % per unit volume of chromic acid
Method 4	Immerse in molten sodium hydroxide	Method 3	Mo-0.5 Ti alloy a. Degrease 10 min. in trichloroethylene
Method 5	Immerse in molten sodium hydride		b. Immerse in commercial alkaline cleaner 2-3 min.
			c. Rinse in cold water
			d. Buff and vapor blast
			e. Repeat step (b)
			f. Rinse in cold water
			g. Electropolish in 80% H_2SO_4 at 120 °F (54 °C) with 8-12 amp
			h. Repeat st (c)
Tantalum		**Columbium**	
Method 1	a. Shot blast tantalum surface	Method 1	a. Degrease
	b. Follow by immersion into HCl solution to remove embedded iron particles		b. Immerse in commercial alkaline cleaner 5-10 min.
	c. Immerse in glass cleaning solution composed of 95% H_2SO_4, 4.5% HNO_3, 0.5% HF and Cr_2O_3 (equivalent to 18.8 g/1)		c. Rinse with water
			d. Immerse in 35-40% HNO_3 for 2-5 min. at room temperature
			e. Rinse with tap water, followed by distilled water
			f. Force air dry

alloying element raises the recrystallization temperature, allowing greater use of different brazing filler metals. The recrystallization temperature of six different alloys are: Mo unalloyed — 2150 °F (1177 °C); Mo-13Cb — 2200 °F (1204 °C); Mo-15Ti — 2450 °F (1343 °C); Mo-1V — 2150 °F (1177 °C); Mo-0.5Ti — 2450 °F (1343 °C); Mo-0.5Ti-0.07Zr — 2700 °F (1482 °C). Molybdenum has application in the electric and electronic industries (e.g., electronic tubes, electrical contacts, supports for transducers and transistors), the aerospace industry (nozzles, leading edges, radiation shields), the nuclear industry (heat exchangers, support grids), and in the chemical, glass, and thermal spraying industries.

Filler Metals

A wide variety of brazing filler metals may be used to join molybdenum; the choice is made on the basis of application. The brazing temperature ranges from 1200 to 3500 °F (649-1927 °C) as in the case of tungsten, and many of the alloys listed in Table 21.1 may be considered equally acceptable for molybdenum. Each filler metal should be evaluated for the particular end use. The service temperature requirement in many cases dictates the brazing filler metal selected. However, consideration must be given to the effect of brazing temperature on the base metal properties, specifically recrystallization. When brazing above the recrystallization temperature, time should be kept as short as possible, but when high temperature service is not required, copper base and silver base filler metals may be used. For electronic parts and other nonstructural applications requiring higher temperatures, copper-gold, nickel-gold, and nickel-copper filler metals can be used. Pure metals may be used at still higher temperatures.

Joint Design

The joint design and clearance requirements are similar to those for tungsten. Reference to Table 6.1 should be made after a brazing filler metal has been selected. Molybdenum has an extremely low coefficient of thermal expansion, and this property should be considered in the design of brazed joints, particularly when molybdenum is joined to other metals. Usually, joint clearances in the range of 0.002-0.005 in. (0.05-0.13 mm) at the brazing temperature are suitable. The selection of a fixture material should also be made with care and should be based on this same property.

Processes and Equipment

Torches, controlled atmosphere furnaces, vacuum furnaces, and induction and resistance heating equipment can be used to braze molybdenum. Oxyacetylene torch brazing can be accomplished with silver base and copper base filler metals and an appropriate flux.

Precleaning and Surface Preparation

Surface oxide removal before brazing is mandatory. The cleaning operation should be performed immediately before brazing to prevent contamination from accumulating. Degreasing should be used to remove oil, fingerprints, and grease — mechanical and chemical cleaning have been found satisfactory. Sand blasting, liquid abrasive cleaning, or abrasion may be used on simple parts to remove oxide films, but chemical cleaning is more satisfactory, especially for complex assemblies. For heavy oxide films, molten salt baths, such as (1) 70% sodium hydroxide and 30% sodium nitrite at 500-700 °F (260-371 °C) and (2) commercial martempering salt (mixture of sodium and potassium nitrates) operating at 700 °F (371 °C) have given good results. The first cleaning bath should be controlled closely, since it attacks molybdenum, but no gross attack has been noted with the second bath. Light surface oxide films should be removed after the salt bath by an appropriate treatment. Electrolytic etchants may be used to remove surface oxides on simple parts; however, grain boundary attack of the molybdenum can be severe. Chemical etchants are the most popular cleaning method. Three successful methods are given in Table 21.2.

Fluxes and Atmospheres

When utilizing an oxyacetylene torch, fair protection may be obtained by using a combination of fluxes, a commercial borate base or silver brazing flux (AWS types 3A and 3B), plus a high temperature flux containing calcium fluoride. The temperature range over which the fluxes are active is between 1050 and 2600 °F (566 and 1427 °C). The molybdenum is first coated with the commercial silver brazing flux and then the high temperature flux is applied. The silver brazing flux is active at the lower end of the active temperature range; the high temperature flux then takes over and is active up to 2600 °F. Purified, dry hydrogen and inert-gas (helium and argon) atmospheres have been found suitable for brazing molybdenum. For brazing pure molybdenum the purity of the hydrogen atmosphere is not critical.

A dew point of 80 °F (27 °C) can be tolerated in hydrogen when reducing molybdenum oxide. A low dew point (−50 °F [−46 °C] or lower) at 2200 °F (1204 °C) is required when brazing titanium-bearing molybdenum alloys. Vacuum furnaces have been used to braze molybdenum; however, precautions should be taken in selecting a proper brazing filler metal that will not volatilize during the brazing cycle. Pressures of less than 1μ (0.15 Pa) are desirable when brazing molybdenum alloys containing titanium.

Assembly, Procedures, and Techniques

Heating cycles should be short, above 2000 °F (1093 °C), to avoid the recrystallization of molybdenum. Another problem to be considered when brazing is the formation of intermetallic compounds between molybdenum and filler metals. Recrystallization occurs by exceeding the recrystallization temperature of molybdenum or lowering the recrystallization temperature by the diffusion of elements of the filler metal, such as nickel, or both. These problems are not important for low temperature service; however, they do manifest themselves at elevated temperatures. Palladium base filler metals in combination with base metals having high recrystallization temperatures (Mo-0.5Ti) have been used to minimize these conditions. Chromium, plated on molybdenum as a barrier layer, has been used to prevent the formation of intermetallic compounds. Service conditions influence the selection of the proper brazing filler metal and brazing cycle. Vacuum tube applications cannot tolerate outgassing or volatilization of elements; therefore, fluxes and brazing filler metals containing cadmium and zinc are not employed. Most high temperature brazing filler metals are as oxidation resistant as the base metal. Those filler metals having very high melting points are most suitable for coating applications. Each application where a coating is to be used should receive trial tests to determine the brazing filler metal coating compatibility.

TANTALUM AND COLUMBIUM

Base Metal

Available in several forms, tantalum and its alloys are midway between tungsten and molybdenum in density and melting points. Tantalum can be worked easily at room temperature. Its thermal conductivity is one-fourth that of molybdenum, its coefficient of expansion is one-third greater, and its elevated-temperature strength is low compared with tungsten and molybdenum. Its corrosion resistance is unusually good in most commercial combinations of acids, except in hot sulfuric acid solutions. Pure tantalum recrystallizes at approximately 2200 °F (1204 °C) depending upon the amount of cold work present. The addition of alloying elements, such as tungsten (Ta-30Cb-10W), raises the recrystallization temperature to 3000 °F (1649 °C). Other tantalum alloys (Ta-20Ti-5A1) achieve a reduction in recrystallization temperature to 1850 °F (1010 °C) by adding titanium. Tantalum has been used as a fixture material because of its low coefficient of expansion, high temperature properties, and ease of working at room temperature. Several tantalum alloys are commercially available.

Columbium is very similar to tantalum, and several alloys are available in the arc-cast and wrought condition. Columbium has the lowest melting point of all the refractory metals covered, the lowest modulus of elasticity and thermal conductivity, and the highest thermal expansion. It also has the lowest strength and lowest density of refractory metals. Columbium's ductile-to-brittle transition temperature ranges from −150 to −250 °F (−101 to −157 °C). This metal also has the low thermal neutron capture cross section required for nuclear applications. Its high melting point warrants its use at temperatures above the maximum service temperatures of the iron base, nickel base, and cobalt base metals, and it has excellent ductility and fabricability. Pure columbium has a recrystallization temperature range of 1800-2000 °F (982-1093 °C). This temperature increases with the addition of alloying elements, i.e., Cb-33Ta-1Zr — 2200 °F (1204 °C); Cb-15W-1Zr-5Ti — 2500 °F (1371 °C). Strengthening mechanisms, such as solid solution and dispersion hardening, are used in designing columbium alloys. Other alloys available are Cb-15W-5Mo-1Zr, Cb-10Ti-10Mo, and Cb-7Ti-0.8Zr.

Filler Metals

Filler metals are chosen on the basis of the application intended. Nickel base filler metals (such as the nickel-chromium-silicon alloys) have been used to braze tantalum and columbium. Tantalum forms a homogeneous solid solution with nickel up to a tantalum content of 36%; the liquidus is reduced from 2640 to 2460 °F (1449 to 1349 °C). These filler metals are satisfactory for service temperatures below 1800 °F (982 °C). Copper-gold alloys having less than 40% gold can be used as filler metals, but gold in amounts between 40 and 90% tends to form age hardening compounds which are also brittle. Although silver base filler metals have been used to join tantalum and columbium, they are not recommended because of a tendency to embrittle the base metals. Copper-tin, gold-nickel, gold-copper, and copper-titanium filler metals have also been used in brazing tantalum and columbium. In addition, pure filler metals, such as titanium, platinum, and palladium are used to join columbium. Literally hundreds of prospective filler metals have been evaluated for nuclear applications for columbium.

Joint Design

The joint designs and clearances are similar to the recommendations of Table 6.1.

Precleaning and Surface Preparation

Tantalum can be cleaned by both mechanical and chemical methods. Hot chromic acid (glass cleaning solution) is quite satisfactory, but hot caustic cleaning solutions will attack the metal and should not be used. Prior to the chromic acid cleaning, tantalum can be blast cleaned; this procedure, however, should be followed by an immersion in a hydrochloric acid solution to dissolve the iron particles. The glass cleaning solution then becomes more effective. Abrasion and other usual mechanical cleaning methods have been found acceptable. Tantalum has a tenacious oxide film that reforms immediately on exposure to air or vapor after any cleaning treatment. One method of preparing tantalum and columbium prior to brazing is to electroplate either copper or nickel onto an acid cleaned surface. The composite is then subjected to a diffusion treatment at 2200 °F (1204 °C) and 10^{-3} torr pressure. The deposits are bonded to the tantalum or columbium by diffusion, and in the case of copper melting actually occurs. Brazing may then be accomplished by using the plated surfaces as a base.

Fluxes and Atmospheres

Only inert-gas (argon and helium), controlled atmospheres and a vacuum are used in joining tantalum and columbium. Atmospheres containing those constituents that enbrittle tantalum, such as oxygen, nitrogen, and hydrogen, should be avoided. Brazing may be performed in air using suitable fluxes for the particular filler metal used. However, both tantalum and columbium will require a protective coating, such as nickel or copper electroplate.

Assembly, Procedures, and Techniques

Special techniques are necessary to satisfactorily braze tantalum and columbium. All gases that have any reactivity must be removed: not only oxygen, but carbon monoxide, ammonia, hydrogen, nitrogen, and carbon dioxide must be eliminated. Tantalum forms oxides, carbides, nitrides, and hydrides very readily with these gases, and loss of ductility ensues. At high temperatures tantalum and columbium should be protected from oxidation. One method is to electroplate the surfaces with copper or nickel; it is, therefore, necessary for the filler metal to be compatible with the plating.

DISSIMILAR METALS BRAZING

Molybdenum has been brazed to tungsten with a manganese-copper filler metal and to AISI 310 stainless steel sheet. Molybdenum has also been furnace brazed to tungsten, using manganese-nickel-cobalt and nickel-chromium-silicon-boron filler metals in an argon atmosphere. Joints between molybdenum and graphite have been effected in hydrogen, helium, and argon, using a gold-nickel-chromium filler metal.

Tungsten has been brazed to graphite and molybdenum using manganese-nickel-cobalt, nickel-chromium-silicon-boron, and nickel-chromium-silicon-molybdenum filler metals in argon and hydrogen atmospheres. Tungsten has been brazed to tantalum with the same filler metals, using an argon atmosphere. Molybdenum has been brazed to porous tungsten, using a manganese-nickel-cobalt filler metal in argon and hydrogen atmospheres.

TYPICAL APPLICATIONS

Figure 21.1 shows a rocket nozzle assembly fabricated from tungsten washer blanks brazed to each other, using a high temperature manganese-base filler metal.

Figure 21.2 shows a photomicrograph of a tungsten brazed joint, using a nickel-chromium-silicon-boron filler metal.

Figure 21.3 shows a gas turbine guide of arc-cast molybdenum brazed with a nickel base filler metal in a hydrogen atmosphere.

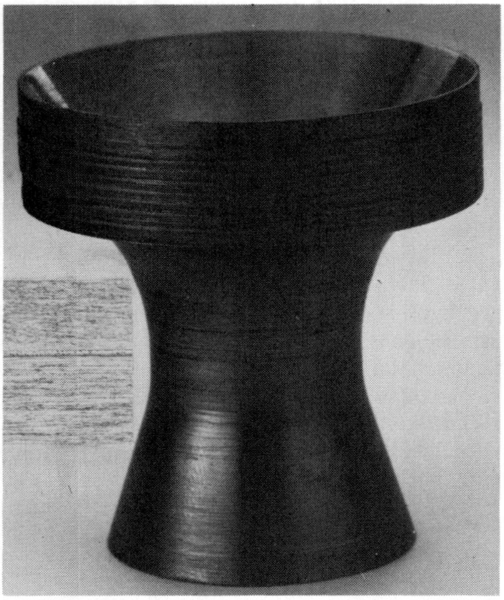

Fig. 21.1 — Brazed tungsten nozzle assembly. Inset shows closeup of joint, 100 × (reduced by 40% on reproduction).

Fig. 21.2 — Tungsten brazed to tungsten, 100 × (reduced by 35% on reproduction).

Fig. 21.3—Brazed molybdenum gas turbine guide.

Fig. 21.4 — Tungsten brazed to Mo-0.5Ti alloy, 100 × (reduced by 35% on reproduction).

Fig. 21.5 — Tungsten brazed to graphite, 500 × (reduced by 50% on reproduction).

Figure 21.4 shows a brazed joint between tungsten and Mo-0.5Ti made using a manganese base filler metal.

Figure 21.5 shows a tungsten-graphite brazed joint, using a nickel base filler metal.

SELECTED READING

Albom, M. J. 1962. Diffusion bonding tungsten. *Welding Journal* 41 (11), Research Supplement, pp. 491-s to 502-s.

Barth, S. D. 1595. *The fabrication of tungsten.* Defense Metals Information Center Report 115.

Bertossa, R. C., and Rau, S. 1959. Heat extractive brazed bimentals show promise for missile and industrial applications. *Welding Journal* 38 (7), Research Supplement, pp. 273-s to 281-s.

Briggs, J. Z. *A new look at joining molybdenum.* Climax Molybdenum Co.

Bruckhart, W. C. 1959. Fabrication of refractory metals. *Metal Industry* (6 February 1959).

Fox, C. W.; Gilliland, R. G.; and Slaughter, G. M. 1963. Development of alloys for brazing columbium. *Welding Journal* 42 (12), Research Supplement, pp. 535-s to 540-s.

Fox, F. G. 1957. Welding and brazing refractory metals. In *Steel processing and conversion.*

Freedman, A. H., and Mikus, E.B.

1964. Brazing of columbium D36 honeycomb structures. *Welding Journal* 43 (9), Research Supplement, pp. 385-s to 391-s.

1966. High remelt temperature brazing of columbium honeycomb structures. *Welding Journal* 45 (6), Research Supplement, pp. 258-s to 265-s.

Fugardi, J., and Zambrow, J. L. 1959. *The cladding and welding of stainless steel to molybdenum and niobium.* WADC Tech. Report 58-624

Gilliland, R. G. 1966. *Joining of tungsten.* USAEC Report ORNL-TM-1606. Oak Ridge: Oak Ridge National Laboratory.

Gilliland, R. G., and Slaughter, G. M. 1969. The development of brazing filler metals for high temperature service. *Welding Journal* 48 (10), Research Supplement, pp. 463-s to 468-s.

Hendricks, J. W.; Cole, N. C.; and Slaughter, G. M. 1972. Compatibility of brazed joints with potassium and vacuum. *Welding Journal* 51 (7), Research Supplement, pp. 329-s to 336-s.

Jacobson, M. I., and Martin, D. C. 1955. Brazing molybdenum for high temperature service. *Welding Journal* 34 (2), Research Supplement, pp. 65-s to 74-s.

Monroe, R. E. 1960. *The joining of tungsten.* Defense Metals Information Center Memo 74.

Neiman, J. T., and Sopher, R. P. 1955. *Brazing tungsten to molybdenum.* AEDC-TV-55-41, BMI.

Pattee, H. E., and Evans, R. M. 1961. *Brazing for high temperature service.* Defense Metals Information Center Report 149.

Robbins, W. P. 1967. *Brazing superalloys and refractory metals.* AEC Research and Development Report no. MLM-1322. Miamisburg: Mount Laboratory, Monsanto Research Corporation. Contract no. AT-33-1-gen-53.

Rosebury, F. 1951. *Tube laboratory manual.* Research Laboratory of Electronics, Massachusetts Institute of Technology.

Scott, M. H., and Knowlson, P. M. 1963. The welding and brazing of the refractory metals niobium, tantalum, molybdenum, and tungsten. *Journal of the Less Common Metals* 5: 205-44.

Stone, L. H.; Freedman, A. H.; and Milkus, E. B.

1966. High-remelt temperature brazing of tantalum using a diffusion sink concept. *Metals Engineering Quarterly* 6 (1): 35-42.

1967. Brazing alloys and techniques for tantalum honeycomb structures. *Welding Journal* 46 (8), Research Supplement, pp. 343-s to 350-s.

Weare, N. E., and Monroe, R. E. 1959. *Welding and brazing of molybdenum.* Defense Metals Information Center Report 108.

CHAPTER 22

PRECIOUS METAL CONTACTS

Precious metal contacts are used extensively in the manufacture of electrical equipment to provide the means for opening and closing electrical circuits. These metals are desirable because of good electrical and thermal conductivity, and for resistance to corrosion; other requirements may be high melting temperatures, suitable behavior under arcing conditions, and freedom from sticking. The contacts vary in size and are attached to a base metal support and conductor.

Although other methods of attachment are sometimes used, brazing is the most common choice. It provides a joint with good thermal and electrical conductivity and is adaptable for joining many combinations of contacts and base metals. The production brazing of electrical contacts is a highly specialized field where the quantity to be produced may greatly influence the selection of materials and techniques.

CONTACT MATERIALS

The precious metal combinations used for electrical contacts may be categorized as follows:

1. Platinum, gold, palladium, and their alloys. These metals and their alloys may also contain iridium, ruthenium, rhodium, silver, nickel, or copper. They are used where oxidation resistance, corrosion resistance, and high melting temperatures are required. Some alloys with high hardness and wear resistance are used in sliding contact applications.

2. Silver and silver alloys. The alloying elements may be palladium, copper, nickel, indium, or cadmium. These contribute to improved properties, such as increased hardness, corrosion resistance, and wear resistance; they are often used for economic reasons and for their good conductivity.

3. Sintered metals. Silver or copper is used with nickel, iron, tungsten, or molybdenum. Resistance to pitting under the arc and resistance to sticking are the advantages of this group.

4. Sintered silver and nonmetallic materials. Compositions of silver with cadmium oxide or graphite will minimize sticking of the contacts. Silver-cadmium oxide has become a major contact material because of its relatively low cost and excellent performance in medium power switch gear.

BRAZING FILLER METALS

Silver (BAg) and precious metal (BAu) filler metals are used for the brazing of contacts to the holders. The lower range of brazing temperatures is preferable unless elevated service temperatures are to be encountered. The filler metal is usually preplaced in the form of washers or discs between the contact and holder, but rings and wire forms are also used. Contacts are often supplied preclad with the brazing filler metal for convenience in assembling. A filler metal containing phosphorus will provide better wetting of contact materials containing molybdenum and those of the silver-cadmium oxide type. However, care must be taken to prevent flow of these filler metals onto the face of the contact by controlling the heating pattern and by controlling the excess usage of flux and filler metals.

Fig. 22.1—Brazed electrical contacts.

JOINT DESIGN

The joint design for contact assemblies follows the same principles used for other brazed joints. Contacts that are punched from sheet or sheared from strip should not be distorted by such operations. The brazing surfaces of contacts formed in this way or by heading, coining, or forging must be flat to provide good contact with flat surfaces on the supports. Curved surfaces or any radii to be brazed must be matched to provide the proper capillary for brazing.

PROCESSES AND EQUIPMENT

Most methods of heating may be used for brazing precious metal contacts (see chapter 1). Manual torch brazing is frequently used for the larger contact assemblies and where production requirements are limited. Induction heating is adaptable to many production needs. Resistance brazing can be done with resistance welding machines, though less current and more time are required than for resistance welding — for example, heating times may vary from 1 to 6 s for brazing. Resistance brazing may also include the use of carbon blocks and low voltage transformers.

Furnace brazing is often used where a large number of contact assemblies are brazed at one time or where more than one contact is to be brazed on each assembly. A contact assembly that is self-positioning is desirable for furnace brazing; otherwise, suitable fixtures are required to position the work (see chapter 8). Volatile elements, such as zinc, should be avoided since they will contaminate the surfaces of precious metal contacts during long brazing cycles.

PRECLEANING AND SURFACE PREPARATION

As in all brazing, best results are obtained when the parts are precleaned and this cleanliness is maintained throughout all subsequent operations. All surface oil should be removed before brazing.

Most silver-nonmetallic contacts are generally supplied with a backing of fine silver to facilitate brazing, and this is the recommended approach. In an alternate method, the cadmium oxide is sometimes removed from the surface to be brazed by chemical leaching. For silver-graphite contacts, the graphite may be removed by firing the contact at elevated temperatures (about 1500 °F [816 °C]) in an oxidizing atmosphere (air). The surfaces that will serve as the contacts should be protected when cleaning these materials to avoid removing the nonmetallic constituents. The treated surface is then suitable for brazing, although coining may be advisable to compact the area from which the nonmetallics have been removed. Silver-backed contacts are easier to clean and to braze and are preferred over the other types where economically possible.

FLUXES AND ATMOSPHERES

Fluxes are used when brazing in air or in an atmosphere furnace if zinc or cadmium is present in the base metal or filler metal. Zinc and cadmium generate highly toxic fumes (consult chapter 29). Controlled atmosphere brazing, without flux, is used extensively in production because it eliminates the need for postbraze cleaning.

ASSEMBLY, PROCEDURES, AND TECHNIQUES

The heating method should bring both members to the brazing temperature simultaneously. If the contact is relatively small compared with the holder, it may be desirable to avoid heating the contact directly by allowing it to become heated by conduction. Some pressure should be applied to the contact to seat it as the brazing filler metal melts and flows.

The dissolution of the precious metal by the brazing filler metal should be avoided. Dissolution is minimized by controlling the amount of filler metal used, avoiding excessive heating, limiting the time at brazing temperature, and striving for uniform heat distribution. It may be necessary or desirable to avoid annealing the contact holders or supports to maintain the rigidity of these members. For such applications, heating must be localized in the joint area. Torch, induction, or resistance brazing may be adjusted to obtain the required localized heating effect. More expensive but more satisfactory is the use of holders made from age hardenable copper alloys, which retain hardness after brazing.

POSTBRAZE OPERATIONS

Any of the applicable postbraze cleaning methods discussed in chapter 10 may be used to remove residual flux.

INSPECTION

In most cases, the inspection of brazed precious metal contacts is done visually as discussed in chapter 11. Recently, inspection by ultrasonic techniques has been used successfully.

TYPICAL APPLICATION

Figure 22.1 shows a representative group of brazed contacts.

CHAPTER 23

CARBIDES

Carbides are extensively used as cutting tools and where wear resistance is an important property. Frequently, the carbide part is relatively small in size and is present as an insert that must be attached to a larger, heavier backing material.

Several techniques for fastening the carbides have been developed. The three major techniques are (1) brazing, (2) cementing, and (3) mechanical fastening. This chapter is limited to the discussion of the use of brazing for attaching the carbides. Brazing was one of the first successful methods of mounting carbides to steel or other base alloys. This method is quite satisfactory for small area, short length joints and, through the use of certain design principles, can be applied satisfactorily to larger joints.

BASE METALS

The hard carbides vary widely in composition, but for brazing purposes the following general classifications can be made:

1. Tungsten carbides with a cobalt binder. The cobalt binder may vary from 3 to 25% depending upon the use but normally will be on the order of 6 to 10%.
2. Tungsten carbides mixed with moderate percentages of carbides of titanium, tantalum, or columbium and a metallic cobalt binder.
3. Titanium carbides or tantalum carbides mixed with tungsten carbides and metallic nickel or cobalt binders.
4. Chromium carbides mixed with metallic nickel or cobalt binders.
5. Various carbides that have been impregnated with hardenable tool steels in order to combine the properties of the steels with the carbides.

Hard carbides are joined to back up materials whose composition and properties are dictated by the expected application of the part. Generally, the metals are iron base, i.e., steel, martensitic and austenitic stainless steel, etc. In addition, special, high temperature applications have utilized nickel base and cobalt base materials.

FILLER METALS

Although it is possible to use any of the BAg-1 through BAg-7 filler metals, those which contain nickel (BAg-3 and BAg-4) are generally recommended because nickel improves wettability. The RBCuZn-D and BCu filler metals also have been used, particularly where a postbraze heat treatment is required. The 85Ag-15Mn and 85Cu-15Mn filler metals are used where the braze is subject to elevated temperatures in service and also for wetting the titanium base or chromium base carbides.

Tungsten base carbides are generally readily wetted by the BAg and BCu filler metals. However, titanium base carbides are somewhat more difficult to wet. Where it is necessary to mount titanium carbides by brazing, the joint must either be made in an inert or vacuum atmosphere, or the surfaces must be specially treated by the carbide manufacturer, or by nickel plating. In each case the filler metals just mentioned will then wet the titanium carbide surface.

In specifying a filler metal for a given job, first consideration should be given to the temperature range of anticipated application. This and the other considerations of corrosion and mechanical properties will dictate the alloy composition selected, the brazing temperature required, the equipment to be used, and the joining atmosphere. Generally speaking, the filler metals mentioned previously are considered for the majority of applications which require simple equipment, fluxes, and brazing temperatures below 1800 °F (982 °C). The most common filler metals used to join hard carbides to steel in this bonding temperature range have been the BAg-3 and BAg-4 metals. The joints so produced have tensile strengths of approximately 70 000 psi (483 MPa), although joints of considerably higher strengths can be made between hard materials with thin joint clearances (see chapter 6).

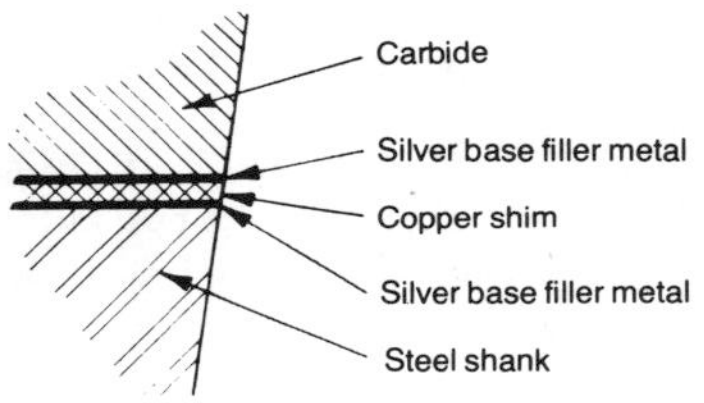

Fig. 23.1—Sandwich braze assembly method.

For high temperature applications pure copper (BCu) may be used as a filler metal, with resultant good wetting. The BCu filler metal retains practically all its strength up to a temperature of 1000 °F (538 °C). The use of the BCu as a filler metal generally is limited to high production operations since it requires a hydrogen atmosphere furnace for best results.

The carbides, in general, are not wet as readily by the filler metals as are most base metals; thus it is preferable, when possible, to preplace shims in the joint rather than to face feed the filler metal in the form of wire. For larger surfaces, shims having a core of copper or nickel with a coating of filler metal on both surfaces are frequently used. In the sandwich-type joint, the soft core of the shim has the function of relaxing stresses set up by differences in thermal expansion between the carbide and the base metal. The core of the shim is generally about 50% of the total thickness with 25% thickness of filler metal on each side (Fig. 23.1). Shims normally used may vary in thickness from 0.010 to 0.025 in. (0.25 to 0.64 mm). The thickness of the shim should be increased as the area of the joint is increased. There is no hard and fast rule to determine the proper thickness, and this can only be determined by experience.

JOINT DESIGN

Carbides have thermal expansion coefficients approximately one-third to one-half those of steel and other base metals to which they are attached, and this factor must be considered in the design of carbide joints. A carbide core can be readily brazed into a steel ring or cylinder — although the steel expands away from the carbide on heating; the joint on cooling will be under compression and is not likely to crack (see Fig. 23.1). If the combination is reversed, insufficient clearance will result at brazing temperature, and the expanding steel, which expands more than the carbide, may cause a stress buildup, resulting in the formation of stresses in the carbide, which may cause it to crack. Sufficient clearance at room temperature to permit the steel to expand at the brazing temperature without stressing the carbide will tend to overstress and crack the filler metal during cooling.

Small carbide pieces are usually brazed directly to an outer surface or in a groove machined to fit the piece. In the latter case, the carbide may be brazed on several surfaces (see Figs. 23.2 and 23.3). With carbides over 0.5 in. (12.7 mm) long, cored shims are used to absorb the contraction stresses (Fig. 23.1); with pieces longer than 3 in. (76.2 mm) it may be necessary to cut the carbide into segments (Fig. 23.4). The size and shape of the carbide to be brazed must be considered when selecting the filler metal form. The filler metal joining the ends of the segments will serve to relieve the excessive stresses that would occur in a single length. Strains may often be relieved or reduced on a poorly proportioned assembly by brazing only on one surface (Figs. 23.5 and 23.6). Using a stop-off paint or a relief gap will reduce flow and prevent wetting on this surface.

Another method of brazing a long length of carbide is to simultaneously braze an opposing carbide piece on the other side of the holder to balance the stress (Fig. 23.7). This method of overcoming curvature provides not only a counter strained assembly, but one with two wear

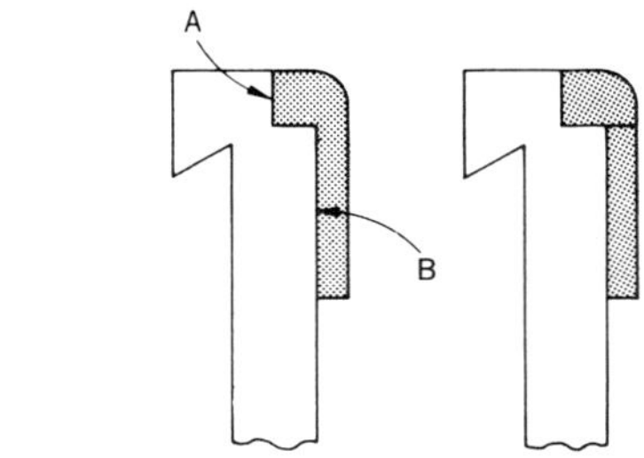

Fig. 23.2—Single piece of carbide cannot be brazed properly to both **a** and **b** surfaces. Two-piece construction permits each to seat properly.

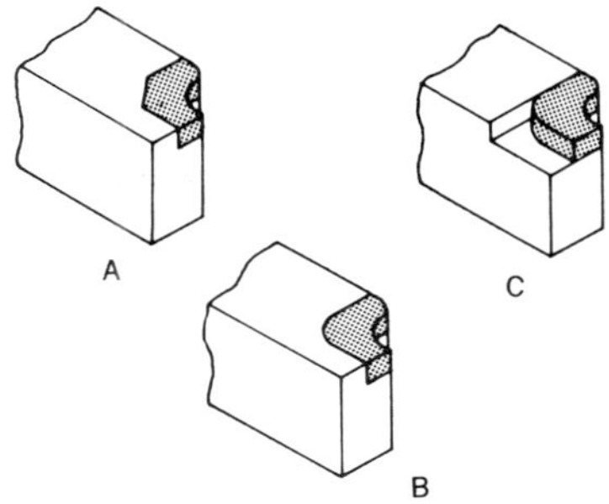

Fig. 23.3—Tip cannot fit all three shoulder surfaces at **a.** The redesign to nonconflicting shoulders at **b** permits proper braze. Removing the steel shoulder as at **c** and brazing only on bottom and end of tip improves strength of assembly.

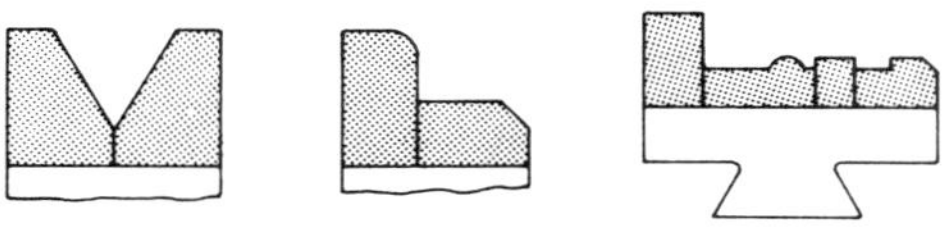

Fig. 23.4—Use of multiple inserts prevents braze strain cracks.

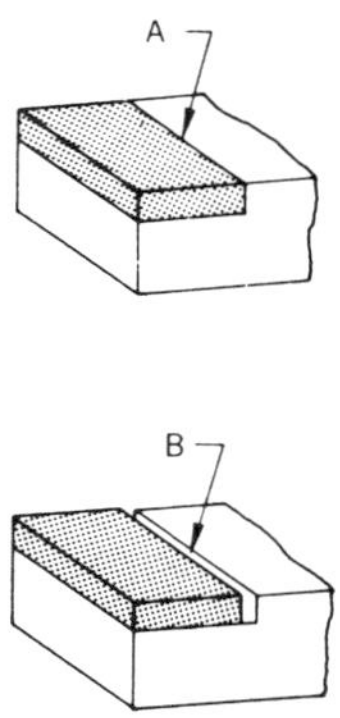

Fig. 23.5—Use of chalklike paint to prevent braze, **a,** or else relief slot, **b,** reduces brazing strain.

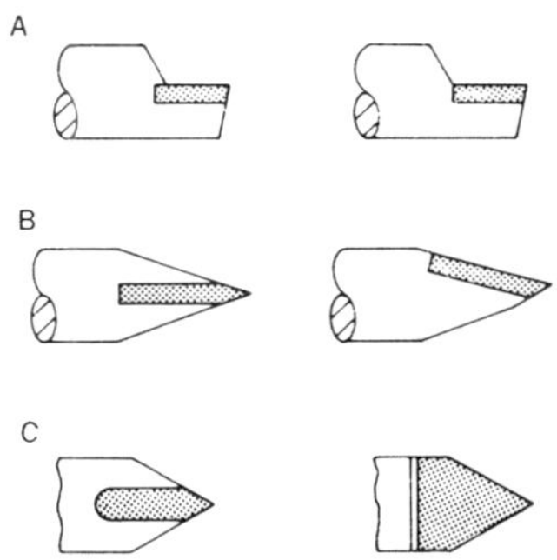

Fig. 23.6—Avoid enclosing carbide in a steel slot; redesign for single braze surface.

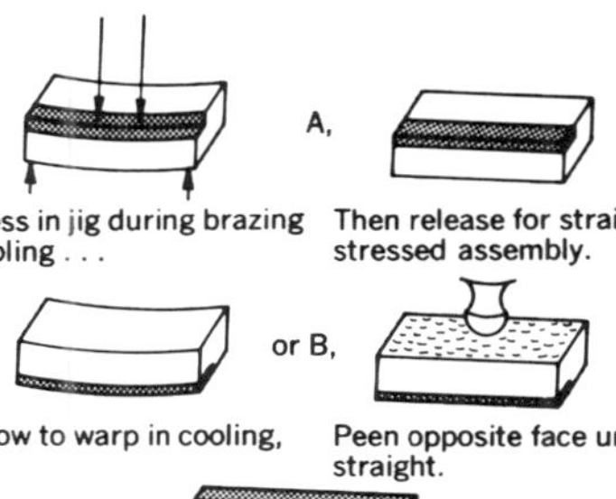

Fig. 23.7—Three methods of counterstraining to overcome curvature of assembly due to brazing strains.

faces that can be indexed to double the life of the part. Other techniques used to produce a similar effect include peening of the steel surface opposite the braze joint and low temperature (400 °F [204 °C]) soaking, which provides a degree of stress relief.

The overall effect of joint thickness or clearance has been discussed in chapter 6. As mentioned earlier, however, this dimension is particularly helpful in carbide joints for improving the tensile strength. The thicker the braze layer, the more readily it can absorb thermal strain and shock loading; this is especially important when large carbide pieces are being brazed.

When the brazed joint is stressed, it tends to elongate prior to failure. The usual reduction in area normally accompanying elongation is restrained in the joint by the closely spaced hard faces of the steel and carbide on either side of the filler metal. Because of this restraint elongation is minimized, resulting in stresses several times that of the tensile strength of the filler metal as the joint thickness is reduced. Table 23.1 illustrates these effects for the shear testing of steel (4340/ tungsten carbide joints brazed with pure copper). Note the significant fracture shear elongation in these 0.5 × 0.25 in. (12.7 × 6.4 mm) joints. These data exhibit the behavior of reduced strength as joint clearance increases; however, the increasing joint clearance suggests improved loading and thermal strain capacities at larger joint spacings.

EQUIPMENT

The torch and induction brazing processes are most commonly used with all but the BCu and copper-manganese alloy filler metals. These filler metals are generally used in controlled atmosphere furnaces or, less frequently, in induction heating apparatus arranged for use in a hydrogen atmosphere.

PRECLEANING AND SURFACE PREPARATION

Much of the difficulty in brazing carbides is the result of improper cleaning. The carbide surfaces should be grit blasted or ground on a silicon carbide or diamond wheel to remove any surface carbon enrichment, as such a surface is not readily wet by the filler metal. The usual precaution of degreasing the surface prior to brazing also should be taken. Occasionally some of the more difficult-to-wet carbides, such as titanium carbide, are prepared by plating or painting on a paste prepared of copper oxide or nickel oxide which is then fired in a reducing atmosphere to fuse the copper or nickel onto the surface. This surface is readily wet by the common filler metals. It is usual, when these surfaces are required, for the carbide supplier to prepare the carbide for brazing.

FLUXES AND ATMOSPHERES

As noted previously, most carbide brazing utilizing BCu and copper-manganese filler metals is done in a controlled atmosphere furnace, usually with a thin flux coating. Flux is always used with the other filler metals. The AWS flux type 3A is the most widely used, but for some of the difficult-to-wet carbides or for the higher melting filler metals a special flux may be required. The recommendations of the supplier of the filler metal should be followed in these cases.

Table 23.1—Properties of tungsten carbide/steel joints brazed with copper (BCu), tested in shear

Joint clearance		Ultimate engineering shear strength		Fracture engineering shear strength		Fracture shear elongation (total)	
in.	mm	psi	MPa	psi	MPa	in.	mm
0.000	0.00	26 700	184	26 700	184	0.010	0.25
0.005	0.13	28 800	199	25 750	178	0.021	0.53
0.010	0.25	33 625	232	22 800	157	0.049	1.24
0.015	0.38	31 300	216	23 500	162	0.060	1.52
0.020	0.51	33 200	229	19 400	134	0.066	1.68
0.025	0.64	29 700	205	18 450	127	0.082	2.08

BRAZING TECHNIQUES

The assembly of a carbide joint offers no unusual features, and the general procedure outlined in chapter 8 is applicable. The carbide insert is usually the lightest part of the assembly, and for this reason the heavier holder should be preheated almost to the brazing temperature before heat is applied directly to the carbide. If this is not done, the carbide may be overheated and oxidized before the holder is hot enough to be wet by the filler metal. Once the entire assembly is brought to the brazing temperature, it is advisable, if possible, to move the carbide slightly. The rubbing action aids in the wetting of the carbide surface and the pressure squeezes out excess filler metal and flux as well as removing undesirable dirt and gas holes.

POSTBRAZE OPERATIONS

After brazing, the assembly should be allowed to air-cool slowly, and quenching should be avoided until the assembly is close to room temperature. This permits relief of stresses in the joint and reduces the danger of cracking the carbide or the joint. The residual flux may be removed from the assembly by washing in hot water or commercial flux-removal compounds, followed by pickling in a solution suitable for removing the oxide coating from the base metal holder. Nitric acid solutions should be avoided because they rapidly attack the brazing filler metal.

INSPECTION

Inspection usually consists of close scrutiny of the carbide and the joint for cracks. This inspection should be made before and after any final grinding operations, as cracking of the carbide may be caused by local overheating and overstressing during the finishing. A test for proper wetting of the carbide may be made by rapping the holder sharply. A poorly brazed carbide insert will pop off. Usually sound joints result if proper wetting and flow of the filler metal within the joint is observed during the brazing operation, that is, if filletting is observed for joints that have filler metals preplaced or in face fed joints, filler metal is noted at the opposite side of the joint from feeding. Good joints normally result. These actions also suggest that the flux has been flushed out of the joint area.

CHAPTER 24

CERAMICS AND GRAPHITE

CERAMICS

The increasing use of ceramic-to-ceramic and ceramic-to-metal joints in industrial and developmental applications is due to the unique combination of properties of ceramic materials (Refs. 1-3). The large usage of ceramic-to-metal joints in vacuum tubes (see also chapter 26) in the electronics industry stems from the following properties: (a) ceramic tubes can be outgassed at higher temperatures than glass tubes, (b) ceramic tubes can withstand higher temperatures than glass tubes of similar dimensions, (c) ceramic tubes are mechanically stronger and less sensitive to thermal shock than glass tubes, (d) ceramic components can be ground to the precise tolerances required for vacuum tube construction, and (e) ceramic materials have very low electrical losses at high frequencies. A photograph of one of these tubes appears in Fig. 24.1. Figure 24.2 shows another large use of ceramic-to-metal joints in the electrical industry: a lead-through device where electrical leads are insulated from another metal.

Because of their inertness in many corrosive environments, ceramics are used as seals in fuel cells and other devices that convert chemical, nuclear, or thermionic energy to electricity. Ceramics are also used as friction materials for brakes, clutches, and other energy-absorbing devices, coatings for nuclear fuel particles, constituents in high temperature adhesives, radomes to enclose antennae, and ablative materials.

Glass-to-metal seals have been made for many years in the vacuum tube industry, and the experience thus obtained gives a general insight into the problem of fabricating ceramic-to-metal joints. More recently, the fabrication of refractory tipped tools, vacuum tubes, and various experimental devices has added to the available knowledge and techniques.

Ceramic materials are inherently difficult to wet with conventional brazing filler metals. Most of these filler metals merely ball up at the joint with little or no wetting occurring. When bonding does occur, it can be either mechanical or chemical. The strength of a mechanical bond can be attributed to interlocking particles or penetration into surface pores and voids, whereas the chemical bond derives strength from material transfer between the filler metal and the base material. Discussions of the bonding mechanism can be found in the literature (Refs. 4-6).

Another basic problem in brazing ceramics stems from the differences in thermal expansion between the base material and the brazing filler metal itself, and, in the case of ceramic-to-metal joints, between the two base materials to be joined. In addition, ceramics are poor conductors of heat, which means that it takes them longer to reach equilibrium temperature than it does metals. Both of these factors may lead to cracking in the joint. Since ceramics generally have low tensile and shear strengths, crack propagation occurs at relatively low stresses when compared to metals. In addition, the low ductilities permit very little distribution of the stresses set up by stress raisers.

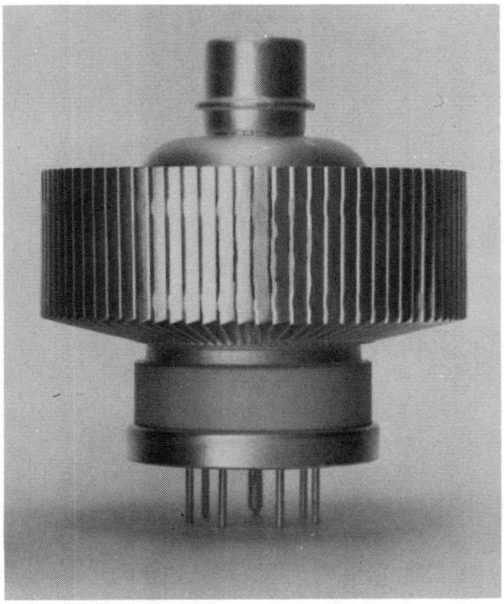

Fig. 24.1—Ceramic vacuum tube.

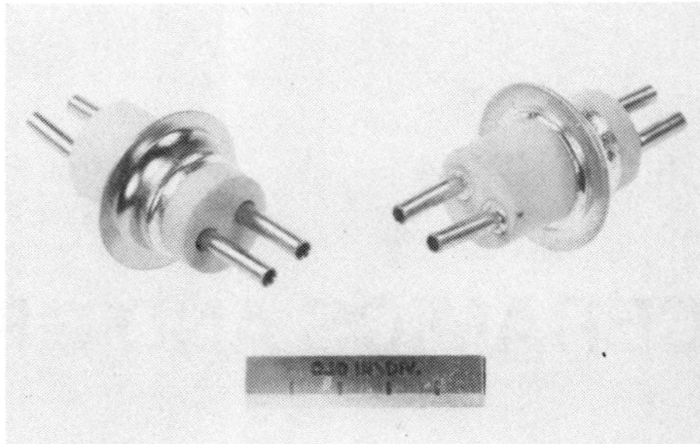

Fig. 24.2—Insulated electrical leads containing ceramic-to-metal joints.

Base Materials

Alumina, zirconia, magnesia, forsterite (Mg_2SiO_4), beryllia, and thoria are among those ceramic materials which can be joined by brazing.

Filler Metals

In those brazing processes where the ceramic is premetallized to facilitate wetting, filler metals such as copper (BCu-1), silver-copper (BAg-8), gold-copper (BAu-8), and other noble-metal filler metals can be used. The commercial filler metal used for brazing non-metallized ceramics is silver-copper-clad or nickel-clad titanium wire. Developments during the past few years have formulated additional filler metals containing active metals, such as titanium or zirconium (Refs. 7, 8). These filler metals were required primarily in the aerospace and nuclear industries to maintain satisfactory properties at high temperatures or in highly corrosive environments. Some of these are now commercial, e.g., Ti-Zr-Be, while others such as Ti-V-Zr, Zr-V-Cb, Ti-V-Be, and Ti-V-Cr are still experimental.

Joint Design

Ceramics are inherently brittle materials and are much stronger in compression than in tension. In comparision to most metals, they have small expansion coefficients; as a result, many types of joints have been developed to accommodate joint members with different expansion coefficients. A frequent solution to the problem of differing coefficients of thermal expansion in ceramic-to-metal brazes is to use a low expansion intermediate metal, such as molybdenum or titanium, adjacent to the ceramic. This intermediate metal will yield and absorb the thermal expansion stresses instead of transmitting these stresses to the ceramic material.

Several typical joint designs are shown in Fig. 24.3.

Processes and Equipment

The brazing of ceramics is usually carried out in high purity, inert-atmosphere furnaces or in vacuum furnaces. High frequency induction heating can be used for brazing ceramics to metals. In induction brazing of ceramics to themselves, it is necessary to interpose a graphite or metal tube (susceptor) between the inductor and the work. This tube then becomes the heat source.

Precleaning and Surface Preparation

The inherent porosity associated with many ceramic bodies necessitates the use of very strict cleaning procedures before brazing (Ref. 3). They are usually fired in air at a temperature from 1472 to 1832 °F (800 to 1000 °C) to permit outgassing. Suitable alkaline cleaning solutions may be used; these are followed by immersion in dilute nitric acid and subsequent rinsing in a neutralizing solution. Trichlorethylene should be avoided as a cleaning agent as it may leave a conductive coating. Care should also be taken to remove particles embedded in ceramics during grinding operations.

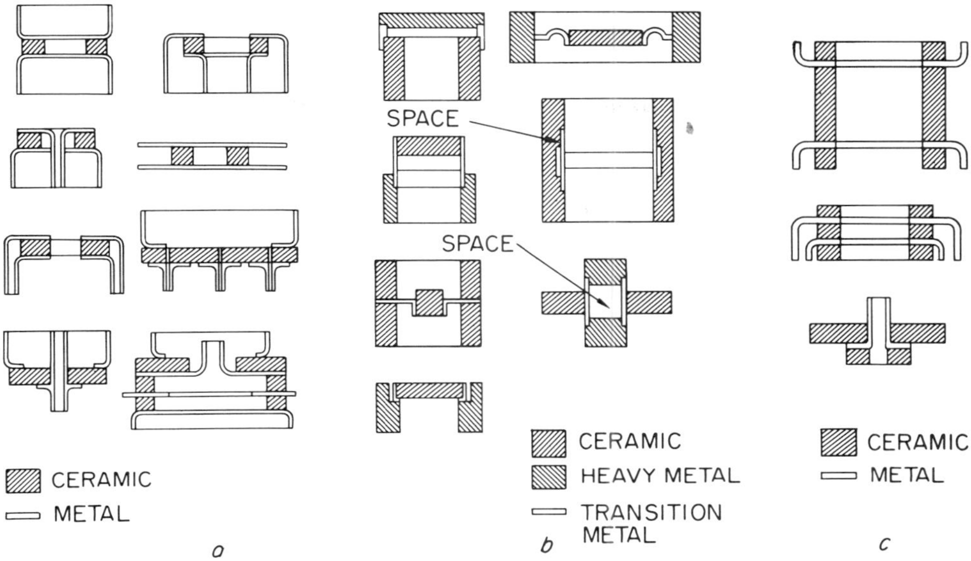

Fig. 24.3—Ceramic-to-metal joint configurations (Ref. 9). **a,** Butt and lap seal joint designs; **b,** Joint designs for transition to thick all-metal members; **c,** Backup of ductile metal seal with blank ceramic.

Fluxes and Atmospheres

Where premetallized ceramics are to be brazed, the operation can be carried out in high purity inert-gas, hydrogen, or vacuum atmospheres. If the ceramic is to be brazed directly, i.e., without being premetallized, a vacuum atmosphere is preferred.

Assembly, Procedures, and Techniques of Joining

Several techniques for joining ceramics have been reported in the literature (Refs. 1, 3). Some of these procedures do not incorporate brazing and, consequently, are not discussed in this section.

The sintered metal powder technique is a widely used brazing method. The process requires several steps to produce a joint: (1) the firing of metal powder held in suspension on the ceramic, (2) a plating or deposition of a thin copper or nickel film, and (3) brazing by conventional methods to make the ceramic-to-metal joint.

The affinity of titanium and zirconium for ceramics is the principle behind the active alloy process. Highly active titanium or zirconium can be made available at the ceramic-metal interface by hydride decomposition of a preplaced powder slurry on the ceramic surface. The reaction of the titanium or zirconium with the ceramic, or with additional metals placed at the interface, forms an intimate bond. In some cases, the titanium or zirconium is merely painted on the ceramic surface and placed in contact with a suitable filler metal and the base metal to which the joint is to be made. This process has an advantage over some others in that only one firing operation is required.

The affinity of reactive and refractory metals for ceramics is also the basis for the direct brazing approach. In this case, active metals are incorporated as one or more of the constituents in the filler metal, which is placed at the joint as it is in conventional brazing. An interesting means of promoting flow of the brazing filler metal involves vapor-deposited coatings of titanium, zirconium, or other metals, on the ceramic sub-

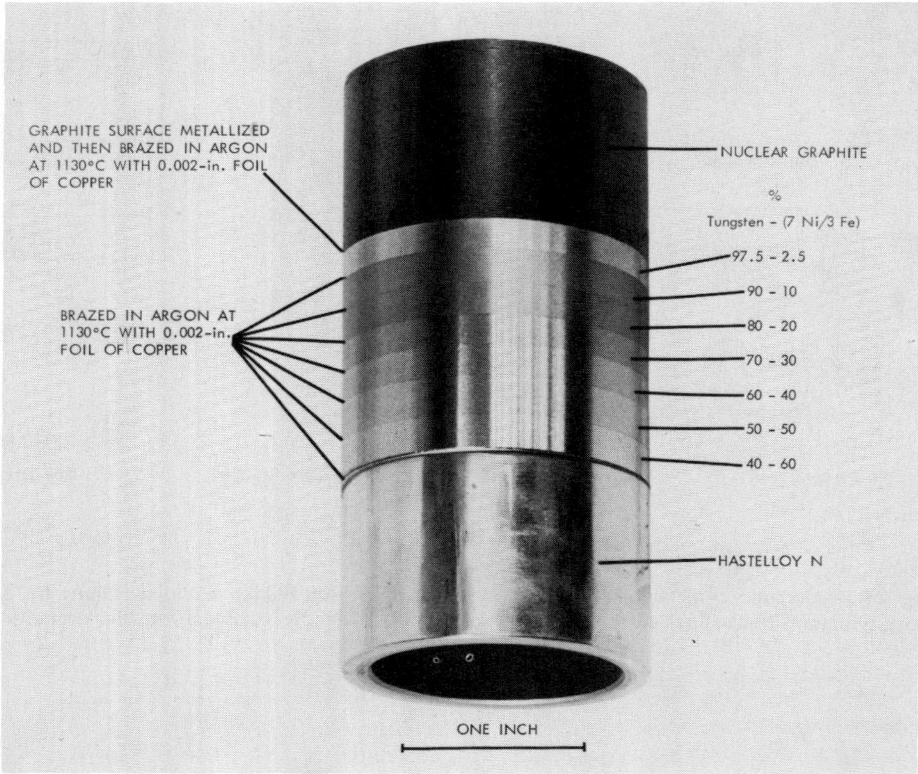

Fig. 24.4—Seven-segment tubular transition joint.

a

b

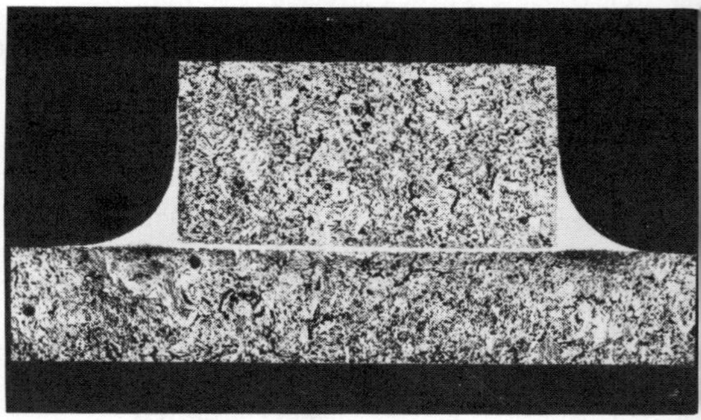

Fig. 24.5—Graphite-to-graphite joint brazed with Ti-Zr-Be filler metal. Part **b** is a photomacrograph of a cross section of **a.**

strate. Electron beam heating can provide a unique means for producing the metal vapor.

Typical Applications

The largest application for ceramic-to-metal seals lies in the electron tube industry. Many different types of ceramic tubes have been developed, and vacuum-tight seals are needed in all of them. A variety of other applications make important use of ceramic-to-metal or ceramic-to-ceramic seals. Electrical lead-through wires can be insulated from the main chamber body with ceramic insulators. The nuclear industry also has several applications, including experimental irradiation capsules and thermocouple attachments. Vacuum switch gear and high temperature connectors are other applications.

GRAPHITE

The joining problems associated with brazing graphite are very similar to those encountered with the ceramic materials previously mentioned (Refs. 2, 10). Compared to metals, wetting is difficult to achieve and the differences in coefficients of thermal expansion between graphite and conventional structural materials are pronounced.

Base Materials

Graphite is available in a number of grades, differing in grain size, density, purity, and other properties. These different grades usually possess different brazing characteristics.

Significant progress in producing graphite with consistent properties has been made in response to the requirements of the aerospace and nuclear industries. Among the most important advances is the development of pyrolytic graphite, which consists of highly oriented planes of graphite molecules closely stacked in a laminated structure. The bonding forces between atoms in a given plane are much stronger than the bonding forces between planes; consequently, the properties of pyrolytic graphite are highly directional. These differences in properties must be considered in designing a structure that incorporates this product; similarly, the dependence of properties on the crystal orientation of graphite has a profound effect on the design of a graphite-to-metal joint.

A transition member provides a useful means for attaching a high expansion metal to a low expansion graphite. Figure 24.4 shows novel joint design (Ref. 11).

Filler Metals

Metals that have strong tendencies to form carbides have been found to wet graphite when they are molten. A commercial filler metal frequently used for brazing graphite is silver-copper-clad titanium wire. Graphite is also readily wet by molybdenum disilicide, titanium, and zirconium. In recent years, the requirements of the aerospace and nuclear industries have resulted in the development of several additional filler metals. In general these metals have incorporated substantial quantities of carbide-forming elements. They include 48 Ti-48 Zr-4Be, 49 Ti-49 Cu-2Be, 35 Au-35 Ni-30 Mo, nickel-clad titanium, Ti-Cr, 60 Au-10 Ni-30 Ta, 54 Ti-21 V-25Cr, 43Ti-42Zr-15 Ge, and 47Ti-48Zr-5Nb. A photograph showing joints brazed with the Ti-Zr-Be filler metal is shown in Fig. 24.5.

Typical Applications

To date, graphite has been brazed for only a very limited number of applications. The nuclear industry has used brazing as a means of sealing graphite-clad fuel elements and for fabricating a variety of experimental assemblies. The rocket and missile industry has also used brazing to advantage in the fabrication of certain components.

REFERENCES

1. Pattee, H. E. 1972. *Joining ceramics to metals and other materials.* Welding Research Council Bulletin no. 178.
2. Pattee, H. E.; Evans, R. M.; and Monroe, R. E. 1968. *Joining ceramics and graphite to other materials.* NASA SP-5052.
3. Kohl, W. H. 1960. *Materials and techniques for electron tubes,* pp. 470-518. New York: Reinhold Publishing Co.
4. School of Ceramics, Rutgers University. 1954. *Ceramic Age* (February-September): Symposium on ceramics and ceramic-metal seals, 21 April 1953.
5. Van Houten, G. R. 1959. Survey of ceramic-to-metal bonding. *Bulletin of American Ceramic Society* 38: 301-7.
6. Lewin, G., and Mark, R. 1959. Theory of dissimilar tubular seals of glass, ceramics and

metals for critical applications. In *1958 Vacuum Symposium Transactions*. American Vacuum Society.

7. Fox, C. W., and Slaughter, G. M. 1964. Brazing of ceramics. *Welding Journal* 43 (7): 591-97.
8. Hanks, G. S.; Kirby, R. S.; and LaMotte, J.D. *Ceramic-to-metal brazes for use at 1000 °C*. LA-DC-7159. Los Alamos Scientific Laboratory.
9. Johnson, C. W. 1976. *Manual of metal-to-ceramic sealing techniques*. Publication no. NA-27-0001. Great Neck: Sperry Gyroscope Co.
10. Donnelly, R. G., and Slaughter, G.M. 1962. Brazing of graphite. *Welding Journal* 41 (5): 461-69.
11. Hammond, J. P., and Slaughter, G. M. 1971. Bonding graphite to metal with transition pieces. *Welding Journal,* January 1971.

CHAPTER 25

HONEYCOMB STRUCTURE BRAZING

Notable among high efficiency advanced aerospace structures are brazed honeycomb sandwich assemblies. Since 1952, this remarkably lightweight, temperature resistant type of construction has been successfully applied to critical-use aircraft and space vehicles.

DEFINITION

The all-metal honeycomb sandwich is an anisotropic, integrated sheet metal assembly. Specifically, this type of structure is comprised of two thin facings of high strength material strongly bonded to a high strength foil-gage core. The core is sandwiched between the facings and utilizes a suitable brazing filler metal to join all faying surfaces.

Facings are the prime load-bearing members. With proper design, complete surface stabilization permits ultimate compression strengths to be realized even in thin gage materials. The core performs the vital function of providing essentially continuous support to the facings to prevent buckling while at the same time transmitting shear forces. Furthermore, excellent stiffness, vibration dampening, thermal, acoustical, and insulation properties are inherent, and the moment of inertia can be very large for a given structural weight. The favorable properties of brazed sandwich are maintained even at elevated temperatures.

Although the basic honeycomb sandwich construction consists of foil gage metal core and thin sheet facings, it is necessary to incorporate edge close-outs, ribs, densified core, massive inserts, or doublers. An optimum design can therefore become complex with attendant difficulties in manufacturing, assembly, and brazing.

MATERIALS

Numerous metals and alloys have been used in brazed honeycomb panels. Corrosion resistant alloys predominate; however, as titanium manufacturing technology advances, applications for this efficient material will increase substantially. For fuselage and wing panels, extensive use has been made of several precipitation hardening stainless steels, such as 17-7PH, PH 15-7Mo, and PH 14-8Mo. Nickel base and cobalt base alloys suitable for high temperature exposure include Inconel 600, 625, and 718; Hastelloy X, Haynes Alloy 25, and RE 41. Other structural metals that have found usage include aluminum alloys, beryllium, steels, dispersion strengthened superalloys, and the refractory metals.

The choice of filler metal depends upon the sandwich base metal and the particular operating environment. Some brazing filler metals are highly oxidation resistant and exhibit strength levels approaching base metal strength up to 2000 °F (1093 °C). In this category are certain nickel base and cobalt base, chromium-silicon and boron-containing fillers; AWS BNi-1, BNi-2, BNi-4, and BNi-5 are representative. Intermediate temperatures are accommodated by nickel-manganese-copper, nickel-phosphorus (BNi-6, BNi-7), nickel-palladium-silver, silver-copper-palladium, and gold filler metals such as BAu-1, BAu-2, BAu-3, and BAu-4. At service temperatures below about 1000 °F (538 °C) silver base filler metals of the AWS BAg-19 and BAg-8a types are widely used. Aluminum 3003 brazing of titanium base honeycomb sandwich allows operation up to about 600 °F (316 °C). This filler metal has been successfully applied to large and complex titanium brazements.

Selection of a proper filler metal is of prime importance. Critical factors to be considered are base metal characteristics, strength level required, operational environment, base metal interaction with braze metal versus exposure duration, assembly size and subassembly approach (if used), filler metal form (foil, powder, plating) and quantity, brazing facility type, brazing cycle, heat treat cycle, and quality control, braze inspection methods applicable.

MANUFACTURING

Stringent dimensional and processing controls must be exercised throughout the honeycomb sandwich panel manufacturing operation. Individual components are generally machined or formed to within several thousandths of an inch of net dimensions. Electrical discharge and electrochemical machining are favored for surfacing stainless steel foil cores. Mating edge members, inserts, and internal doublers must be precisely fitted.

Preliminary assembly of all components is generally made in the brazing support to check out the tolerance accumulation. Adjustments are made to obtain proper fits. When the preliminary assembly is satisfactory, the components are dismantled, cleaned, and reassembled. Final assembly generally includes tack welding the various components into position. Shear ties from core to edge members and inserts may be accomplished by minute resistance tack welds.

Cover sheets or facings are frequently held in place by small foil tabs. During preparation for titanium panel brazing, the facings are sometimes fusion welded or seam welded directly to the close-out or edge members. In this way, it is possible to produce a hermetically tight brazing retort in situ while ensuring integrity of the panel. See Fig. 25.1. This desirable approach also allows evacuation and purging of the sandwich and ensures brazing conditions with accurate control of internal gas pressure.

BRAZING

Several different brazing techniques are in use today. All provide a purified inert, reducing, or vacuum atmosphere. The method of heating and fixturing varies:

1. Sealed retort, conventional furnace (gas or electric)

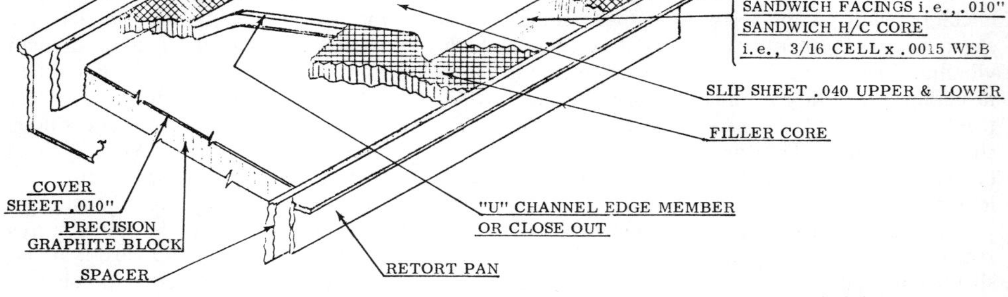

Fig. 25.1— Retort assembly for brazing of precision honeycomb sandwich.

Fig. 25.2—B-70 all-honeycomb wing surfaces.

2. Blanket, electrical strip heaters
3. Radiant lamp, infrared quartz bulb
4. Vacuum, cold wall, resistance heaters

The muffle or panel itself is instrumented with thermocouples to accurately measure and afford close temperature control during heating, brazing, cooling, and heat treatment if required. Frequently the configuration of the part or tooling necessitates more than one zone of control to assure uniform temperature distribution. The brazing temperature selected and the time at temperature are critical parameters. Experience, possibly subscale runs or a full scale dry run will serve to define these conditions for a given honeycomb sandwich brazement. The effects of base metal(brazing filler metal diffusion interaction and of gravity upon the liquid filler metal can be deleterious. Close temperature-time control and furnace uniformity are paramount considerations.

NONDESTRUCTIVE INSPECTION

Inspection and evaluation of brazed panels are usually accomplished by radiography. Each panel is examined for continuity, size of fillets, and lap joint integrity. Experience is required to interpret these films and determine whether the flow has been uniform throughout or whether the flow is excessive is some areas and deficient in others. Each panel supplied is accompanied by its own X-ray picture as a permanent record.

Ultrasonic inspection is used in conjunction with radiographic techniques; C-scan type equipment is favored. New advances are being made in the use of holography, which promises to be a highly sensitive and practical tool. Automated inspection also appears feasible with this

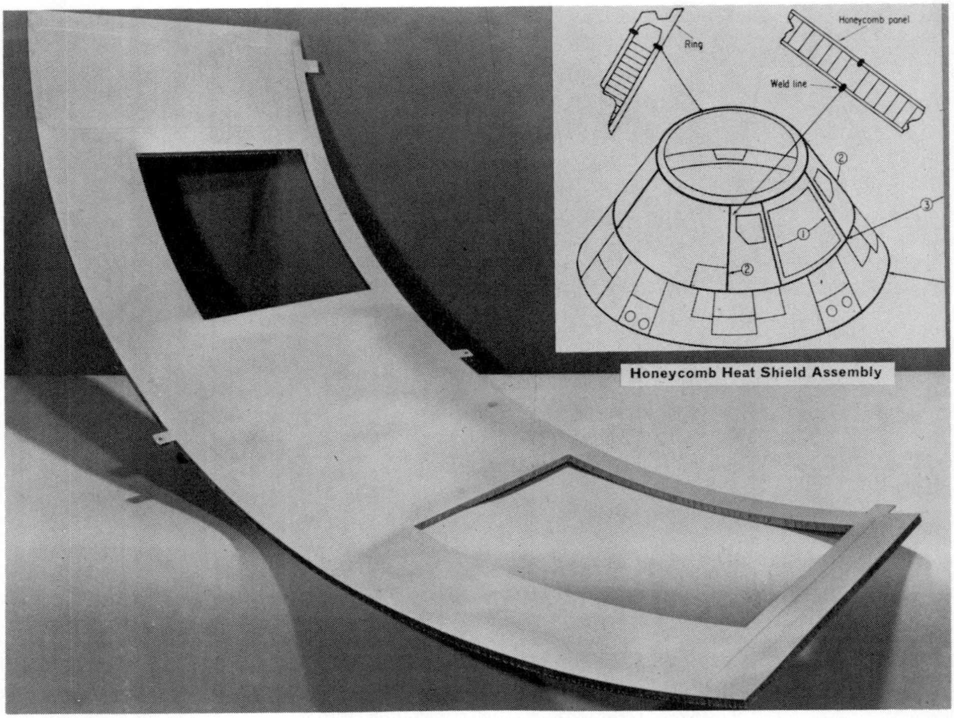

Fig. 25.3—Apollo CM sketch.

method, while acoustic emission research indicates yet another emerging technique that may be applicable.

APPLICATIONS

In the mid 1950s, the supersonic B-58 bomber was the first production requirement for brazed all-metal honeycomb sandwich. Approximately 1200 ft^2 (111.5 m^2) of engine nacelle and wing paneling was fabricated from thin gage 17-7PH stainless steel. Over 100 ship sets resulted from this major development and production program.

The U.S. Air Force's XB-70 Mach 3 airplane utilized honeycomb sandwich extensively. About 20 000 ft^2 (1858 m^2) of brazed stainless steel panels were used for fuselage, wing, and tail surfaces. Figure 25.2 shows brazed sandwich sections being fitted to the 250-ton (227 000 kg) machine during construction. Materials were predominantly PH 15-7Mo, an advanced, higher strength stainless alloy as compared to 17-7PH.

Twenty Command Modules (CM) incorporating brazed stainless steel honeycomb for the outer cone shell and aft heat shield were produced for the Apollo program. One manufacturer fabricated building-block panels; another welded these panels into the CM vehicle. Figure 25.3 shows various brazed sandwich elements that combine to form the high strength, lightweight, truncated cone structure approximately 13 ft (4 m) in diameter by 10 ft (3 m) high. The ingenious weld joint designs made integration of the sandwich subassemblies practical and reliable. Noteworthy is the use of two opposing gas tungsten arc welding heads operating simultaneously on front and back honeycomb sandwich facings.

Cores, facings, and attachments for the CM panels are of PH 14-8Mo stainless steel. After brazing, subzero cooling, then aging, minimum specification mechanical properties are

Ultimate tensile strength	190 000 psi (1310 MPa)
0.2% Yield strength	175 000 psi (1210 MPa)
Elongation	3%

Fig. 25.4—Command module mating prior to girth welding.

Chemical milling techniques were extensively used to thin panel facings to 0.008 in. (0.2 mm) while maintaining a "picture frame" of full sheet thickness to permit high strengths as butt welded.

The CM cone structure typically comprises brazed honeycomb panels 0.5 in. (12.7 mm) in thickness; the aft heat shield's four quadrant-shaped panels utilize 2 in. (50.8 mm) deep, relatively high density core. The aft shield was welded to the cone (Fig. 25.4). Braze process specifications, tooling, and methods used were essentially those developed during the billion dollar B-70 program. Brazing filler metal sheet was a commercial product.

For core-to-facing joints, the silver base 7.5Cu-5.5In-2.2Pd-0.2Li filler metal is used in a nickel matrix composite (80/20); solidus is 1400 °F (760 °C), liquidus 1615 °F (879 °C). For flat faying surfaces, such as internal and external doublers and inserts, the silver base filler metal 62.5Ag-32.5Cu-5Ni is specified. This filler metal is very fluid, whereas the nickel matrix or wick-containing composite is sluggish.

Panel brazing is prescribed as 1685 ± 25 °F (918 ± 14 °C) for 15-20 min at temperature in argon. Cooling of PH 14-8 brazements in the heated precision fused silica tooling dies proceeds to 1525 °F (829 °C), held for 2 hours, then cooled to −140 °F (−96 °C) for 4 hours to effect transformation. Aging treatment is next accomplished at 1072 ± 15 °F (578 ± 8 °C) for 1 hour.

Brazed sandwich construction design of the U.S. supersonic transport was largely based on manufacturing technology established during the aforementioned programs. As a result of NASA-sponsored development, brazed titanium alloy honeycomb sandwich was also incorporated as a high efficiency structure (see chapter 20).

Today's advanced commercial jetliners are utilizing sandwich brazements. A noteworthy application involves hot structures for sound attenuation treatment. Fan ducts, exhaust ducts, splitters, and center body structure represent important areas where honeycomb sandwich with perforated facing sheet is being used successfully to quiet noise generated by turbojet engine. Optimum results require proper facing sheet and honeycomb core cell design.

Figure 25.5 illustrates one example of acoustic sandwich structural component. This 9-ft (2.7 m) long jet engine exhaust plug is used on each power plant for a jumbo commercial jet. Hot gases up to 1400 °F (760 °C) flow over the contoured Inconel 625 plug which supports the thrust reverser linkage and controls the nozzle flow area. Weight savings of 100 lb (45.4 kg) per unit were achieved by designing a brazed honeycomb sandwich construction. Over 1000 exhaust nozzle plugs manufactured attest to the efficiency of these extraordinary brazements.

HONEYCOMB SEALS

Open face honeycomb sandwich can be brazed for high temperature corrosion resistant applications, and the development of these efficient air-seal structures has made major contributions to the evolution of large, lightweight, turbojet and turbofan engines. Over $200 million in labyrinth and blade tip types of brazed honeycomb seals have been manufactured (seal types, Fig. 25.6). The unique attribute of such honeycomb shrouds is that the core structure provides an effective air seal and will tolerate actual interference or rub between rotating and stationary components without damaging the turbine/compressor rotors. Consequently, engines can be assembled tighter and cases made lighter because distortion and rubbing can be accommodated.

A small segmented rub-ring seal is shown in Fig. 25.7. Both 0.125 in. (3.2 mm) cell core and

Fig. 25.5—Sound attenuation plug for jumbo commercial jet.

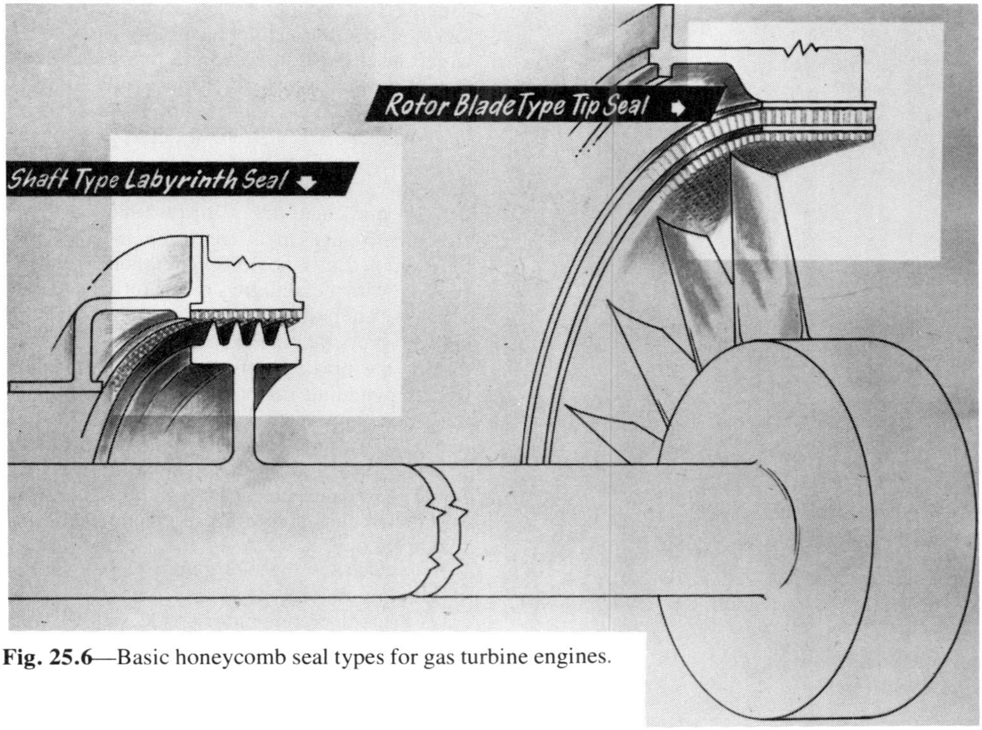

Fig. 25.6—Basic honeycomb seal types for gas turbine engines.

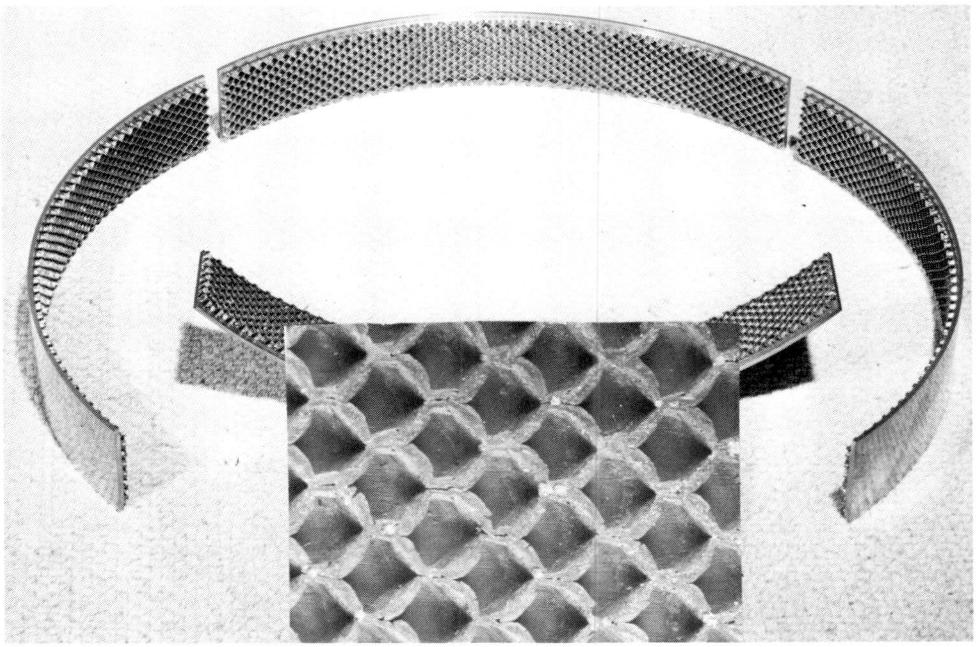

Fig. 25.7—Inconel 600 blade tip seal.

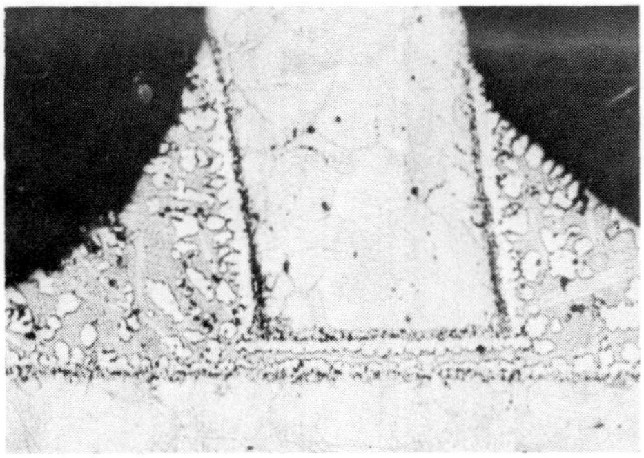

a

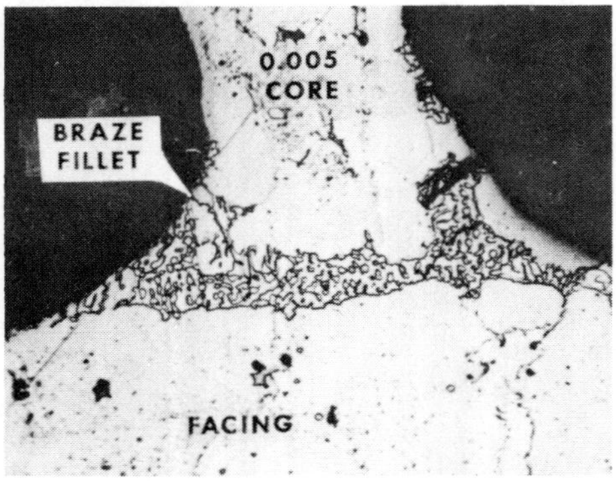

b

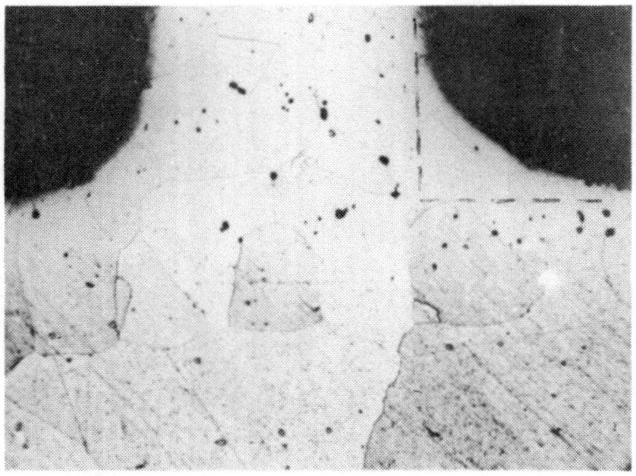

c

Fig. 25.8—Braze diffusion sequence. Core is 0.005 in. (0.013 mm) Inconel; facing is 0.1 in. (2.5 mm) Inconel; filler metal is BNi-4, high temperature, corrosion resistant. Approx. 275×. **a,** Brazed for 1 min at 1950 °F (1066 °C); **b,** Brazed for 10 min at 1950 °F (1066 °C); **c,** Soaked for 5 h at 1950 °F (1066 °C). Reduced by 30% on reproduction.

backing are Inconel 600 brazed joined with nickel-silicon-boron filler metal (BNi-4). The enlarged inset illustrates a severe interference rub caused by the turbine rotor blades operating at 1750 °F (954 °C) gas temperature. Efficacy of the deformable/abradable core structure is evident. Welded core cells used are: 0.031 in. (0.8 mm), 0.046 in. (1.2 mm), 0.062 in. (1.6 mm), 0.093 in. (2.4 mm), and 0.125 in. (3.2 mm). Microstructural changes in the core-to-facing joint may be examined in Fig. 25.8. The phenomena associated with metallurgical diffusion between filler metal and base metal account for outstanding strength properties at temperatures that may even exceed the brazing temperature. Homogenization is assuming important manufacturing significance in production of honeycomb sandwich hardware by such techniques known as liquid-interface diffusion, diffusion alloying, transient liquid phase, and eutectic brazing. The American Welding Society defines these phenomena as diffusion brazing.

Aircraft gas turbines rely heavily on a variety of brazing processes. Honeycomb-type seals cover the temperature range up to 2100 °F (1150 °C); at these temperatures, abradable, porous fillers are being used in conjunction with open face honeycomb sandwich-supporting structure. Developmental seals are being tested with gas temperatures up to 3000 °F (1650 °C).

Typical applications for honeycomb sandwich brazing are:

Aircraft skins	Empennage
Wing ribs	Flight control surfaces
Leading edges	Tail cones
Trailing edges	Access doors
Wing flaps	Cowling
Acoustical structure	Fan vanes
Fan ducts	Thrust reversers
Exhaust ducts	Dive brakes
Inlet ramp	Electronic cold plates
Engine nacelles	Microwave reflectors

SELECTED READINGS

Elred, S., and Lovell, D. T. 1972. *Development of aluminum brazed titanium honeycomb sandwich*. Contract DOT-FA-SS-71-12.

Filippi, F. J. 1958. Qualitative analysis of brazed sandwich. Paper read at meeting of Society for Nondestructive Testing, 27 October 1958, in Cleveland.

Kramer, Bruce E. 1972. Brazed titanium honeycomb core sandwich. Paper read at Third Air Force Metalworking Conference, WESTEC, 14 March 1972.

Long, John V., and Cremer, G. D. 1954. High temperature all-metal sandwich structures. Paper read at SAE Aeronautic Meeting, 5 October 1954, in Los Angeles.

Long, John V., and Cremer, G. D. 1967. Brazing bonanza. Paper read at ASM National Metal Congress, 17 October 1967.

Schwartz, M. 1967. *Modern metal joining techniques*. John Wiley and Sons.

Schwartz, M. 1973. *Brazed honeycomb structures*. Welding Research Council Interpretative Report no. 182.

Setterholm, V. C., and Kuenza, Edward. 1956. *Performance of stainless steel construction at high temperatures*, Part 1. WADC Tech. Report no. 55-417. ASTIA Document no. AD97288.

CHAPTER 26

ELECTRON TUBE AND VACUUM EQUIPMENT BRAZING

Brazing of structures for vacuum tubes and other high vacuum devices has required the development of highly refined procedures, furnace equipment, and high purity, low vapor pressure brazing filler metals. Vacuum tubes, of necessity, are operated at very low pressures (10^{-6} to 10^{-8} torr [10^{-4} to 10^{-6} Pa]), and they must maintain this very low pressure for the thousands of hours of their useful life. High vacuum devices and specialized equipment can tolerate no condition which inhibits the ability to secure and maintain extremely low pressures. As a further complication, vacuum tubes and other vacuum devices are heated at temperatures up to 932 °F (500 °C) for extended periods during their gas evacuation cycle ("bake out") to drive out the gases entrapped in their metal structures. Vacuum tubes operate at elevated temperatures and are constructed of widely dissimilar materials; hence, the brazed joints must also withstand differential thermal expansion stresses. The foregoing discussion indicates that the brazing of components for electron tubes, vacuum devices, and equipment should only be undertaken after a thorough analysis of the problem. Consultation with persons experienced in this special brazing field is advised.

BASE METALS AND BRAZING FILLER METALS

Copper, nickel, stainless steels, copper-nickel alloys, iron-nickel-cobalt alloys, molybdenum, tungsten, and alumina ceramics are commonly used materials for constructing vacuum tubes, special vacuum devices, and vacuum equipment. While the need to join such a wide range of materials has stimulated the development of many new brazing filler metal compositions, the number of possible filler metals available for brazing of vacuum systems is reduced considerably by the necessity of using only low vapor pressure elements in their composition. This limitation, for all practical purposes, restricts the selection of filler metals to those containing elements of platinum, copper, silver, gold, palladium, nickel, indium, tin, and gallium. All other elements have either too high a melting point or too high a vapor pressure. Consequently, brazing filler metals must be completely free of zinc, cadmium, lithium, and phosphorus.

The necessity for completely vacuum-tight joints, rather than just the mechanically strong and reasonably air tight or liquid-tight joints normally expected of a braze, complicates brazing procedure on vacuum systems. Materials, such as fluxes, which are a great aid to flowing a molten metal onto a solid metal, cannot normally be used since their subsequent removal is questionable and their presence, in even trace amounts, in a vacuum system is very objectionable. Consequently, vacuum system components are generally brazed in hydrogen atmospheres, or where this atmosphere is objectionable, in helium, argon, or vacuum.

Voids in a brazed joint are to be avoided, not only because they impair joint strength, but more importantly, because they contain gases that if released during the operation of a vacuum device could seriously impair its performance.

Metals to be joined by brazing must be clean

Table 26.1—Composition of brazing filler metals specially produced for the brazing of vacuum devices and equipment

AWS classification	Filler metal	Liquidus, C	Solidus, C	Nominal composition, % Pd	Au	Ag	Cu	Ni	Sn	In
	Pt	1768	1768							
	Pd	1552	1552	100						
	Pd-35Co	1235	1230	65						
	Pd-Au	1240	1200	8	92					
	Au-Pd-Ni	1121	1102	25	50			25		
BAu-3	Au-Cu-Ni	1030	1000		35		62	3		
	Au-Cu	1010	990		35		65			
BAu-1	Au-Cu	1005	985		37.5		62.5			
	Au-Cu	970	955		50		50			
BAu-4	Au-Ni	950	950		82			18		
	Ag-Pd-Cu	950	900	25		54	21			
	Au-Cu-Ni	925	910		81.5		16.5	2		
	Ag-Pd-Cu	900	850	15		65	20			
	Au-Ag-Cu	895	885		75	5	20			
	Ag-Pd-Cu	852	824	10		58	32			
	Au-Ag-Cu	845	835		60	20	20			
	Ag-Cu clad 4.5% Ti	850	830			68.8	26.7			
	Ag-Pd-Cu	810	807	5		68	27			
	Ag-Cu-Ni	795	780			71.15	28.1	.75		
BAg-8	Ag-Cu	780	780			72	28			
	Ag-Cu-Sn	760	743			68	27		5	
BAg-18	Ag-Cu-Sn	718	602			60	30		10	
	Ag-Cu-In	705	630			61.5	24			14.5

NOTE: Impurity limits are Zn $<$ 0.001%; Cd $<$ 0.001%; Pb $<$ 0.002%; P $<$ 0.002%; C $<$ 0.005%. All other impurities not specified having a vapor pressure higher than 10^{-7} torr at 500 °C, are limited to 0.002% each. Impurities having a vapor pressure lower than 10^{-7} torr at 500 °C are limited to a total of 0.05%.

and essentially free of oxides as must, of course, the brazing filler metal. Brazing filler metals for vacuum devices must be produced with far more care than conventional filler metals, not only to prevent inclusion of high vapor pressure elements, but also to keep mechanically included impurities such as dirt and carbon to a minimum. These mechanically incorporated impurities are generally picked up during ingot casting and in the subsequent wire drawing or sheet rolling procedures as a result of the entrapment of organic lubricants used in these processes. The entrapped organic compound is converted to carbon during the numerous annealing treatments required to reduce the original ingot to its final wire or sheet form. Dirt, if present, is pressed into the metal during rolling or wire drawing reductions. On melting a filler metal so contaminated, the carbon and dirt generally rise to the surface of the molten filler metals where they may not be too objectionable; however, foreign materials, because of the possibility of their adversely affecting electron emission and evolving gases when heated, are always to be avoided in vacuum systems. Of more immediate danger is the possibility that the dirt and carbon may remain within the braze joint and form a gas pocket and possibly a "leaky" joint.

Brazing filler metals for vacuum devices should not only be free of the aforementioned impurities, but also have a low oxygen and metal oxide content. Oxides are particularly objectionable when brazing in a hydrogen atmosphere since they will react with the hydrogen to form water vapor. When such a reaction takes place while the filler metal is molten, the rapidly expanding water vapor splatters the molten alloy into areas adjacent to the braze joint where their presence is often very undesirable. The choice of a particular brazing filler metal depends on the top operating temperatures of parts to be joined, the vapor pressure of the filler metal at these temperatures, the number and requirements of subsequent brazes to be made on the assembly,

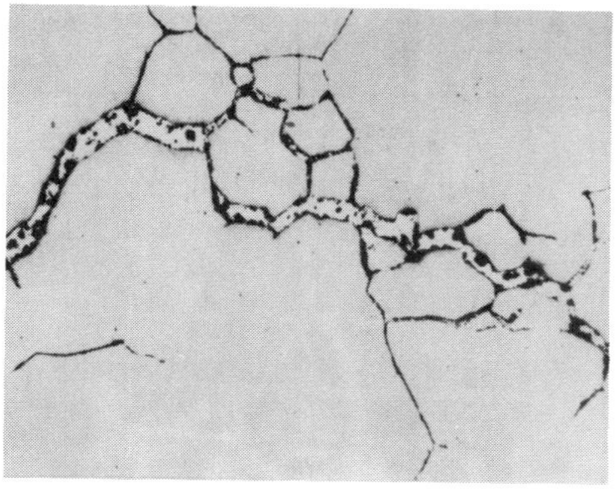

Fig. 26.1—Penetration of silver into grain boundaries of Kovar brazed with copper silver eutectic. Such penetration is eliminated by adding palladium to the braze filler composition, or by nickelplating the Kovar being joined. Both techniques provide a diffusion barrier to the copper silver filler. 100 × (reduced by 35% on reproduction).

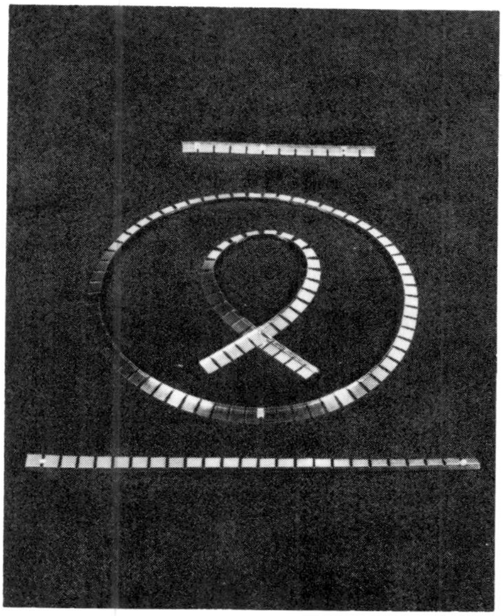

Fig. 26.2—Method of braze filler application used by vacuum equipment manufacturers to make large diameter preforms without the need to stamp them out of sheet, thereby eliminating tooling and large dropouts of precious metal fillers.

and the properties of the materials to be joined. Brazing filler metals are available ranging from the high melting point platinum and platinum alloys for brazing of tungsten and molybdenum parts through pure copper down to a silver-copper-indium alloy having a liquidus of 1301 °F (705 °C). The vapor pressures of these filler metals usually vary inversely as their melting points.

Good copper to copper, copper to nickel, and nickel to nickel braze joints can be made with the silver-copper eutectic as well as with various gold-copper and gold-copper-nickel alloys. Economic considerations would dictate the use of the silver-copper eutectic filler metal for these applications; however, since the gold-copper and gold-copper-nickel filler metal have considerably lower vapor pressures, vacuum system performance may dictate the use of the latter, more costly filler metals. Since the liquidus and solidus of the gold filler metals is higher than the silver copper, they are also often used for applications in which a component is best formed by two or more successive brazing steps. In such cases, the solidus of the first filler metal used must occur at a higher temperature than the liquidus of the next filler metal so that the first braze joint does not soften and deform during the second brazing cycle. The choice between the various gold-copper filler metals available is normally dictated by their brazing temperature range since all compositions flow about equally well on copper, nickel, Kovar-type alloys, and nickel plated stainless steels. The copper-gold-nickel and the 82% gold, 18% nickel filler metals are, however, generally preferred for brazing of Kovar and stainless alloys since they flow better on these iron-bearing materials and produce a stronger braze joint than do the binary gold-copper filler metals.

The iron-nickel-cobalt (Kovar-Rodar) and nickel-iron alloys are widely used for sealing to glass and brazing to metallized alumina ceramic structures because they closely match the thermal expansions of these insulating materials. While the silver-copper eutectic filler metal will wet and flow well on nickel-iron and Kovar-type alloys, it is somewhat risky to use because of its tendency to penetrate the grain boundaries of these alloys, when they are under stress, and cause them to crack. (See Fig. 26.1.) The stress corrosion cracking of nickel-iron and Kovar-type alloys by the silver-copper filler metals can be minimized

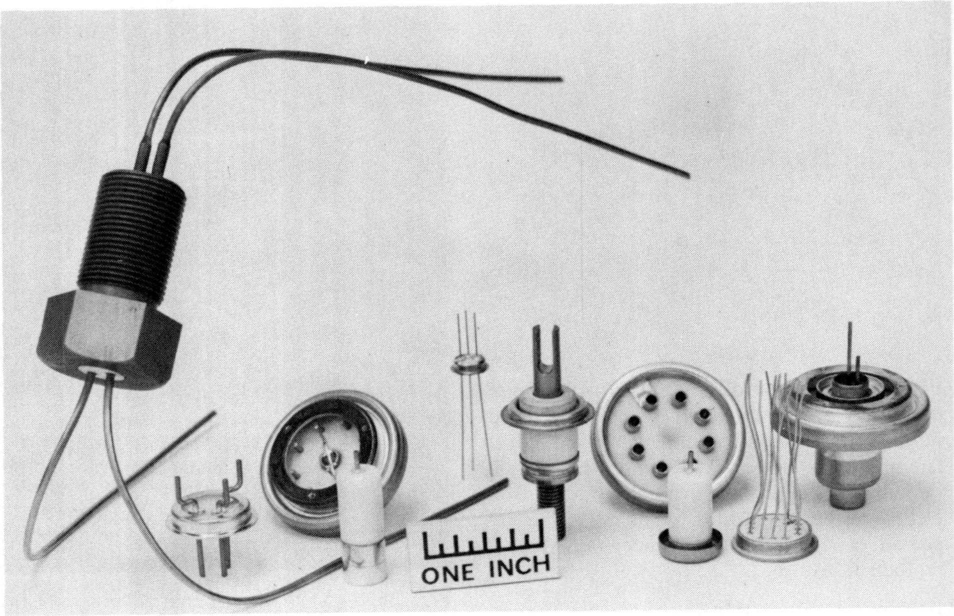

Fig. 26.3—Miscellaneous metal-to-ceramic assemblies for vacuum applications and feed-throughs for vacuum. Filler metals consist of copper silver eutectic, silver-copper-gold, gold-nickel, and OFHC copper.

by stress relieving the formed parts prior to brazing. Electroplating these alloys with nickel also appears to reduce stress corrosion by the silver filler metals. When these alloys are brazed to a higher expansion metal such as copper, care must be exercised in joint design to prevent restressing them during the brazing cycle. It has been found that additions of as little as 5% palladium to the silver-copper eutectic filler metal will greatly inhibit the stress corrosion effect on Kovar-type alloys. Filler metals are now available with 5, 10, 15, and 25% palladium additions to silver-copper. Since they are not too costly and far safer to use, the 5% and 10% palladium-bearing silver-copper filler metals are being increasingly favored over the silver-copper eutectic for brazing of nickel-iron and Kovar-type alloy structures.

When it becomes necessary to make a vacuum tight braze joint between two metals with widely different coefficients of thermal expansion, such as copper to Kovar or molybdenum, a difficult brazing problem arises. The use of any braze filler metal with a relatively large difference between its solidus and liquidus temperatures should be avoided. Any differential movement between parts being joined during the period in which the filler metal is in a liquid-solid, hence, in a very weak state, will open up cracks in the body of the filler metal. To minimize this problem select a filler metal with little or no temperature differential between its solidus and liquidus.

BRAZING OF REFRACTORY METALS

Molybdenum, tungsten, and tantalum are extensively used in electron tube and other vacuum device construction and are often brazed to each other or to other metals. (See also chapter 21.) Tungsten and molybdenum are readily brazed with platinum and platinum alloys and palladium-gold-nickel filler metals. The choice between these available filler metals is generally determined by the operating temperature of parts to be brazed; thus, tungsten electron tube support rods can be brazed with either pure platinum or a platinum-4% tungsten alloy. Molybdenum vacuum tube ''plate'' structures are also commonly brazed with these filler metals. The tungsten contacts of vacuum switches are brazed to copper conductors with the 82Au-18Ni filler metal. Tantalum is readily wet by copper-gold, gold-nickel, palladium-nickel, and palladium-gold-nickel al-

Fig. 26.4—Operator loading a metal section of large vacuum tube in bell type brazing furnace.

Fig. 26.5—Section of electron accelerator mounted in brazing fixture. White sections are alumina; braze filler is copper silver eutectic.

loys; however, tantalum is most often brazed in a vacuum since it reacts with hydrogen.

AVAILABLE FORMS OF BRAZING FILLER METALS

Vacuum tube brazing filler metals (Table 26.1) are readily available from their manufacturers in the form of wire of any diameter, any thickness of foil down to 0.0003 in. (0.008 mm), powder, and preforms such as wire rings, stampings, and formed wire shapes. Extrudables, pastes, and ''tapes'' made of filler metal powders mixed with appropriate organic vehicles are also available; however, they are seldom used in vacuum device brazing because of the undesirable residual carbon left by the decomposition of the organic vehicle. A serrated strip can also be bent around virtually any configuration; the advantage of this form of filler metal is that the user can produce preforms of the size and shape he requires (see Fig. 26.2).

BRAZING OF CERAMIC TO METAL STRUCTURES

Vacuum tight ceramic structures, generally of alumina, are widely used in electron tube vacuum switches and vacuum equipment, having largely replaced glass as an electrical insulator in these devices. Ceramic shapes formed to very close dimensional tolerances are far stronger than glass and can be obtained with a high bond strength metal coating in selected areas to which one can braze a metal part. Alumina ceramic parts serve as electrical insulators and structural members in electron tube, vacuum switch, and vacuum equipment construction. Alumina ceramics are most commonly metallized by sintering a coating of a molybdenum-manganese powder mixture at around 2730 °F (1500 °C) in hydrogen; the sintered moly-manganese coating is then generally nickel plated and resintered at around 1830 °F (1000 °C). The nickel coating is applied simply to allow a greater choice of filler metals for brazing the part.

Ceramic to metal joints must be carefully designed to avoid failure of the brittle ceramic as a result of differential thermal expansion stresses set up in the brazing process. The most commonly used metals are the Kovar-type alloys because their thermal expansion more closely matches that of alumina ceramics; however, even when using these alloys, the metal section is designed to yield at a stress level below that which will rupture the ceramic. When for other considerations Kovar-type alloys cannot be used, such ductile metals as copper and occasionally copper nickel are used. While these metals have a considerable expansion mismatch with alumina, the section joined to the ceramic by brazing is always designed to yield before the ceramic. Ideally, the brazing filler metals used must also have a low yield point; hence, the silver-copper eutectic and the 5% palladium-silver-copper filler metals are most commonly used. Should higher melting point and lower vapor pressure filler metals become necessary, the gold-copper filler metals are indicated.

Ceramics are brazed directly to metals, without first metallizing the ceramics, by use of filler metals containing one or more of the reactive metals, such as zirconium or titanium. Some common filler metals for this application are silver-copper eutectic clad titanium and nickel-clad titanium. These clad filler metals contain about 5% titanium and must be brazed in vacuum or inert atmospheres since the titanium will react with hydrogen. See also chapter 24.

TYPICAL APPLICATIONS

Figures 26.1-26.5 show typical applications of electron tube and vacuum equipment brazing.

CHAPTER 27

BRAZING PROVISIONS OF THE ASME BOILER AND PRESSURE VESSEL CODE

The ASME Boiler and Pressure Vessel Code is designed to establish rules of safety governing design, fabrication, and inspection, during construction, of boilers and unfired pressure vessels. The ASME Boiler and Pressure Vessel Code Committee is charged with the preparation of these rules and their interpretations when questions arise regarding intent. In formulating the rules, the Committee considers the needs of users, manufacturers, and inspectors of pressure vessels. The Code is published in several sections as follows:

- Section I Power Boilers
- Section II Material Specifications
  - Part A-Ferrous
  - Part B-Nonferrous
  - Part C-Welding Rods, Electrodes, Filler Metals (includes brazing filler metals)
- Section III Nuclear Power Plant Components
- Section IV Heating Boilers
- Section V Nondestructive Examination
- Section VI Recommended Rules for Care and Operation of Heating Boilers
- Section VII Recommended Rules for Care of Power Boilers
- Section VIII Pressure Vessels
  - Division 1
  - Division 2-Alternate Rules
- Section IX Welding Qualifications (includes brazing qualifications)
- Section X Fiberglass-Reinforced Plastic Pressure Vessels
- Section XI Rules for Inservice Inspection of Nuclear Reactor Coolant Systems

The individual sections were prepared by subcommittees of the main committee. Specific rules with respect to welding and brazing may appear in the individual sections as deemed necessary by the various subcommittees. Welding and brazing qualifications are set forth in Section IX and are applicable to all sections of the Code.

Provisions for brazing are in Section VIII of the Code. A complete revision was prepared and appeared in the 1962 edition; at the same time, a new Part C, Requirements for Brazing Ferrous and Nonferrous Materials, was added to Section IX. The entire Code is republished every three years; in the intervals, corrections, revisions, and additions appear in semiannual addenda. New materials, processes, and applications are covered by the "Code Case" procedures, as needed. Each new edition incorporates all the changes accomplished by the addenda and the Code Cases of the intervening period. It is impractical to reproduce Section IX, or the other sections, in this book; however, the salient points are reviewed here for the benefit of those persons who would operate under the Code rules. It should be pointed out that those who propose to produce products under the brazing requirements of the Code should obtain and use Sections VIII and IX and such other sections as may be applicable to the particular area of activity in which the persons are engaged.

Base metals used in brazed constructions must conform to Section II, Material Specifications: Part A for Ferrous Materials and Part B for Nonferrous Materials. While these specifications are similar to the ASTM specifications carrying the same numbers, there are, occasionally, significant differences in some of the specifications.

Such differences are noted under the titles of the individual specifications in Section II; therefore, it is important to work with the Code versions. A further limitation is imposed on the use of base metals: only those base metals for which allowable stress values have been established may be used.

In the tabulation of the Code sections just given, it will be noted that Section II, Part C, relates to Welding Rods, Electrodes, and Filler Metals (including brazing filler metals). When the AWS Filler Metal Specifications were separated from the ASTM specifications, an agreement was reached between ASME and AWS for the inclusion of the AWS Filler Metal Specifications in Section II, as Part C. The Code versions of these specifications are essentially identical with the AWS versions. As is the case with the ASTM specifications in Parts A and B, any deviations are noted under the title. The Code identifies the filler metal specifications with the prefix "SF," followed by the AWS designation. Thus, AWS A5.8 becomes SFA5.8, Brazing Filler Metals. Further, under the title it is noted that SFA5.8 is identical with A5.8, except that Classification BMg has been deleted. Since there is no provision for the use of magnesium base metals in the Code, it follows that there is no need for the magnesium brazing filler metals.

There are no specifications for fluxes or atmospheres under the Code requirements; however, the successful qualification of a brazing procedure in turn qualifies the specific flux or atmosphere used. Responsibility for the strength of the brazed joints rests with the designer, who must determine "from suitable tests or from past experience that the specific brazing filler metal

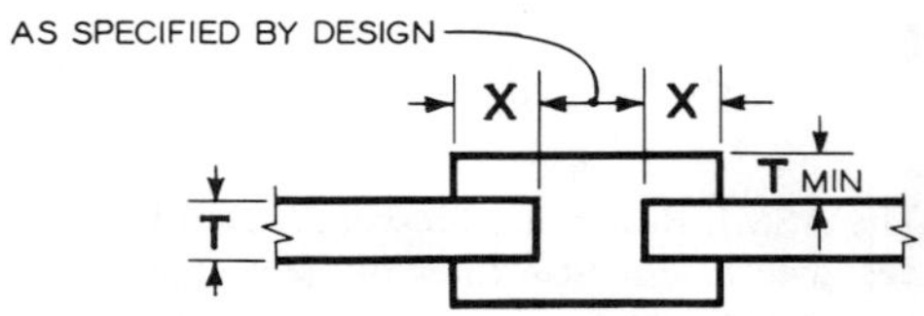

Fig. 27.1—Rabbet joint.

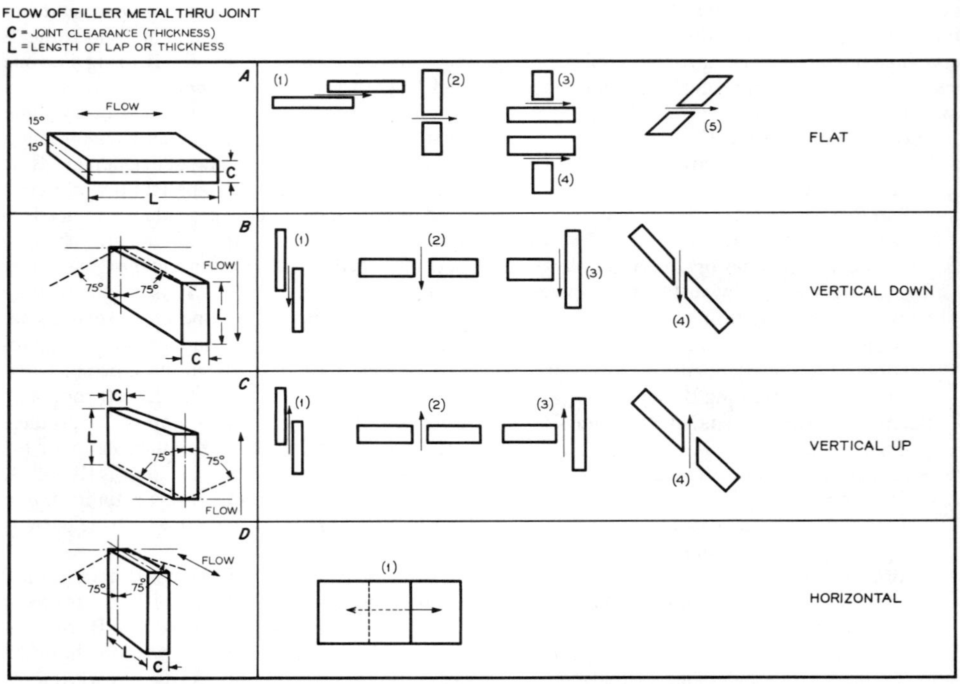

Fig. 27.2—Brazing positions showing flow of filler metal.

selected can produce a joint which will have adequate strength over the operating temperature range.'' The strength of brazed joints must also equal that of the base metal or the weaker of the two base metals in dissimilar metal joints. A joint efficiency factor is used in strength calculations. A factor of 1.0 applies when a visual examination assures that the brazing filler metal has penetrated the entire joint; when a joint is of such configuration and location that it cannot be inspected from both sides then a joint efficiency factor of 0.5 applies.

In conformance with the principles of the Code, as set forth in Section IX, the matter of brazing qualifications is treated in two parts, i.e., procedure qualification and performance qualification. Performance qualification includes qualification of brazers as well as those who operate automatic and semiautomatic equipment. Before a procedure qualification can be undertaken it is necessary to prepare a Procedure Specification, responsibility for the development, recording, and qualification of which rests with the contractor. The procedure must list all variables that might influence the results. The general categories of items covered are base metal or metals, brazing filler metals, brazing temperature, brazing process, brazing flux or atmosphere, flow direction, and joint design. These are essential variables, any change in which requires requalification of the procedure. Within these general categories there are several specific items relating to the brazing process to be used. Other, nonessential variables may be changed without the necessity for requalification, provided the details are recorded on the qualified procedure. This procedure is then subjected to qualification testing under the jurisdiction of a qualified inspector. The types of tests used in the procedure qualification depend upon the type and thickness of the joint involved in the procedure. Thus a tension test is required for butt, scarf, lap, and rabbet joints. (A drawing of the rabbet joint is shown in Fig. 27.1.) A peel test is used for lap joints, a bend test on butt and scarf joints, and a sectioning test on rabbet joints. For those special joint configurations that do not lend themselves to any of these tests, a workmanship sample is used, which may be a complete assembly or a typical component of an assembly. This test is used to supplement such of the other tests as may be applicable.

The brazer who performs the work in a procedure qualification test is automatically qualified, assuming that the procedure passes the necessary tests. Other brazers are required to pass the appropriate performance qualification test, using the qualified procedure. For checking the acceptance of performance test specimens, three tests discussed previously are used, i.e., the peel test, the sectioning test, and the workmanship sample.

In conducting both the procedure and the performance qualification tests, it is necessary to identify the positions of brazing. These positions differ significantly from those used for welding, and it is therefore necessary to provide a new system of identifying positions.

Figure 27.2, a reproduction of Fig. QB-461 as published in the Code, was specifically prepared for use in identifying brazing positions under the Code. The left column shows segments of brazing filler metal in a joint, but without any of the joint details; the right column shows typical joint configurations. Filler metal flow is indicated by means of arrows. The purpose of these illustrations is to define the flow of the brazing filler metal.

Figure 27.2 is as basic to brazing as is the corresponding illustration for welding positions. Note, however, that the descriptions of the positions, i.e., flat, vertical down, vertical up, and horizontal, while similar in wording, are different in intent, in that the brazing positions are predicated upon the flow of the molten filler metal.

Brazing has been used in the fabrication of unfired pressure vessels, particularly for heat exchangers. It is currently finding other areas of use. The rules of the Code, reflecting modern practices, permit further extension of these applications. Revisions to these rules will be made as advances in the science and art of brazing dictate.

CHAPTER 28

PIPE AND TUBE

It would be difficult to say when man first used metal tubular goods to convey materials from one location to another. Remnants of copper pipe show that metal was used to convey water as early as 3000 B.C., and certainly by the time of the Roman Empire the use of metals to convey water and other materials was well established. At this time, too, tubular members were probably first joined to various alloys, and the basic joining method has changed little since, as for the most part it is still a manual operation. Increased demands and service conditions in today's technology, however, have brought about an array of material requirements and the development of new alloys, as well as advanced joining techniques.

Brazed joints are economical and efficient and are easily and quickly made by skilled personnel with the correct equipment and materials. Brazed joints can readily be made in aluminum alloys, copper and copper alloys, steel, stainless steel, and almost all other metals, which will be stronger than the base metal material. Brazed joints can usually be made in those materials that will be serviceable at temperatures compatible with the end use temperature of the base metal. Most of the pipe and tube materials that can be soldered can also be brazed, and when service conditions are too severe to allow soldered joints, brazed joints can very often be used satisfactorily.

A wide variety of filler metals and fluxes for brazing are available, and most of them are readily applicable to pipe and tube. A filler metal for use with tube or pipe material should be chosen from the charts in the chapters concerned with the various base metals. In joints where joint clearances cannot be held to close tolerance it is advantageous to use a brazing filler metal that has a melting range wide enough to be easily workable, with enough fluidity to flow into the capillary space and provide good wetting action.

Most brazed joints in pipe and tubing do not need as much depth of socket insertion (lap) as do soldered joints. Ordinarily a depth of tube insertion three times the thickness of the tube will make a joint as strong as or stronger than the tube. Attempting to draw brazing filler metal to the base of a socket designed for soldering will often result in serious overheating.

CUTTING AND SIZING

Pipe or tubing should be cut square to seat evenly on the shoulder at the bottom of the socket, and all uneven or burred surfaces formed by the cutting operation should be removed. Hack saws or abrasive cutoff discs are suitable for cutting lengths of pipe and tubing; smaller size tubing, such as copper water service tubing, can be cut with a roller cutter. During any cutting operation, the tube ends should not be so severely clamped as to cause distortion in the wall diameter. Care should be exercised to ensure that the ends for joining are true and round. Both fittings and tubings are sized to provide joint clearances for capillary flow of the filler metals and joint strength.

Tables 28.1 and 28.2 show the standard dimensions of copper tube and copper and brass socket-type fittings. Tables 28.3 and 28.4 pro-

Table 28.1—Sizes and weights of copper water tube

Nominal size		Outside diameter Types K-L-M-DWV		Inside diameter Type K		Type L		Type M		Type DWV	
in.	mm	in.	mm	in.	mm	in.	mm	in.	mm	in.	mm
¼	6.4	0.375	9.5	0.305	7.75	0.315	8.00	0.325	8.00	—	—
⅜	9.5	0.500	12.7	0.402	10.21	0.430	10.92	0.450	11.43	—	—
½	12.7	0.625	15.9	0.527	13.39	0.545	13.84	0.569	14.45	—	—
⅝	15.9	0.750	19.1	0.652	16.56	0.666	16.92	0.690	17.53	—	—
¾	19.1	0.875	22.2	0.745	18.92	0.785	19.94	0.811	20.60	—	—
1	25.4	1.125	28.58	0.995	25.27	1.025	26.04	1.055	26.80	—	—
1¼	31.8	1.375	34.93	1.245	31.62	1.265	32.13	1.291	32.79	1.295	32.89
1½	38.2	1.625	41.28	1.481	37.62	1.505	38.23	1.527	38.79	1.541	39.14
2	50.8	2.125	53.98	1.959	49.76	1.985	50.42	2.09	51.03	2.041	51.84
2½	63.5	2.625	66.68	2.435	61.85	2.465	62.61	2.495	63.37	—	—
3	76.2	3.125	79.38	2.907	73.84	2.945	74.80	2.981	75.72	3.035	77.09
3½	88.9	3.625	92.08	3.385	85.98	3.425	87.00	3.459	87.86	—	—
4	101.6	4.125	104.78	3.857	97.97	3.905	99.19	3.935	99.94	4.009	101.83
5	127	5.125	130.18	4.805	122.05	4.875	123.83	4.907	124.64	4.981	126.52
6	152	6.125	155.58	5.741	145.82	5.845	148.46	5.881	149.38	5.959	151.36
8	203	8.125	206.38	7.583	192.61	7.725	196.22	7.785	197.74	7.907	200.84
10	254	10.125	257.18	9.449	240.00	9.625	244.48	9.701	246.41	—	—
12	305	12.125	307.98	11.315	287.40	11.565	293.75	11.617	295.07	—	—

*Slight variations from these weights must be expected in practice.

Table 28.2—Dimensional data, solder, brazed joint fitting ends

Nominal size		Tolerances—solder joint fittings: Male end (Fitting connection) O D max		O D min		Female end (solder cup) I D max		I D min	
in.	mm	in.	mm	in.	mm	in.	mm	in.	mm
¼	6.4	0.376	9.55	0.374	9.50	0.380	9.65	0.378	9.60
⅜	9.5	0.501	12.73	0.499	12.67	0.505	12.83	0.503	12.78
½	12.7	0.626	15.90	0.624	15.85	0.630	16.00	0.628	15.95
⅝	15.9	0.751	19.08	0.749	20.19	0.755	19.18	0.753	19.13
¾	19.1	0.876	22.25	0.874	22.20	0.880	22.35	0.878	22.30
1	25.4	1.1265	28.61	1.1235	28.54	0.1305	28.71	1.1285	28.66
1¼	31.8	1.3765	34.96	1.3735	34.89	1.3805	35.06	1.3785	35.01
1½	38.2	1.627	41.33	1.623	41.22	1.6315	41.44	1.629	41.38
2	50.8	2.127	54.03	2.123	53.92	2.1315	54.14	2.129	54.08
2½	63.5	2.627	66.73	2.623	66.62	2.6315	66.84	2.629	66.78
3	76.2	3.127	79.43	3.123	79.32	3.1315	79.54	3.129	79.48
3½	88.9	3.627	92.13	3.263	92.02	3.632	92.25	3.629	92.18
4	101.6	4.127	104.8	4.123	104.7	5.132	130.4	4.129	104.9
5	127	5.127	130.2	5.123	130.1	5.132	130.4	5.129	130.3
6	152	6.127	155.6	6.123	155.5	6.132	155.8	6.129	155.7
8	203	8.127	206.4	8.123	206.3	8.132	206.6	8.129	206.5
10	254	10.127	257.2	10.119	257.0	10.132	257.4	10.129	257.3
12	305	12.127	308.0	12.119	307.8	12.132	308.2	12.129	308.1

*For general plumbing use.
†For copper drainage use.

Wall thickness								Weight*							
Type K		Type L		Type M		Type DWV		Type K		Type L		Type M		Type DWV	
in.	mm	in.	mm.	in.	mm	in.	mm	lb/ft	kg/m	lb/ft	kg/m	lb/ft	kg/m	lb/ft	kg/m
0.035	0.89	0.030	0.76	0.025	0.064	—	—	0.145	0.216	0.126	0.187	0.106	0.158	—	—
0.049	1.24	0.035	0.89	0.025	0.064	—	—	0.269	0.400	0.198	0.295	0.145	0.216	—	—
0.49	1.24	0.040	1.02	0.028	0.71	—	—	0.344	0.512	0.285	0.424	0.204	0.304	—	—
0.49	1.24	0.042	1.07	0.030	0.76	—	—	0.418	0.622	0.362	0.539	0.263	0.391	—	—
0.065	1.65	0.045	1.14	0.032	0.81	—	—	0.641	0.954	0.455	0.677	0.328	0.488	—	—
0.065	1.65	0.050	1.27	0.035	0.89	—	—	0.839	1.25	0.665	0.975	0.465	0.692	—	—
0.065	1.65	0.055	1.40	0.042	1.07	0.040	1.02	1.04	1.55	0.884	1.32	0.682	1.01	0.650	0.
0.072	1.83	0.060	1.52	0.049	1.24	0.042	1.07	1.36	2.02	1.14	1.70	0.940	1.40	0.809	1.
0.083	2.11	0.070	1.78	0.058	1.47	0.42	1.07	2.06	3.07	1.75	2.60	1.46	2.17	1.07	1.
0.095	2.41	0.080	2.03	0.065	1.65	—	—	2.93	4.36	2.48	3.70	2.03	3.02	—	—
0.109	2.77	0.090	2.29	0.072	1.83	0.045	1.14	4.00	5.95	3.33	4.96	2.68	3.99	1.69	2.
0.120	3.05	0.100	2.54	0.083	2.11	—	—	5.12	7.62	4.29	6.38	3.58	5.33	—	—
0.134	3.40	0.110	2.79	0.095	2.41	0.058	1.47	6.51	9.69	5.38	8.01	4.66	6.93	2.87	4.
0.160	4.06	0.125	3.18	0.109	2.77	0.072	1.83	9.67	14.4	7.61	11.3	6.66	9.91	4.43	6.
0.192	4.88	0.140	3.56	0.122	3.10	0.083	2.11	13.9	20.7	10.2	15.2	8.92	13.3	6.10	9.
0.271	6.88	0.200	5.08	0.170	4.32	0.109	2.77	25.9	38.5	19.3	28.7	16.5	24.6	10.6	15.
0.338	8.59	0.250	6.35	0.212	5.38	—	—	40.3	60.0	30.1	44.8	25.6	38.1	—	—
0.405	10.29	0.280	7.11	0.254	6.45	—	—	57.8	86.0	40.4	60.1	36.7	54.6	—	—

Solder joint fittings; cast bronze, wrought copper				Solder joint fittings†							
Fitting end length		Solder cup length		Fitting end length				Solder cup depth			
				Wrought drainage		Cast drainage		Wrought drainage		Cast drainage	
in.	mm	in.	mm	in.	mm	in.	mm	in.	mm	in.	mm
3/8	9.5	5/16	7.9	—	—	—	—	—	—	—	—
7/16	11.1	3/8	9.5	—	—	—	—	—	—	—	—
9/16	14.3	1/2	12.7	—	—	—	—	—	—	—	—
11/16	17.5	5/8	15.9	—	—	—	—	—	—	—	—
13/16	20.6	3/4	19.1	—	—	—	—	—	—	—	—
31/32	24.6	29/32	23.0	—	—	—	—	—	—	—	—
1 1/32	26.2	31/32	24.6	11/16	17.5	0.44	11.2	5/8	15.9	0.38	9.7
1 5/32	29.4	1 3/32	27.8	3/4	19.1	0.50	12.7	11/16	17.5	0.44	11.2
1 13/32	35.7	1 11/32	34.1	13/16	20.6	0.56	14.2	3/4	19.1	0.50	12.7
1 17/32	38.9	1 15/32	37.3	—	—	—	—	—	—	—	—
1 23/32	43.7	1 21/32	42.1	1	25.4	0.81	20.6	15/16	23.8	0.75	19.1
1 31/32	50.0	1 29/32	48.4	—	—	—	—	—	—	—	—
2 7/32	56.4	2 5/32	54.8	1 1/4	31.8	1.06	26.9	1 3/16	30.2	1.00	25.4
23/32	69.1	2 21/32	67.5	—	—	1.31	33.3	—	—	1.25	31.8
3 5/32	80.2	3 3/32	78.6	—	—	1.62	41.1	—	—	1.50	38.2
4 1/32	102.4	3 31/32	100.8	—	—	2.12	53.8	—	—	2.00	50.8
4 1/8	104.8	4	101.6	—	—	—	—	—	—	—	—
4 5/8	117.5	4 1/2	114.3	—	—	—	—	—	—	—	—

Table 28.3—Steel schedule 40

Pipe size		Outside diameter		Inside diameter		Wall thickness	
in.	mm	in.	mm	in.	mm	in.	mm
⅛	3.2	0.405	10.29	0.307	7.80	0.049	1.24
¼	6.4	0.540	13.72	0.410	10.41	0.065	1.65
⅜	9.5	0.675	17.15	0.545	13.84	0.065	1.65
½	12.7	0.840	21.34	0.675	17.12	0.083	2.11
¾	19.1	1.050	26.67	0.884	22.45	0.083	2.11
1	25.4	1.315	33.40	1.097	27.86	0.109	2.77
1¼	31.8	1.660	42.16	1.442	36.63	0.109	2.77
1½	38.2	1.900	48.26	1.682	42.72	0.109	2.77
2	50.8	2.375	60.33	2.157	54.79	0.109	2.77
2½	63.5	2.875	73.03	2.635	66.93	0.120	3.05
3	76.2	3.500	88.90	3.260	82.80	0.120	3.05
3½	88.9	4.000	101.6	3.760	95.50	0.120	3.05
4	101	4.500	114.3	4.260	108.2	0.120	3.05
5	127	5.563	141.3	5.295	134.5	0.134	3.40
6	152	6.625	168.3	6.357	161.5	0.134	3.40
8	203	8.625	219.1	8.239	211.6	0.148	3.76

Table 28.4—Stainless steel schedule 5

Pipe Size		Outside diameter		Inside diameter		Wall thickness	
in.	mm	in.	mm	in.	mm	in.	mm
½	12.7	0.840	21.34	0.710	18.03	0.065	1.65
¾	19.1	1.050	26.67	0.920	23.37	0.065	1.65
1	25.4	1.315	33.40	1.185	30.10	0.065	1.65
1¼	31.8	1.660	42.16	1.530	38.86	0.065	1.65
1½	38.2	1.900	48.26	1.770	44.96	0.065	1.65
2	50.8	2.375	60.33	2.245	57.02	0.065	1.65
2½	63.5	2.875	73.03	2.709	68.81	0.083	2.11
3	76.2	3.500	88.90	3.334	84.68	0.083	2.11
3½	88.9	4.000	101.6	3.834	97.38	0.083	2.11
4	101.6	4.500	114.3	4.334	110.1	0.083	2.11

vide the dimensions for steel pipe, and Table 28.5 those for threadless copper pipe. Roundness tolerances are given in Table 28.6. The dimensions given in these tables generally allow about 0.004 in. (0.10 mm) clearance between the outside wall of the tubing and the inner wall of the solder/braze cup. The overall difference in diameter between the male end (fitting connection) and the female end (solder cup) is about 0.009 in. (0.23 mm). Usually a clearance of 0.004 in. (0.10 mm) will assure a sound metallurgical joint. Clearances larger than about 0.012 in. (0.30 mm) provide less dependable joints, whereas joint clearances that are too tight may be weak from poor filler metal penetration or flux inclusions.

After cutting, the tube is reamed to remove burrs or uneven edges that would interfere with the flow of the filler metal or the alignment of the mating parts. A half-round file or other device will remove these surface burrs and irregularities; however, care should be exercised to prevent distortion or increasing of the joint clearance by overreaming of the fitting sockets.

CLEANING

The ends of tubing or pipe as well as the internal surfaces of fittings must be thoroughly cleaned to provide surfaces that allow for adherence, wetting (alloying), and distribution of the filler metals at the joint interfaces. All traces of dirt, grease, lacquers, or oxides on the base metals must be removed. Degreasing with organic solvents will often remove various oils, but these materials will not ordinarily be effective in removing oxides or organic coatings applied to provide oxidation resistant surfaces. Cleaning

Table 28.5—Dimensions and weights and tolerances in diameter and wall thickness for copper threadless pipe (TP) sizes

Standard pipe size in.	Nominal dimensions: Outside diameter		Inside diameter		Wall thickness		Cross-sectional area of bore		Nominal weight lb/ft (kg/m)		Tolerances: Average outside diameter* all minus		Wall thickness plus or minus	
	in.	mm	in.	mm	in.	mm	in^2	cm^2	lb/ft	kg/m	in.	mm	in.	mm
¼	0.540	13.7	0.410	10.4	0.065	1.65	0.132	0.852	0.376	0.559	0.004	0.10	0.0035	0.089
⅜	0.675	17.1	0.545	13.8	0.065	1.65	0.233	1.50	0.483	0.719	0.004	0.10	0.004	0.10
½	0.810	21.3	0.710	18.0	0.065	1.65	0.396	2.55	0.613	0.912	0.005	0.13	0.004	0.10
¾	1.050	26.7	0.920	23.4	0.065	1.65	0.665	4.29	0.780	1.16	0.005	0.13	0.004	0.10
1	1.315	33.4	1.185	30.1	0.065	1.65	1.10	7.10	0.989	1.47	0.005	0.13	0.004	0.10
1¼	1.660	42.2	1.530	38.9	0.065	1.65	1.84	11.9	1.26	1.87	0.006	0.15	0.004	0.10
1½	1.900	48.3	1.770	45.0	0.065	1.65	2.46	15.9	1.45	2.16	0.006	0.15	0.004	0.10
2	2.375	60.3	2.245	57.0	0.065	1.65	3.96	25.5	1.83	2.72	0.007	0.18	0.006	0.15
2½	2.875	73.0	2.745	69.7	0.065	1.65	5.92	38.2	2.22	3.30	0.007	0.18	0.006	0.15
3	3.500	88.9	3.334	84.7	0.083	2.11	8.73	56.3	3.45	5.13	0.008	0.20	0.007	0.18
3½	4.000	102	3.810	96.8	0.095	2.41	11.4	73.5	4.52	6.73	0.008	0.20	0.007	0.18
4	4.500	114	4.286	109	0.107	2.72	14.4	92.9	5.72	8.51	0.010	0.25	0.009	0.23
5	5.562	141	5.298	135	0.132	3.40	22.0	142	8.73	13.0	0.012	0.30	0.010	0.25
6	6.625	168	6.309	160	0.158	4.01	31.3	202	12.4	18.5	0.014	0.36	0.010	0.25
8	8.625	219	8.215	209	0.205	5.21	53.0	342	21.0	31.2	0.018	0.46	0.014	0.36
10	10.750	273	10.238	260	0.256	6.50	82.3	531	32.7	48.7	0.018	0.46	0.016	0.41
12	12.750	324	12.124	308	0.313	7.95	115	742	47.4	70.5	0.018	0.46	0.020	0.51

*The average outside diameter of a tube is the average of the maximum and minimum outside diameters, as determined at any one cross section of the tube.

Table 28.6—Roundness tolerances

t/d, Ratio of nominal wall thickness to nominal outside diameter	Diameter roundness tolerances, % of nominal outside diameter (expressed to the nearest 0.001 in.) (nearest 0.01 mm)*
0.01 to 0.03, incl	1.5
Over 0.03 to 0.05, incl	1.0
Over 0.05 to 0.10, incl	0.8
Over 0.10	0.7

*The deviation from roundness is measured as the difference between major and minor outside diameters as determined at any one cross section of the tube.

should be performed so that only the surface contaminants are removed without excessive loss of the base metals; this is best accomplished by lightly abrading the tube ends and solder/braze cups with small compatible wire brushes, steel wool, or fine grades (00) of abrasive papers or cloth. Embedding of abrasive particles in the joint surfaces should be avoided, and any material resulting from the mechanical cleaning operation must be removed.

Although mechanical cleaning of steel or copper pipe and tubing is usually adequate, the use of mineral acids or alkaline cleaners is often required to prepare surfaces for brazing. Nickel, aluminum, and their alloys, as well as stainless steel and cast iron, usually require active chemical cleaning before joining. Thorough washing of chemically cleaned parts is necessary to remove any residues resulting from the operation. Suitable chemical cleaning compounds are given in chapter 10.

FLUXES AND FLUXING OPERATIONS

Fluxes are usually required for all manual brazing of tubing or pipe, though not when brazing filler metals containing copper phosphorus (BCuP) are used in joining copper tubing, or when brazing in controlled atmospheres or under vacuum conditions. Flux must be used if the copper tube is being joined to a cast brass or bronze component. One purpose of a flux is to prevent oxidation of joint surfaces both after cleaning and during the brazing process. Fluxes are not intended to be used as cleaning compounds to remove heavy oxide films, but they should be sufficiently active to remove thin oxide films that may develop after cleaning. In addition, fluxes should dissolve or absorb any metallic oxide formed on joint surfaces during the joining operation.

It is important that flux be applied to the joint as soon as possible after cleaning. A small brush or a clean cloth will assist in coating the joint interfaces, and it is essential that joints be brazed soon after they are fluxed. One hour is the usual limit for brazing to follow flux application, but a shorter interval is desirable.

Fluxes used in brazing pipe or tubing usually fall under AWS brazing flux classifications 3A, 3B, or 4. When choosing a brazing flux, the proprietary mixtures recommended by the manufacturer of the filler metal being used are usually most satisfactory. Flux activity and stability depend upon the extent and type of oxides present, the heating method employed, and the temperature of the heated parts. Fluxes are characterized by their high fluidity at the temperature at which the filler metals melt; prolonged heating or overheating of fluxes will usually destroy their effectiveness. All surfaces to be wetted by the filler metals should be coated with flux, but application of flux to the inside surfaces of pipe should be avoided. Flux inclusions in the joint can often be minimized by preplacing a ring of the brazing filler metal at the seat of the pipe joint. Upon heating the filler metal from the ring will flow outward into the joint, displacing the flux ahead of it.

ASSEMBLY

Before any joining operation, the assembly should be carefully aligned and adequately supported. Misalignment will affect the joint by altering the clearance between the tube and the fittings. Vertical pipe runs should be plumb to prevent any inconsistencies in joint spacing. Supports and fixtures are an excellent means for allowing for expansion and contraction of the assembly during the joint operation and while in service. Use rings, straps, and clamps to help maintain alignment of the joints.

APPLYING HEAT AND FILLER METALS

All the preliminary steps for preparing the joint surfaces must be followed before any attempt is made to heat the joints and flow the alloy to effect a bond. Since most of the pipe and tubing assemblies are brazed with gas-fired torches, we will emphasize the heating of joints by torches fueled with acetylene, propane, natural gas, or other gases in the presence of oxygen. The amount of heat required for the size of the pipe or tubing being joined will determine the selection of torch tips. The tip should be of a size to develop a low velocity bulbous flame which will heat rapidly and evenly. This flame should be neutral or only slightly reducing. A neutral flame has a smooth, even inner cone, as large as possible without giving the end of the cone a feathered appearance. To adjust an acetylene flame to its neutral state, light a flame that has an excess of acetylene, then reduce the flow of the gas, or increase the oxygen flow very slowly until the cone just starts to feather.

The flame should be directed so that it wraps itself around small diameter pipe or fittings. If the pipe is of large diameter (i.e., greater than 4 in. [101.6 mm]) multiple tip torches and even additional heating torches may be required. Usually the pipe or tube is heated before the fittings, and after a short heating period the flame is directed alternately from the pipe and then to the fitting. This procedure will bring the fittings and pipe up to equal heat. Holding the flame at one location on the fitting or pipe will cause localized overheating, excessive drying of the flux, distortion of the pipe, and possible cracking of the fittings. Overheating is the most common single cause of poor brazed joints. The inner cone tip of the flame should not impinge directly on the shoulder of the tube and fitting or into the socket of the fitting. The sharp point of the inner cone of the flame should not touch the metal surfaces, and the flame should be kept in motion so that heat is distributed evenly to as large an area as possible.

The correct temperature for making brazed pipe joints can be tested by touching the filler metal to the heated junction between the pipe and fitting and observing the flow of the filler metal. The flux should be very fluid on both fittings and pipe. Brazing fluxes usually melt into a clear liquid at brazing temperatures. When the filler metal begins to melt, it is fed into the capillary space with a gentle pressure. At the same time, the torch is moved around the fitting with the filler metal being fed to the joint just behind the path of flame. Heat should not be applied directly to the filler metal. If the torch flames are directed to the heel or bottom of the socket of the fitting, the filler metal will distribute throughout the joint.

Opinions differ as to whether or not a fillet is

necessary or desirable on pipe joints. Often a full annular fillet indicates good filler metal penetration and distribution of the joint; however, generous fillets at the bottom of the assembly may be caused by accumulation of filler metal that has solidified after flowing over relatively cold metal. Heating the bottom of the fitting will largely obviate this condition.

Induction or furnace heating is sometimes used to braze pipe and tubing assemblies. Usually the filler metal is preplaced in or adjacent to the fluxed joint. Alternately, brazing pastes may be used wherein the joint surfaces are painted with mixtures of flux and powdered filler metals before joining.

FILLER METALS

Filler metal used in brazing of pipe and tubing must be metallurgically compatible with the base metals they are intended to join. In addition, they must provide adequate strength to the joint under specific service conditions. Brazing filler metals commonly used in joining pipe and tubing are presented in Table 28.3. Brazing filler metals used in pipe assembly preferably should allow for a melting range, that is, the temperature at which the filler metal starts to melt to the temperature at which it is fully liquid. The melting range is useful in providing distribution of the filler metal over the surfaces to be joined during the joining process.

POSTBRAZE CLEANING OPERATIONS

After the filler metal has solidified, the remaining flux and residues can be removed by wiping with a wet cloth or by wet brushing. Cast fittings should be allowed to cool naturally to some extent before applying swabs to the joints. Since most brazing fluxes are hygroscopic and many are active corrosive agents they should be removed. Where aluminum alloys are used, it is essential that flux residues be removed.

INSPECTION

Though brazed joints are very rarely examined or inspected by nondestructive testing methods

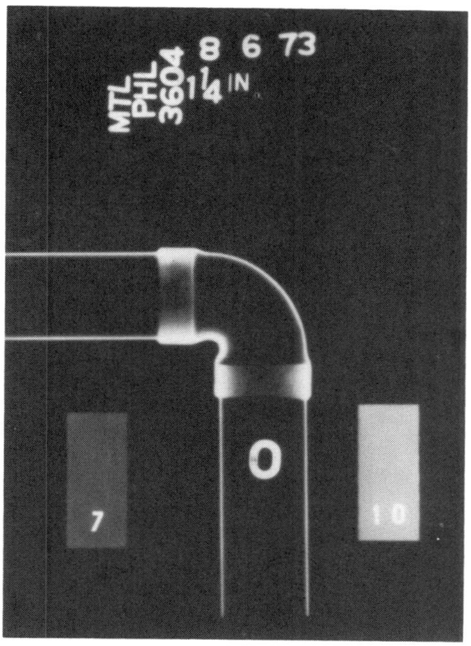

Fig. 28.1—Copper tube brazed to wrought copper fitting joint in 1.25 in. (31.8 mm) with 15Ag-5P-80Cu filler metal shows minor voids.

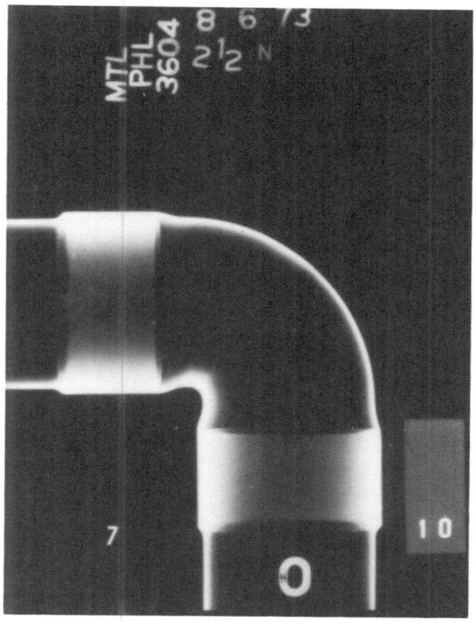

Fig. 28.2—Copper tube brazed to wrought copper fitting joint in 2.5 in. (63.5 mm) with 50Ag-Cu-Cd-Zn filler metal shows a nearly perfect joint.

other than visual, it is possible to conduct both radiographic and ultrasonic inspections. Ultrasonic inspection is fairly quick and easy, but does require a highly skilled and experienced operator, and it usually does not give a permanent record. Radiography also requires an able operator and a longer time for inspection, but it does give a permanent record. Figures 28.1, 28.2, 28.3 show brazed pipe joints that were inspected by radiograph.

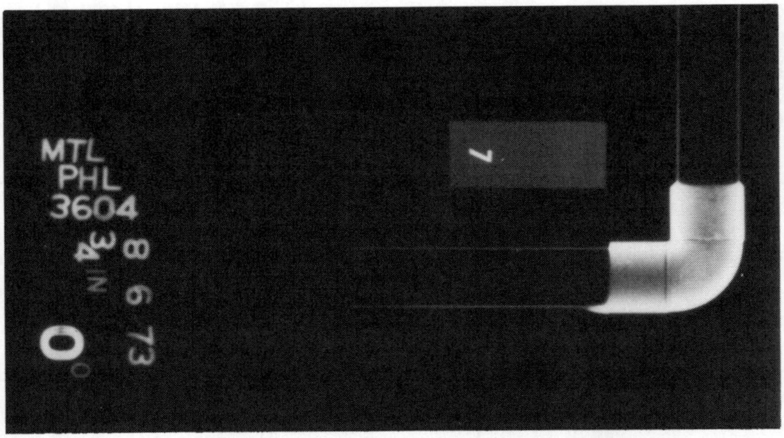

Fig. 28.3— Copper tube brazed to cast brass fitting in 0.75 in. (19.05 mm) with 50Ag-Cu-Cd-Zn filler metal showing a very good joint and indicating the greater difficulty of radiographing brazed joints in castings as contrasted with wrought fittings (BAg-1a).

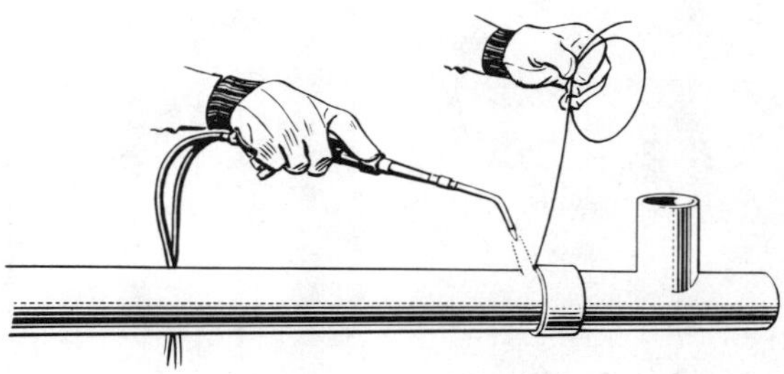

Fig. 28.4 —Brazing pipe.

CHAPTER 29

SAFETY AND HEALTH PROTECTION

Brazing is subject to the provisions of the American National Standard Z49.1, Safety in Welding and Cutting. Although for the most part the requirements of this standard are applicable to brazing, there are certain aspects peculiar to brazing. It is necessary, therefore, to establish these areas of special application.

In the past ANSI Z49.1 has been a voluntary standard, sponsored and prepared under the auspices of the AWS in accordance with the procedures established by ANSI. In 1970 the U.S. Congress passed the Williams-Steiger Occupational Safety and Health Act (OSHA). This act applies to virtually all industrial operations, including welding, cutting, brazing, and soldering. Since the federal government did not have available safety standards in all areas, the Department of Labor adopted many of the standards developed in industry, notably those of ANSI. American National Standard Z49.1 was adopted in this manner and has thus become the law for safety in welding, cutting, brazing, and soldering. It is also published by the Department of Labor, Occupational Safety and Health Administration, as Occupational Safety Standards, Chapter XVII, Part 1910, Subpart Q.

In the earlier, voluntary, Z49.1, there were mandatory rules and recommended practices. When the standard became law it was necessary to separate these two functions. The mandatory phases become the law; the safety recommendations are merely advisory — they appear as such in the latest edition of ANSI Z49.1, but cannot be enforced as law. This latest edition provided for the separation of functions.

With this new function for ANSI Z49.1 it becomes necessary for users of the brazing processes to follow strictly the provisions of this standard. In a review or abstract of the standard, such as provided here, it is not possible to cover every detail; thus, those persons working with the brazing processes should obtain and follow the complete standard.

Care must be exercised in the handling of hot objects. Brazers must wear clothing, including shoes, that will afford protection against falling objects, heat, and other hazards incidental to the welding and brazing processes. Goggles or spectacles, as dictated by the brazing operations, are necessary.

Some brazing filler metals contain toxic materials, such as cadmium or beryllium; most fluxes contain fluorides and other fluorine compounds. The use of such filler metals and fluxes requires special precautions as detailed in the standard.

VENTILATION

The standard establishes the following basis for ventilation:

> The dimensions of the working space with special regard to ceiling height and the presence of obstructions. The number of brazers, plus any welders or cutters. The

evolution of hazardous fumes, gases, or dusts all related to the base and filler metals in use. Location of the brazer's breathing zone with respect to the rising plume of fumes.

This is supplemented by general conditions such as atmospheric conditions; the heat generated; the presence of volatile solvents.

It is a further requirement that local exhaust or general ventilating systems be provided, capable of keeping the amount of toxic fumes, gases, or dusts below the acceptable concentrations as established by Toxic Dusts and Gases, ANSI Z37, or the latest threshold limit values (TLV) of the American Conference of Governmental Industrial Hygienists, or the exposure limits as published by Public Law 91-596 (Occupational Safety and Health Act of 1970). Compliance with this requirement must be determined by sampling of the atmosphere in the work area. The samples must reflect the exposure of the persons involved. Natural ventilation is considered to be acceptable if an atmospheric survey demonstrates that these limits or restrictions are not exceeded.

Mechanical ventilation must be provided when brazing is performed under the following conditions:

In working areas of less than 10 000 ft^3 (284 m^3) per brazer.

In a work area with a ceiling height less than 16 ft (5 m).

In confined spaces or where the brazing space contains partitions, balconies, or other structural barriers to the extent that they significantly obstruct cross ventilation.

The standard lists rates for general ventilation, and provides for mechanical local exhaust ventilation along with the appropriate value for the performance of the equipment. A footnote warns that higher rates for ventilation may be necessary "When brazing with cadmium bearing or other toxic materials..."

When brazing is performed in confined spaces, other, special requirements are enforced. As a basic provision, adequate ventilation must be provided to prevent the accumulation of toxic materials or possible oxygen deficiency for the brazer and persons working with him. When adequate ventilation cannot be supplied, then it is necessary to use air-supplied respirators or hose masks as approved by the U.S. Bureau of Mines, National Institute of Occupational Safety and Health (NIOSH), or another approving authority. If, however, the area is immediately hazardous to life, then hose masks, hose masks with blowers, or self-contained breathing apparatus (respirators) are required. Again these devices must be of types approved as noted above. It is a further requirement, under these circumstances, that an attendant be stationed outside to ensure the safety of those working inside. Oxygen must never be used for ventilation purposes.

SPECIAL PRECAUTIONS

Cadmium

Cadmium is present as an element in some brazing filler metals, and may also be present as a plating on base metals. It is essential that all working near sources of cadmium avoid the inhalation of the fumes. Plated objects, when heated, will produce greater volumes of fumes than will brazing on unplated metals.

Brazing filler metals containing cadmium in significant amount must carry the following notice on tags, boxes, or other containers:

WARNING
CONTAINS CADMIUM
POISONOUS FUMES MAY
BE FORMED ON HEATING

Do not breathe fumes. Use only with adequate ventilation such as fume collectors, exhaust ventilators, or air-supplied respirators. See ANSI Z49.1. If chest pain, cough, or fever develops after use, call physician immediately. Keep children away when using.

When brazing is performed, indoors or in confined spaces, on cadmium-bearing or cadmium-coated base metals, it is necessary to use local exhaust ventilation or air-supplied respirators unless atmospheric tests under the most adverse conditions establish that the exposure of the workers is within the acceptable limits defined in the ventilation section. Outdoors such operations require the use of respiratory protective equipment, such as fume respirators. All respiratory equipment must be of approved types as previously described. Brazing with cadmium-bearing filler metals requires the use of ventilation in the manner described under ventilation. In this connection note that cadmium-free filler metals are available (see chapter 3).

Fluoride

Brazing fluxes containing fluoride must carry the following warning label on all tags, boxes, or other containers:

CAUTION
CONTAINS FLUORIDES
This flux when heated gives off fumes that may irritate eyes, nose, and throat.

1. Avoid fumes — use only in well ventilated spaces.
2. Avoid contact of flux with eyes or skin.
3. Do not take internally.
4. Keep out of reach of children.

When brazing in confined spaces, with fluoride fluxes, one must have ventilation or use respiratory equipment described under ventilation. In open spaces, where air samples taken in the breathing zone show fluorides below the weighted threshold limit values, no special ventilation may be necessary. Furnace brazing with fluoride-bearing fluxes requires regular exhaust of fumes generated to prevent attack of exposed metals. The discharge of such exhaust may be required to meet local ecological standards. Dip brazing using fluorine-bearing flux bath will also require suitable exhaust systems.

Beryllium

Beryllium is used as a melting point depressant in certain brazing filler metals (e.g., 48Ti-48Zr-4Be). Beryllium may also be present in small amounts in magnesium filler metals, and to a lesser extent in aluminum brazing filler metals. Beryllium and beryllium-copper base metals are also brazed.

When brazing indoors, outdoors, or in confined spaces with beryllium-containing base or filler metals, both local exhaust ventilation and air-supplied respirators must be used unless atmospheric tests under the most adverse conditions have established that the exposure of the workers is within acceptable concentrations as given in the ventilation section. A further requirement is that workers in the immediate vicinity of the brazing operation be similarly protected, as necessary. And, it is prudent practice to observe adequate cleanliness, so as to avoid absorption of the material orally or through skin openings.

Zinc

Zinc may be found as a base metal coating or as a constituent of base or filler metals. It melts at a low temperature and oxidizes readily, producing zinc oxide fumes. These fumes may cause an illness known as "zinc chills" or "zinc fume fever." While the health hazard from zinc fumes is considered to be less than that of the other fumes here listed, it is still necessary to provide adequate ventilation. The standard requires that indoor brazing involving zinc-bearing base or filler metals be done as described in the ventilation section. Similarly, if the work involves confined spaces the previous discussions apply.

Other Materials of Toxic Significance

Other metals are specifically listed in the standard, e.g., lead and mercury, together with the appropriate controls. However, these metals are not normally encountered in brazing, and therefore are not detailed here. If they are to be used in brazing operations, the rules of the standard will apply. Under this general heading the standard also lists a number of elements that might be encountered in base metals, filler metals, or fluxes. The control of such fume sources is by means of atmospheric sampling and maintenance of the element below the time weighted threshold limit value.

Cleaning Compounds

It is imperative that proper ventilation be provided in areas where cleaning operations are being conducted. The use of carbon tetrachloride is generally prohibited: its presence, even in small quantities and at any temperature, is a serious cumulative health hazard. Trichloroethylene and tetrachlorethylene vapors, if in contact with a hot object or in the vicinity of an electric arc, become dissociated, generating free halogens and phosgene, which are toxic. The vapors of these cleaning solvents are heavier than air and can travel long distances from cleaning vats.

Some solvents, such as benzene and gasoline, are flammable as well as toxic. Consequently they are not often used, and when they are, proper protection against the possibility of fire must be taken.

Before using a chemical cleaner it is necessary to determine if any of its constituents are flammable or produce toxic fumes. A well designed and properly located hood with an exhaust system will effectively remove fumes from the cleaning and pickling tank area. Caution must be exercised in the handling and use of alkalies and acids. Attention is directed to the instruction sheets provided by the Manufacturing Chemists

Association, which indicate the preferred methods of handling the various acids and alkalies. The standard warns of such conditions and requires that safety precautions, such as manufacturers' instructions be followed; also, that degreasing or other cleaning operations involving chlorinated hydrocarbons be so located that no vapors from these operations will reach or be drawn into the atmosphere surrounding molten weld metal. The standard also warns against permitting fumes from trichloroethylene and perchlorethylene in welding atmospheres in which ultraviolet radiation is present.

Chlorinated solvents of the types discussed for cleaning operations may also be used in preparing binder for powdered brazing filler metals, and in spray guns. The precautions detailed in the foregoing apply equally for these operations and must be followed.

Brazing Atmospheres

American National Standard Z49.1 has specific requirements with respect to the handling of flammable furnace atmospheres, as defined in chapter 4, AWS Brazing Atmospheres, Type Numbers 2 through 7. It is required that such atmospheres be purged to remove oxygen and air by one of the following methods.

1. Below 1400 °F (760 °C) purging shall reduce the oxygen below the explosive level by:
   a. Evacuating to a pressure below 5 mm mercury and backfilling to working pressure.
   b. Purging with a neutral or inert atmosphere for 4 to 6 volumes and subsequently with 4 to 6 volumes of the flammable atmosphere.
2. Above 1400 °F (760 °C), heated furnaces or retorts may be purged with the flammable gas atmosphere with the atmosphere entering the 1400 °F (760 °C) chamber area. Under these conditions, the air will be progressively consumed, thus eliminating any explosive condition.

It is further required that adequate area ventilation be provided to exhaust and discharge outdoors all explosive or toxic gases that may emanate from such furnace purging and brazing operations. Local ecological regulations should be consulted before exhausting outdoors.

This section of the standard also provides for relaxing the above requirements in terms of advances in furnaces and atmospheres, provided such changes can be demonstrated to be safe operations.

Asphyxiation

The use of gases, whether active or inert, carries with it another potential source of hazard — asphyxiation. The standard therefore requires that confined spaces be adequately ventilated to prevent possible oxygen deficiency. However, in brazing this deficiency is not always recognized. Nitrogen, normally an inert gas at normal temperatures, can produce noxious oxides at elevated temperatures; argon is inert at all temperatures. Both of these gases are used in brazing principally because of their inertness. Although they are inert, their very presence in high concentrations will lead to conditions of asphyxiation. Therefore, do not permit anyone to enter a bell, retort, or vacuum furnace that has been flushed with or still contains an inert atmosphere. If it is necessary to enter such an area, provide for flushing out the inert atmosphere, replacing it with respirable air. If this is not possible, then persons entering should be provided with suitable air-supplied respirators.

Dip Brazing

In dip brazing it is extremely important that the parts immersed in the baths be completely dry. The presence of moisture on the parts will cause an instantaneous generation of steam, accompanied by explosive violence, which will expel the contents of the dip pot, creating a serious burning hazard. Predrying of the parts will obviate this problem. If supplementary flux is necessary in a dip brazing operation, it must be adequately and properly dried to remove not only moisture, but also water of hydration, so as to avoid explosion hazards.

APPENDIX

Properties of Brazeable Metals and Alloys

NOTE: Yield strengths are reported for the 0.2% offset method unless otherwise specified. Values suffixed YP, PL, or 0.5% signify yield point, proportional limit, or 0.5% extension under load.

Hardness values are Brinell Hardness Numbers unless suffixed R_b, R_c, or R_f (Rockwell B, Rockwell C, or Rockwell F).

[1]Hard temper
[2]As-cast
[3]Age-hardened
[4]Annealed
[5]Half hard temper
[6]Solution heat treated
[7]Artificially aged only
[8]Solution heat treated, then artificially aged
[9]Monel 326 same as Monel except nickel content 55-60%
[10]Heat treated
[11]Water quenched, drawn at 200 °F (93 °C)
[12]Water quenched, drawn at 1200 °F (649 °C)
[13]Normalized at 1200 °F (649 °C)
[14]Water quenched, drawn at 1250 °F (667 °C)
[15]Normalized at 1150 °F (621 °C)
[16]Water quenched, drawn at 1150 °F (621 °C)
[17]Rare earths

Metal	Nominal composition (%)	Condition and temper	Yield strength ksi	Yield strength MPa	Tensile strength ksi	Tensile strength MPa	Elongation in 2 in. (50mm) (%)	Hardness	Density lb/in³	Density 10³ kg/m³	Specific gravity
Aluminum 2S	Al 99+	Annealed -0	5	34	13	90	45	23			
		Cold worked[5] - H14	16	110	17.5	121	20	32	0.098	2.71	2.71
		Cold worked[1] - H18	22	152	24	165	15	44			
Aluminum 3S	Al Bal.	Annealed -0	6	41	16	110	40	28			
	Mn 1.2	Cold worked[5] - H14	19	131	21.5	148	16	40	0.099	2.74	2.73
		Cold worked[1] - H18	26	179	29	200	10	55			
Aluminum 4S	Al Bal.	Annealed -0	10	69	26	179	25	45			
	Mn 1.2	Cold worked[5] - H34	27	186	34	234	12	63	0.098	2.71	2.72
	Mg 1.0	Cold worked[1] - H38	34	234	40	276	6	77			
Aluminum B50S	Al Bal.	Annealed -0	8	55	21	145	24	36			
	Mg 1.2	Cold worked[5] - H34	24	165	27.5	190	8	50	0.097	2.68	2.69
		Cold worked[1] - H38	28	193	31	214	6	57			
Aluminum 52S	Al Bal.	Annealed -0	12	83	27	186	30	45			
	Mg 2.5	Cold worked[5] - H34	31	214	37	255	14	67	0.097	2.68	2.68
	Cr 0.25	Cold worked[1] - H38	36	248	41	283	8	85			
Aluminum 53S	Al Bal.	Annealed -0	8	55	16	110	35	26			
	Si 0.7	Heat treated[6] - T4	20	138	30	138	21	62	0.097	2.68	2.69
	Mg 1.3	Heat treated[7] - T5	27	186	21	145	15	60			
	Cr 0.25	Heat treated[8] - T6	32	221	37	255	13	80			
Aluminum 61S	Al Bal.	Annealed -0	8	55	18	124	30	30			
	Cu 0.25	Heat treated[6] - T4	21	145	35	241	25	65	0.098	2.71	2.70
	Mg 1.0	Heat treated[8] - T6	40	276	45	310	17	95			
	Cr 0.25										
	Si 0.6										
Aluminum 62S	Al Bal.	Annealed -0	6.5	45	17	117	30	28			
	Cu 0.25	Heat treated[6] - T4	21	145	35	241	25	65	0.098	2.71	2.70
	Si 0.6	Heat treated[8] - T6	40	276	45	310	17	95			
	Mg 1.0										
Aluminum 63S	Al Bal.	Heat treated[6] - T42	13	90	22	152	20	42			
	Si 0.4	Heat treated[7] - T5	27	186	21	145	12	60	0.098	2.71	2.70
	Mg 0.7	Heat treated[8] - T6	31	214	35	241	12	73			
Aluminum 43	Al Bal.	Sand cast[2]	8	55	19	131	8	40	0.097	2.68	2.69
	Si 5	Perm. mold cast	9	62	23	159	10	45			
Aluminum A612	Al Bal.	Sand cast[2]	25	172	35	241	5	75	0.102	2.82	2.81
	Cu 0.5										
	Mg 0.7										
	Zn 6.5										
Aluminum C612	Al Bal.	Perm. mold cast	18	124	35	241	8	70	0.103	2.85	2.84
	Cu 0.5										
	Mg 0.35										
	Zn 6.5										
Copper, oxygen free (OF)	Cu 99.92 Min	0.050 mm anneal	10 (.5%)	69	32	221	45	40Rf	0.321	8.89	8.89
		Cold rolled, hard	45 (.5%)	310	50	345	6	90Rf	to	to	to
		Hot rolled	10 (.5%)	69	34	234	45	45Rf	0.323	8.94	8.94
Copper, electrolytic tough pitch (ETP)	Cu 99.9+	0.050 mm anneal	10 (.5%)	69	32	221	45	40Rf	0.321	8.89	8.89
	O 0.04	Cold rolled, hard	45 (.5%)	310	50	345	6	90Rf	to	to	to
		Hot rolled	10 (.5%)	69	34	234	45	45Rf	0.323	8.94	8.94
Copper, deoxidized (DLP)	Cu 99.9+	0.050 mm anneal	10 (.5%)	69	32	221	45	40Rf	0.323	8.94	8.94
	P 0.02										
Commercial bronze, 90%	Cu 90	0.035 mm anneal	12 (.5%)	83	38	262	50	57Rf			
	Zn 10	1/2 Hard	45 (.5%)	310	52	359	15	58Rb	0.318	8.80	8.80
		Hot rolled	10 (.5%)	69	37	255	45	53Rf			
Red brass, 85%	Cu 85	0.035 mm anneal	14 (.5%)	97	41	283	46	63Rf	0.316	8.75	8.75
	Zn 15	1/2 Hard	49 (.5%)	338	57	393	12	65Rb			

Note: Lake copper (silver bearing, fire refined) has essentially the same properties as electrolytic tough pitch copper.

Melting range		Specific heat (32-212 °F) (0-100 °C)		Coeff. of thermal exp. (32-212 °F) 10^{-6} (0-100 °C) 10^{-6}		Thermal conductivity (Room temp.)		Electrical resistivity		Tensile modulus of elasticity	
°F	°C	Btu/lb•°F	J/kg•K	in./in.•°F	mm/mm•°C	Btu•in./h•ft²•°F	W/m•K	ohms•cir mil/ft	ohms•mm²/m	10³ ksi	10³ MPa
1190-1215	643-657	0.22	921	13.1	7.3	1540 1510	222 218	17.6 18.2	0.029 0.030	9.9	68
1190-1210	643-654	0.22	921	12.9	7.2	1335 1105 1075	192 159 155	20.7 25.3 25.9	0.034 0.042 0.043	10	69
1165-1205	629-652	0.22	921	13.3	7.4	1130	169	24.7	0.041	10	69
1160-1205	627-652	0.22	921	13.2	7.3	1330	192	20.7	0.034	10	69
1100-1200	593-649	0.22	921	13.2	7.3	958	138	29.6	0.049	10.1	70
1075-1205	579-652	0.22	921	13.1	7.3	1190 1075 1190 1075	171 155 171 155	23.0 25.9 23.0 25.9	0.038 0.043 0.038 0.043	9.9	68
1080-1205	582-652	0.22	921	13.1	7.3	1190 1075 1075	171 155 155	23.0 25.9 25.9	0.038 0.043 0.043	9.9	68
1080-1205	582-652	0.22	921	12.4	6.9	1190 1070 1070	171 154 154	23.0 23.0	0.038 0.038	9.9	68
1140-1205	616-652	0.22	921	13.0	7.2	1335 1450 1450	192 209 209	20.7 18.8 18.8	0.034 0.031 0.031	9.9	68
1065-1170	574-632	0.22	921	12.3	6.8	1015	146	28	0.047	10.3	71
1105-1195	596-646	0.22	921	13.4	7.4	957	138	29.6	0.049	10.3	71
1120-1190	604-643	0.22	921	13.1	7.3	1103	159	25.9	0.043	10.3	71
1981	1083	0.092	385	9.8	5.4	2762	398	10.3	0.017	17.0	117
1949-1981	1065-1083	0.092	385	9.8	5.4	2712	391	10.3	0.017	17.0	117
1981	1083	0.092	385	9.8	5.4	2352	339	12.2	0.020	17.0	117
1870-1910	1021-1043	0.09	377	10.2	5.7	1303	188	23.5	0.039	17.0	117
1810-1880	988-1027	0.09	377	10.4	5.8	1104	159	28.0	0.047	17.0	117

Metal	Nominal composition (%)	Condition and temper	Yield strength ksi	Yield strength MPa	Tensile strength ksi	Tensile strength MPa	Elongation in 2 in. (50mm) (%)	Hardness	Density lb/in^3	Density 10^3 kg/m^3	Specific gravity
Low brass, 80%	Cu 80 Zn 20	0.035 mm anneal 1/2 hard	15 (.5%) 50 (.5%)	103 345	46 61	317 421	48 18	$66R_f$ $70R_b$	0.313	8.66	8.67
Cartridge brass, 70%	Cu 70 Zn 30	0.025 mm anneal 1/4 hard 1/2 hard	19 (.5%) 40 (.5%) 52 (.5%)	131 276 359	51 54 62	352 372 427	55 43 23	$72R_f$ $55R_b$ $70R_b$	0.308	8.55	8.53
Yellow brass	Cu 65 Zn 35	0.025 mm anneal 1/2 hard Hard	19 (.5%) 50 (.5%) 60 (.5%)	131 345 414	51 61 74	352 421 510	55 23 8	$72R_f$ $70R_b$ $80R_b$	0.306	8.47	8.47
Muntz metal	Cu 60 Zn 40	Hot rolled Cold rolled (1/2 hard)	21 (.5%) 50 (.5%)	145 345	54 70	372 483	45 10	$85R_f$ $75R_b$	0.303	8.39	8.39
Low-leaded brass	Cu 65 Pb 0.5 Zn 34.5	1/2 hard	50 (.5%)	345	61	421	23	$70R_b$	0.306	8.47	8.47
Medium-leaded brass	Cu 65 Pb 1.0 Zn 34	1/2 hard	50 (.5%)	345	61	421	21	$70R_b$	0.306	8.47	8.47
High-leaded brass	Cu 65 Pb 2.0 Zn 33	1/2 hard	50 (.5%)	345	61	421	20	$70R_b$	0.306	8.47	8.47
Leaded muntz metal	Cu 60 Pb 0.6 Zn 39.4	Hot rolled	20 (.5%)	138	54	372	45	$80R_f$	0.304	8.41	8.41
Inhibited admiralty	Cu 71 Zn 28 Sn 1	Hot rolled	18 (.5%)	124	48	331	65	$70R_f$	0.308	8.55	8.53
Aluminum brass	Cu 76 Zn 22 Al 2	0.025 mm annealed	27 (.5%)	186	60	414	55	$77R_f$	0.301	8.33	8.33
Naval brass	Cu 60 Zn 39.25 Sn 0.75	Light anneal 1/2 hard Hot rolled	30 (.5%) 53 (.5%) 25 (.5%)	207 365 172	63 75 55	434 517 379	40 20 50	$60R_b$ $82R_b$ $55R_b$	0.304	8.41	8.41
Free-cutting brass	Cu 61.5 Zn 35.5 Pb 3	1/4 hard (11%)	45 (.5%)	310	56	386	20	$62R_b$	0.307	8.50	8.50
Phosphor bronze, 5% (A)	Cu 95 Sn 5	1/2 hard Hard	55 (.5%) 75 (.5%)	379 517	68 81	469 558	28 10	$78R_b$ $87R_b$	0.320	8.86	8.86
Phosphor bronze, 8% (C)	Cu 92 Sn 8	1/2 hard Hard	55 (.5%) 72 (.5%)	379 496	76 93	524 641	32 10	$84R_b$ $93R_b$	0.318	8.80	8.80
Phosphor bronze, 10% (D)	Cu 90 Sn 10	1/2 hard Hard	— —	— —	83 100	572 690	32 13	$92R_b$ $97R_b$	0.317	8.77	8.78
Phosphor bronze 1.25% (E)	Cu 98.75 Sn 1.25 P Trace	1/2 hard Hard	45 (.5%) 50 (.5%)	310 345	55 65	379 448	16 8	$64R_b$ $75R_b$	0.321	8.89	8.89
Cupro-nickel, 30%	Cu 70 Ni 30	Hot rolled	20 (.5%)	138	55	379	45	$35R_b$	0.323	8.94	8.94
Cupro-nickel, 10%	Cu 89 Ni 10 Fe 1	0.25 mm anneal Light drawn	15 (.5%) 57 (.5%)	103 393	45 60	310 414	40 12	$65R_f$ $72R_b$	0.323	8.94	8.94
Nickel silver, 65-18	Cu 65 Zn 17 Ni 18	0.035 mm anneal 1/2 hard Hard	25 (.5%) 62 (.5%) 74 (.5%)	172 427 510	58 74 85	400 510 586	40 8 3	$85R_f$ $83R_b$ $87R_b$	0.316	8.75	8.73

Melting range		Specific heat (32-212 °F) (0-100 °C)		Coeff. of thermal exp. (32-212 °F) 10^{-6} (0-100 °C) 10^{-6}		Thermal conductivity (Room temp.)		Electrical resistivity		Tensile modulus of elasticity	
°F	°C	Btu/lb•°F	J/kg•K	in./in.•°F	mm/mm•°C	Btu•in./h•ft²•°F	W/m•K	ohms•cir mil/ft	ohms•mm²/m	10^3 ksi	10^3 MPa
1770-1830	966-999	0.09	377	10.6	5.9	972	140	32	0.053	16	110
1680-1750	916-954	0.09	377	11.1	6.2	840	121	37.0	0.061	16.0	110
1660-1710	904-932	0.09	377	11.3	6.3	804	116	38	0.063	15.0	103
1650-1660	899-904	0.09	377	11.6	6.4	852	123	37	0.061	15.0	103
1650-1700	899-927	0.09	377	11.3	6.3	804	116	40	0.066	15.0	103
1630-1700	888-927	0.09	377	11.3	6.3	804	116	40	0.066	15.0	103
1630-1670	888-910	0.09	377	11.3	6.3	804	116	40	0.066	15.0	103
1630-1650	888-899	0.09	377	11.6	6.4	852	123	38	0.063	15.0	103
1650-1720	899-938	0.09	377	11.2	6.2	768	111	42	0.070	16.0	110
1710-1780	932-971	0.09	377	10.3	5.7	696	100	45	0.075	16	110
1630-1650	888-899	0.09	377	11.8	6.6	804	116	40	0.066	15.0	103
1630-1650	888-899	0.09	377	11.4	6.3	804	116	40	0.066	14	97
1750-1920	954-1049	0.09	377	9.9	5.5	564	81	58	0.096	16	110
1620-1880	822-1027	0.09	377	10.1	5.6	432	62	80	0.133	16	110
1550-1830	843-999	0.09	377	10.2	5.7	348	50	94	0.156	16	110
1900-1970	1038-1077	0.09	377	9.9	5.5	1440	208	22	0.037	17	117
2140-2260	1171-1238	0.09	377	9.0	5.0	204	29	225	0.374	22	152
2010-2090	1099-1143	0.09	377	9.3	5.2	312	45	90	0.150	18	124
1960-2030	1071-1110	0.09	377	9.0	5.0	228	33	170	0.283	18	124

Metal	Nominal composition (%)	Condition and temper	Yield strength ksi	Yield strength MPa	Tensile strength ksi	Tensile strength MPa	Elongation in 2 in. (50mm) (%)	Hardness	Density lb/in^3	Density 10^3 kg/m^3	Specific gravity
Nickel silver 55-18	Cu 55 Zn 27 Ni 18	Hard	85 (.5%)	586	100	690	3	$91R_b$	0.314	8.69	8.70
Nickel silver 65-15	Cu 65	0.050 mm anneal	19 (.5%)	131	55	379	42	$73R_f$			
	Zn 20	1/2 hard	62 (.5%)	427	74	510	10	$80R_b$	0.314	8.69	8.70
	Ni 15	Hard	75 (.5%)	517	85	586	3	$87R_b$			
Nickel silver 65-12	Cu 65	0.050 mm anneal	19 (.5%)	131	54	372	45	$73R_f$			
	Zn 23	1/2 hard	60 (.5%)	414	73	503	11	$80R_b$	0.314	8.69	8.69
	Ni 12	Hard	75 (.5%)	517	85	586	4	$89R_b$			
Nickel silver 65-10	Cu 65	0.050 mm anneal	19 (.5%)	131	51	352	46	$71R_f$			
	Zn 25	1/2 hard	60 (.5%)	414	73	503	12	$80R_b$	0.314	8.69	8.69
	Ni 10	Hard	75 (.5%)	517	86	593	4	$89R_b$			
High-silicon bronze, (A)	Cu 94.8	0.035 mm anneal	25 (.5%)	172	60	414	60	$62R_b$			
	min	1/2 hard	45 (.5%)	310	78	538	17	$87R_b$	0.308	8.53	8.53
	Si 3	Hard	58 (.5%)	400	94	648	8	$93R_b$			
		Hot rolled	25 (.5%)	172	64	441	60	$72R_b$			
Aluminum bronze, 5% nickel	Cu 81.5 Al 10 Ni 5 Fe 2.5 Mn 1	Extruded and sized	50 (.5%)	345	110	758	9	$98R_b$	0.273	7.56	7.54
Aluminum bronze, 9.5%	Cu 87.5 Al 9.5 Fe 3	Extruded and sized	40 (.5%)	276	85	586	15	$87R_b$	0.276	7.64	7.65
Aluminum bronze, 8%	Cu 90.5 Al 7 Fe 2.5	As rolled	32 (.5%)	221	72	496	35	$80R_b$	0.279	7.72	7.73
Magnesium, AZ10	Al 1.2 Zn 0.4	As extruded - F	22	152	35	241	10.0		0.063	1.74	1.75
Magnesium, AZ31B	Al 3.0	Cold worked - H24	32	221	42	290	15.0	73			
	Zn 1.0	Annealed - O	22	152	37	255	21.0	56	0.064	1.77	1.77
	Mn 0.4	As extruded - F	29	200	38	262	15.0	49			
Magnesium, KIA	Zr 0.7	Sand cast - F	8	55	26	179	19.0		0.063	14	1.76
Magnesium, MIA	Mn 1.2	Cold worked - H24	29	200	39	269	10.0	56	0.063	1.74	1.76
		Annealed - O	19	131	34	234	18.0	48	0.063	1.74	1.76
Magnesium, ZE10A	Zn 1.2	Cold worked - H24	28	193	38	262	12.0		0.063	1.74	1.76
	RE[17] 0.17	Annealed - O	26	179	33	228	23.0		0.063	1.74	1.76
Nickel (pure)	Ni 99.9	Annealed	8.5	59	46	317	28	85	0.322	8.91	8.91
Nickel (wrought)	Ni 99.0	Annealed	20	138	70	483	40	100			
	min	Hot rolled	25	172	75	517	40	110	0.321	8.89	8.89
		Cold drawn	70	483	95	655	25	170			
		Cold rolled[1]	95	655	105	724	5	210			
Low-carbon nickel	Ni 99.0 min C 0.02 max	Annealed	15	103	60	414	50	90	0.321	8.89	8.89
Nickel (cast)	Ni Rem. Si 2.0 Fe 1.25	Sand cast[2]	25	172	50	345	25	100	0.301	8.33	8.34
"D" Nickel	Ni 95	Annealed	35	241	75	517	40	140			
	Mn 4.5	Hot rolled	50	345	90	621	35	150	0.317	8.77	8.78
		Cold drawn	80	552	100	690	25	190			
"E" Nickel	Ni 98	Annealed	35	241	75	517	40	140			
	Mn 2	Hot rolled	50	345	90	621	35	150	0.320	8.86	8.86
		Cold drawn	80	552	100	690	25	190			

Melting range		Specific heat (32-212 °F) (0-100 °C)		Coeff. of thermal exp. (32-212 °F) 10^{-6} (0-100 °C) 10^{-6}		Thermal conductivity (Room temp.)		Electrical resistivity		Tensile modulus of elasticity	
°F	°C	Btu/lb•°F	J/kg•K	in./in.•°F	mm/mm•°C	Btu•in./h•ft²•°F	W/m•K	ohms•cir mil/ft	ohms•mm²/m	10^3 ksi	10^3 MPa
1930	1054	0.09	377	9.3	5.2	204	29	190	0.316	18	124
1970	1077	0.09	377	9.0	5.0	252	36	150	0.249	18	124
1900	1038	0.09	377	9.0	5.0	276	40	130	0.216	18	124
1870	1021	0.09	377	9.1	5.1	312	45	115	0.191	17.5	121
1780-1880	971-1027	0.09	377	10.0	5.6	252	36	150	0.249	15	103
1890-1910	1032-1043			8.5	4.7	551	79			17	117
1890-1910	1032-1043			9.0	5.0	551	79			17	117
1910-1940	1043-1060			9.0	5.0	609	88			18	124
1170-1180	632-638	0.25	372	14.5	8.1					6.5	45
1050-1160	556-627	0.25	372	14.5	8.1	535	77	55	0.091	6.5	45
1198-1200	648-649	0.25	372	14.5	8.1	840	121	34	0.057	6.5	45
1198-1200	648-649	0.25	372	14.5	8.1	870	125	32	0.053	6.5	45
1198-1200	648-649	0.25	372	14.5	8.1	870	125	32	0.053	6.5	45
1100-1195	593-646	0.25	372	14.5	8.1	900	130	31	0.052	6.5	45
1100-1195	593-646	0.25	372	14.5	8.1	960	138	30	0.050	6.5	45
2650	1454	0.11	164	7.4	4.1	575	83	37	0.061	30	207
2615-2635	1435-1446	0.11	164	7.2	4.0	420	61	57	0.095	30	207
2615-2635	1435-1446	0.11	164	7.2	4.0	420	61	50	0.083	30	207
2540-2600	1393-1427	0.11	164	7.2	4.0	410	59	125	0.208	21.5	148
2500-2600	1371-1427	0.11	164	7.4	4.1	335	48	110	0.183	30	207
2570-2600	1410-1427	0.11	164	7.4	4.1	335	48	85	0.141	30	207

Metal	Nominal composition (%)	Condition and temper	Yield strength ksi	Yield strength MPa	Tensile strength ksi	Tensile strength MPa	Elongation in 2 in. (50mm) (%)	Hardness	Density lb/in³	Density 10³ kg/m³	Specific gravity
"Dura-nickel"	Ni 94	Hot rolled	50	345	105	724	35	180	0.298	8.25	8.26
	Al 4.4	Hot Rolled[3]	130	896	170	1172	15	320			
		Cold drawn	90	621	120	827	25	220			
		Cold drawn[3]	175	1207	210	1448	15	340			
Monel (wrought)[9]	Ni 67	Annealed	35	241	75	517	40	125	0.319	8.83	8.84
	Cu 30	Hot rolled	50	345	90	621	35	150			
		Cold drawn	80	552	100	690	25	190			
		Cold rolled[1]	100	690	110	758	5	240			
Monel (cast)	Ni 63	Sand cast	35	241	80	552	35	135	0.312	8.64	8.63
	Cu 32										
	Si 1.25										
"R" monel	Ni 67	Hot rolled	45	310	85	586	35	145	0.319	8.83	8.84
	Cu 30	Cold drawn	75	517	90	621	25	180			
	S 0.035										
"K" monel	Ni 66	Hot rolled	45	310	100	690	40	160	0.306	8.47	8.47
	Cu 29	Hot rolled[3]	110	758	150	1034	25	280			
	Al 2.75	Cold drawn	85	586	115	793	25	210			
		Cold drawn[3]	130	896	170	1172	20	290			
"H" monel	Ni 65	Sand cast[2]	60	414	100	690	15	210	0.305	8.44	8.48
	Cu 29.5										
	Si 3.0										
"S" monel	Ni 63	Sand cast[4]	70	483	90	621	3	275	0.302	8.36	8.36
	Cu 30	Sand cast[2]	100	690	130	896	2	320			
	Si 4	Sand cast[3]	100	690	130	896	2	350			
Inconel (wrought)	Ni 77.5	Annealed	35	241	85	586	45	150	0.307	8.50	8.51
	Cr 15	Hot rolled	60	414	100	690	35	180			
	Fe 7.5	Cold drawn	90	621	115	793	20	200			
		Cold drawn[1]	110	758	135	931	5	260			
Inconel (cast)	Ni 78.0	Sand cast[2]	40	276	85	586	20	175	0.300	8.30	8.30
	Fe 6.0										
	Cr 13.5										
	Si 2.0										
Inconel "X"	Ni 73	Annealed	50	345	115	793	50	200	0.300	8.30	8.30
	Cr 15.0	Hot rolled[3]	120	827	180	1241	25	360			
	Fe 7.0										
	Ti 2.5										
	Al 0.7										
	Cb 1.0										
Inconel "W"	Ni 73.0	Annealed	45	310	105	724	55	180	0.301	8.33	8.31
	Cr 15.5	Annealed[3]	100	690	150	1034	27	260			
	Fe 7.0										
	Al 0.7										
	Ti 2.5										
80 Ni 20 Cr alloy (wrought)	Ni 78	Cold drawn[4]	60	414	95	655	30	90R_b	0.303	8.39	8.40
	Cr 19	Cold drawn[1]			165	1138	1	102R_b			
60 Ni 15 Cr alloy (wrought)	Ni 61	Hot rolled			110	758					
	Fe Bal.	Cold drawn[4]	60	414	95	655	35	200	0.303	8.39	8.4
	Cr 15.5	Cold drawn[1]			160	1103					
60 Ni 15 Cr alloy (cast)	Ni 60	Cast[2]	45	310	65	448	2	180	0.303	8.39	8.4
	Fe Bal.										
	Cr 14										
"Hastelloy" Alloy A	Ni Bal.	Sand cast[4]	48	331	73	503	12	168	0.318	8.80	8.80
	Mo 21	Hot rolled[4]	44	303	110	758	50	175			
	Fe 19	Investment cast[2]	45	310	76	524	21	173			
"Hastelloy" Alloy B	Ni Bal.	Sand cast[4]	58	400	80	552	8	207	0. 4	9.25	9.24
	Mo 28	Hot rolled[4]	60	414	135	931	50	201			
	Fe 5	Investment cast[2]	61	421	85	586	14	216			

Melting range		Specific heat (32-212 °F) (0-100 °C)		Coeff. of thermal exp. (32-212 °F) 10^{-6} (0-100 °C) 10^{-6}		Thermal conductivity (Room temp.)		Electrical resistivity		Tensile modulus of elasticity	
°F	°C	Btu/lb•°F	J/kg•K	in./in.•°F	mm/mm•°C	Btu•in./h•ft²•°F	W/m•K	ohms•cir mil/ft	ohms•mm²/m	10^3 ksi	10^3 MPa
2615-2635	1435-1446	0.104	435	7.1 7.3	3.9 4.1	128 137	18 20	280 260	0.465 0.432	30	207
2370-2460	1299-1349	0.13	544	7.8	4.3	180	26	290	0.482	26	179
2400-2450	1316-1343	0.13	544	6.8	3.8	180	26	320	0.532	19	131
2370-2460	1299-1349	0.13	544	7.8	4.3	180	26	290	0.482	26	179
2400-2460	1316-1349	0.13	544	7.8	4.3	130	19	350	0.582	20	138
2350-2400	1288-1316	0.13	544	6.8	3.8	180	26	370	0.615	20	138
2300-2350	1260-1288	0.13	544	6.8	3.8	180	26	380	0.632	21	145
2540-2600	1393-1427	0.11	461	6.4	3.6	104	15	590	0.981	31	214
2500-2550	1371-1399	0.11	461	6.4	3.6					23	159
2540-2600	1393-1427	0.105	440	7.6	4.2	102	15	750	1.247	31	214
2540-2600	1393-1427	0.105	440	7.6	4.2	102	15	750	1.247	31	214
2550	1371	0.107	448	7.3	4.1	104	15	650	1.080	31	214
2450-2550	1343-1399	0.107	448	7.6	4.2	94	14	675	1.122	31	214
2450-2550	1343-1399	0.136	569	6.2	3.4	96	14	675	1.122	30.5	210
2370-2425	1299-1329	0.094	394	6.1	3.4	116	17	762	1.266	24 27 —	165 186 —
2410-2460	1321-1349	0.091	381	5.6	3.1	78.5	11	812	1.350	26.5 30.8 28.5	183 212 197

Metal	Nominal composition (%)	Condition and temper	Yield strength ksi	Yield strength MPa	Tensile strength ksi	Tensile strength MPa	Elongation in 2 in. (50mm) (%)	Hardness	Density lb/in³	Density 10³ kg/m³	Specific gravity
"Hastelloy" Alloy C	Ni Bal.	Investment cast[2]	60	414	80	552	7	225			
	Mo 16	Sand cast[4]	58	400	78	538	8	218	0.323	8.94	8.94
	Cr 16	Hot rolled	55	379	130	896	45	211			
	Fe 5										
	W 4										
"Hastelloy" Alloy D	Ni Bal.	Sand cast[4]	118	814	118	814	0-2	390	0.282	7.81	7.80
	Si 10						(in 1 in.)				
	Cu 3						(25 mm)				
"Hastelloy" Alloy X	Cr 22	Cast, solution treated	43	296	77	531	23.0		0.30	8.30	
	Mo 9										
	Fe 20										
	Ni Bal.										
"Illium" G	Ni Bal.	Cast[2]	57	393	68	469	7	185	0.310	8.58	8.58
	Cr 22										
	Fe 6										
	Mo 6										
	Cu 6										
	Mn 1.25										
	Si 0.65										
Rene 41	Cr 19		150	1034	206	1420	14.0	$32R_C$	0.298	8.25	8.3
	Co 11										
	Mo 10										
	Ti 3										
	Al 1.5										
	Fe 2.5										
	Ni Bal.										
J-1500	Cr 19	Age hardened	127	876	177	1220	21.0		0.298	8.25	
	Ni 55										
	Co 10										
	Mo 10										
	Fe										
	Al										
	Ti										
J-1610	Cr 19	Age hardened	148	1020	185	1276	12.0		0.296	8.19	
	Ni 53										
	Co 11										
	Mo 10										
	Fe										
	Al										
	Ti										
Kovar	Ni 29	Annealed	50	345	75	517	25.0	150	0.30	8.30	8.3
	Co 17										
	Fe Bal.										
Multimet (N-155)	Cr 21	Cr 20%	135	931	145	1000	20.0				
	Ni 20	Annealed	58	400	118	814	49.0	194	0.295	8.17	8.2
	Co 20	Investment cast	58	400	101	696	31.0	180			
	Mo 3										
	W 3										
	Cb 1										
Stellite 21	Ni 2.5	As cast	82	565	101	696	8.2	$30R_C$	0.30	8.30	8.3
	Cr 27	Aged	110	758	124	855	1.7	$34R_C$			
	Mo 5.5										
	Fe 2										
	Co Bal.										
Stellite 25 (L-605)	Ni 10	Sheet, annealed	68	469	148	1020	61.0	250	0.33	9.13	9.0
	Cr 20	Sheet, CR 20%	163	1124	192	1324	27.0	$43R_C$			
	W 15										
	Mn 1										
	Si 1										
	Co Bal.										
Tungsten	W	Sintered			18.5	128	1-4	225	0.697	19.29	19.3
		Drawn 15 min.			285	1965		375			

Melting range		Specific heat (32-212 °F) (0-100 °C)		Coeff. of thermal exp. (32-212 °F) 10^{-6} (0-100 °C) 10^{-6}		Thermal conductivity (Room temp.)		Electrical resistivity		Tensile modulus of elasticity	
°F	°C	Btu/lb•°F	J/kg•K	in./in.•°F	mm/mm•°C	Btu•in./h•ft²•°F	W/m•K	ohms•cir mil/ft	ohms•mm²/m	10^3 ksi	10^3 MPa
2320-2380	1271-1304	0.092	385	6.3	3.5	87	13	800	1.330	24.5 26	169 179
2030-2050	1110-1121	0.108	452	6.1	3.4	145	21	680	1.130	28.9	199
2350	1288	0.11	461	7.7	4.3	750	108	710	1.180		
2375	1302	0.105	440	7.5	4.2	84	12	735	1.222	26.8	185
		0.108	452	7.5	4.2	85	12			25.0	172
2250-2500	1232-1371			5.9	3.3	82	12			29.8	205
2250-2475	1232-1357	0.11	461	6.6	3.7					31.6	218
2715	1491	0.16	670	2.3	1.3	133	19	294	0.489	20.0	138
2350-2475	1288-1357	0.104	435	7.9	4.4	101	15	558	0.927	30.0	207
2465	1352			7.8	4.3	100	14	525	0.873	16.0	110
2425-2570	1329-1410	0.09	377	7.2	4.0	90	13	532	0.884	34.0	234
6170 ±35	3410 ±20	0.034	142	2.2	1.2	1160	167	32.9	0.055	50	345

Metal	Nominal composition (%)	Condition and temper	Yield strength ksi	Yield strength MPa	Tensile strength ksi	Tensile strength MPa	Elongation in 2 in. (50mm) (%)	Hardness	Density lb/in³	Density 10³ kg/m³	Specific gravity
Tungsten alloy	Mo 15		88	608	108	745	22.0	265	0.615	17.02	
Molybdenum	Mo	Drawn 14 min.	57-85	393-586	215-240	1482-1655	2-5	155-250	0.368	10.19	10.3
Molybdenum	Ti 0.5	Stress relieved 1600 F	85	586	110	758	35.0	275	0.37	10.24	10.3
M1A ("Dow-metal" M)	Mg Bal. Mn 1.2	Hot rolled Cold rolled Annealed	 28 18	 193 124	33 37 33	228 255 228	12 7 16	 56 48	0.064	1.77	1.76
Silver (pure)	Ag 99.9+	Annealed Cold rolled[1]	12 38	83 262	23 43	159 296	45 6	30 90	0.379	10.49	10.50
Gold (pure)	Au 99.99	Annealed Cold rolled[1]	0 PL 30 PL	0 PL 207 PL	17 32	117 221	45 4	25 60	0.698	19.32	19.32
Platinum (pure)	Pt 99.99	Annealed Cold rolled[1]	10 PL 27 PL	69 PL 186 PL	24 36	165 248	24 2.5	40 100	0.772	21.37	21.40
Platinum (commercial)	Pt 99+	Annealed Cold rolled[1]	6.5 PL 27 PL	45 PL 186 PL	23 37	159 255	29 3.5	57 113	0.772	21.37	21.40
Iridium-Platinum 10%Ir	Pt Bal. Ir 10	Annealed Cold rolled[1]	30 PL 54 PL	207 PL 372 PL	53 82	365 565	27 2.5	110 170	0.778	21.54	21.53
Rhodium platinum 10%Rh	Pt Bal. Rh 10	Annealed Cold rolled[1]	17 PL 56 PL	117 PL 386 PL	47 84	324 579	37 3	65 160	0.742	20.54	20.55
Palladium (commercial)	Pd 99.5	Annealed Cold rolled[1]	5 PL 32 PL	34 PL 221 PL	30 47	207 324	40 1.5	44 106	0.432	11.96	11.98
Palladium (hard)	Pd Bal. Ru 4 Rh 1	Annealed Cold rolled[1]	28 PL 46 PL	193 PL 317 PL	59 72	407 496	23 3		0.432	11.96	11.98
Tantalum	Ta 99.9+	Annealed or unannealed sheet			42 to 178	290 to 1227	45-25 (in 1 in.) (25 mm)	55 to 125	0.60	16.60	16.6
Wrought iron	Fe Bal. Slag 2.5	Hot rolled	30	207	48	331	30 (in 8 in.) (200 mm)	100	0.278	7.70	7.70
Ingot iron	Fe 99.9+	Hot rolled Annealed	29 19	200 131	45 38	310 262	26 45	90 67	0.284	7.86	7.86
Gray cast iron	C 3.1 Si 1.8 Mn 0.7 Fe Bal.	Cast[2]	None	None	32	221	0.5 max.	180	0.260	7.20	7.20
Malleable iron	C 2.5 Si 1 Mn 0.55	Cast[2]	36.5 YP	252 YP	55	379	20	130	0.264	7.31	7.32
"Ni-Tensyl-iron"	C 2.7 Si 1.5 Mn 0.8 Ni 2.3 Cr 0.3 Fe Bal. Mo 0.4	Cast[2] Cast[10]	30 PL 40 PL	207 PL 276 PL	40 100	276 690		220 350	0.260	7.20	7.20

Melting range		Specific heat (32-212 °F) (0-100 °C)		Coeff. of thermal exp. (32-212 °F) 10^{-6} (0-100 °C) 10^{-6}		Thermal conductivity (Room temp.)		Electrical resistivity		Tensile modulus of elasticity	
°F	°C	Btu/lb•°F	J/kg•K	in./in.•°F	mm/mm•°C	Btu•in./h•ft²•°F	W/m•K	ohms•cir mil/ft	ohms•mm²/m	10^3 ksi	10^3 MPa
4760	2627	0.065	272							48.0	331
4750	2621	0.06	251	3.0	1.7	1010	146	31.1	0.050	42	290
1198-1200	648-649	0.25	1047	3.4	1.9	820	118			46.0	317
				14.4	8.0	870	125	30	0.050	6.5	45
1760	960	0.056	234	10.9	6.1	2900	418	9.55	0.016	11	76
1945	1063	0.031	130	7.8	4.3	2000	288	14	0.023	12	83
3225	1774	0.032	134	4.9	2.7	480	69	64	0.106	21	145
3225	1774	0.032	134	4.9	2.7	480	69	72	0.120	22.6	156
3230-3270	1777-1799			4.93	2.74			150	0.249		
3345	1841							114	0.189		
2830	1554	0.058	243	6.5	3.6	490	71	65	0.108	17	117
2860	1571										
5425	2996	0.036	151	3.57	1.98	375	54	90.2	0.150	27	186
2750	1510	0.11	461	6.7	3.7	418	60	70	0.116	29	200
2795	1535	0.108	452	6.5	3.6	494	71	57	0.095	30.1	208
2075	1135			6.0	3.3	310	45	400	0.665	13± 1.5	90± 10
2250	1232	0.122	511	6.6	3.7	435	63	180	0.299	25	172
2150	1177			6.5	3.6	320	46			20± 1.5	138± 10

Metal	Nominal composition (%)	Condition and temper	Yield strength ksi	Yield strength MPa	Tensile strength ksi	Tensile strength MPa	Elongation in 2 in. (50 mm) (%)	Hardness	Density lb/in³	Density 10³ kg/m³	Specific gravity
"Ni-Resist" Type 1	C 2.8 Si 1.5 Mn 1.0 Ni 14.0 Cr 2.0 Cu 7.0 Fe Bal.	Cast[2]			30	207	2	150	0.270	7.47	7.48
"Ni-Resist" Type 2	C 2.8 Si 1.8 Mn 1.3 Ni 20 Cr 2.5 Fe Bal.	Cast[2]			30	207	2	150	0.270	7.47	7.48
Ductile iron (Mg containing)	C 3.4 Si 2.5 Mn 0.7 P 0.1 max Ni 1.5 Mg 0.06 Fe Bal.	Cast[2] Cast[4]	70 55	483 379	100 70	690 483	3 20	245 160	0.26	7.20	7.2
Ductile "Ni-Resist" (Mg containing)	C 2.8 Si 2.5 Mn 1 P 0.01 max Ni 20 Cr 2 Mg 0.1 Fe Bal.	Cast[4]	35	241	60	414	10	175	0.268	7.42	7.4
"Ni-Hard" low-carbon	C 2.7 Si 0.6 Mn 0.5 Ni 4.5 Cr 1.5 Fe Bal.	Sand cast[2] Chill cast[2]					55 75	550 625	0.275	7.61	7.70
"Ni-Hard" high-carbon	C 3.0 Si 0.6 Mn 0.5 Ni 4.0 Cr 1.5 Fe Bal.	Sand cast[2] Chill cast[2]					30 80	600 700	0.275	7.61	7.70
Carbon steel SAE 1020	Fe Bal. Mn 0.45 Si 0.25 C 0.20	Annealed Hot rolled Hardened[11] Hardened[12]	40 45 80 62	276 310 552 427	60 69 104 90	414 476 717 621	35 31 6 22	130 146 205 175	0.284	7.86	7.86
Cast carbon steel	Fe Bal. Mn 0.70 Si 0.40 C 0.30	Cast[4] Cast[13] Cast[14]	40 45 60	276 310 414	72 80 90	496 552 621	26 30 25	140 160 185	0.283	7.83	7.84
Cast alloy steel	Fe Bal. Ni 1.75 Mn 0.80 Cr 0.75 C 0.30 Mo 0.25	Cast[4] Cast[15] Cast[16]	60 95 135	414 655 931	105 120 150	724 827 1034	20 17 13.5	225 260 325	0.284	7.86	7.87
4340	C 0.4 Mn 0.7 Si 0.3 Cr 0.8 Fe Bal.	As rolled Annealed	147 99	1014 683	180 120	1241 827	9.0 17.0	380 240	0.283	7.83	7.7

Melting range		Specific heat (32-212 °F) (0-100 °C)		Coeff. of thermal exp. (32-212 °F) 10^{-6} (0-100 °C) 10^{-6}		Thermal conductivity (Room temp.)		Electrical resistivity		Tensile modulus of elasticity	
°F	°C	Btu/lb•°F	J/kg•K	in./in.•°F	mm/mm•°C	Btu•in./h•ft²•°F	W/m•K	ohms•cir mil/ft	ohms•mm²/m	10^3 ksi	10^3 MPa
2150	1177			10.6	5.9	275	40	1050	1.745	14.5	100
2150	1177			10.3	5.7	275	40	1050	1.745	14.5± 1.5	100± 10
2100	1149			7.5	4.2			360	0.598	25	172
2250	1232			10.4	5.8			610	1.014	18.5	128
2150	1177			4.8	2.7						
2150	1177			4.8	2.7						
2760	1516	0.116	486	6.5	3.6	357	51	101	0.168	30	207
2745	1507	0.107	448	6.7	3.7	400	58	90	0.150	30	207
2745	1507	0.107	448	6.7	3.7					30	207
2740	1504	0.107	448	6.3	3.5	260	37	180	0.299	29.5	203

Metal	Nominal composition (%)	Condition and temper	Yield strength ksi	Yield strength MPa	Tensile strength ksi	Tensile strength MPa	Elongation in 2 in. (50 mm) (%)	Hardness	Density lb/in³	Density 10³ kg/m³	Specific gravity
Vascojet 1000 (H-11)	Cr 5 Mo 1.3 V 0.5 C 0.4 Fe Bal.	Hardened and tempered	215	1482	268	1848	5.0		0.280	7.75	7.7
Stainless steel Type 202	Cr 18 Ni 5 Mn 9 C 0.1	Annealed	55	379	105	724	55.0	90R_b	0.28	7.75	7.83
Stainless steel Type 301	Fe Bal. Cr 17 Ni 7 C 0.11	Annealed C.R. 52%	35 164	241 1131	100 190	690 1310	65 32	160 385	0.29	8.03	8.02
Stainless steel Type 302	Fe Bal. Cr 18 Ni 9 C 0.10	Annealed C.R. 51%	30 156	207 1076	90 177	621 1220	55 5.5	160	0.29	8.03	8.02
Stainless steel Type 304	Fe Bal. Cr 19 Ni 9.0 C 0.08	Annealed C.R. 50%	30 152	207 1048	85 167	586 1151	50 8	160	0.29	8.03	8.02
Stainless steel Type 304L	Cr 19 Ni 9 C 0.03 max	Annealed	30	207	80	552	55.0	76R_b	0.29	8.03	8.02
Cast stainless steel Type 304	Fe Bal. Cr 19 Ni 9 C 0.1	Cast[4]	35	241	79	545	54	131	0.286	7.92	7.92
Stainless steel Type 309	Fe Bal. Cr 23 Ni 13 C 0.20	Annealed C.R. 45%	30 132	207 910	82 144	565 993	50 6	165	0.29	8.03	8.02
Stainless steel Type 310	Fe Bal. Cr 25 Ni 20 C 0.25	Annealed C.R. 50%	40 135	276 931	100 145	690 1000	50 3	165	0.29	8.03	8.02
Stainless steel Type 316	Fe Bal. Cr 17 Ni 12 Mo 2.5 C 0.10	Annealed C.R. 49%	40 136	276 938	90 148	621 1020	50 6	165 275	0.292	8.08	7.94
Cast stainless steel Type 316	Fe Bal. Cr 19 Ni 9 Mo 2.0 C 0.05	Cast[4]	35	241	83	572	60	161	0.286	7.92	7.92
Stainless steel Type 321	Fe Bal. Cr 18 Ni 10 C 0.08 Ti 5 X C	Annealed C.R. 50%	35 141	241 972	85 160	586 1103	50 4.0	160 300	0.292	8.08	7.94
Stainless steel Type 347	Fe Bal. Cr 18 Ni 11 C 0.08 Cb 10 X C	Annealed C.R. 52%	40 144	276 993	90 169	621 1165	50 4.0	160	0.292	8.08	7.94
AM 350	Fe Bal. Cr 16 Ni 4.3 Mo 2.8 Mn 0.8 Si 0.3	Sol. treated and hardened	165	1138	190	1310	13.0	41R_c	0.290	8.03	8.0

Melting range		Specific heat (32-212 °F) (0-100 °C)		Coeff. of thermal exp. (32-212 °F) 10^{-6} (0-100 °C) 10^{-6}		Thermal conductivity (Room temp.)		Electrical resistivity		Tensile modulus of elasticity	
°F	°C	Btu/lb•°F	J/kg•K	in./in.•°F	mm/mm•°C	Btu•in./h•ft²•°F	W/m•K	ohms•cir mil/ft	ohms•mm²/m	10^3 ksi	10^3 MPa
2550-2590	1399-1421	0.12	502	6.1	3.4	198	29			31.0	214
				9.8	5.4			427	0.710		
				8.0	4.4	113	16	435	0.723	29	200
2550-2590	1399-1421	0.12	502	8.0	4.4	113	16	435	0.723	28	193
2550-2650	1399-1454	0.12	502	8.0	4.4	113	16	435	0.723	28	193
2550-2650	1399-1454	0.12	502	8.0	4.4	113	16	435	0.723	28.0	193
2550-2590	1399-1421	0.12	502	8.0	4.4	110	16	420	0.698	29	200
2550-2650	1399-1454	0.12	502	8.0	4.4	108	16	470	0.781	29	200
2550-2650	1399-1454	0.12	502	8.0	4.4	96	14	470	0.781	30	207
2500-2550	1371-1399	0.12	502	8.8	4.9	114	16	445	0.740	29	200
2550-2590	1399-1421	0.12	502	8.0	4.4	110	16	479	0.796	29	200
2550-2600	1399-1427	0.12	502	8.3	4.6	113	16	435	0.723	28	193
2550-2600	1399-1427	0.12	502	8.3	4.6	113	16	435	0.723	28	193
2500-2550	1371-1399	0.120	502	6.8	3.8	114	16	425	0.706	30.4	210

Metal	Nominal composition (%)	Condition and temper	Yield strength ksi	Yield strength MPa	Tensile strength ksi	Tensile strength MPa	Elongation in 2 in. (50 mm) (%)	Hardness	Density lb/in^3	Density 10^3 kg/m^3	Specific gravity
15-7 Mo	Cr 15 Ni 7 Mo 2.5 Al 1.0 Mn 0.7 Fe Bal.	TH 1050 RH 950	200 215	1379 1482	210 240	1448 1655	7.0 6.0	$45R_c$ $48R_c$	0.277	7.68	7.7
17-7 PH	Cr 17 Ni 7 Al 1.3 Mn 0.7 Si 0.7 Fe Bal.	TH 1050 RH 950	150 190	1034 1310	180 210	1241 1448	6.0 6.0	$38R_c$ $44R_c$	0.267	7.39	7.9
Stainless steel Type 405	Cr 13 Al 0.2 C 0.08 max	Annealed	40	276	70	483	30.0	$80R_b$	0.28	7.75	7.75
Stainless steel Type 410	Fe Bal. Cr 12.5 C 0.15	Annealed Heat treated	40 115	276 793	75 150	517 1034	30 15	150 300	0.28	7.75	7.75
Stainless steel Type 414	Fe Bal. Cr 12.5 Ni 1.75 C 0.15	Annealed Heat treated	80 150	552 1034	100 200	690 1379	22 17	217 387	0.28	7.75	7.75
Stainless steel Type 420	Fe Bal. Cr 13 C 0.30	Annealed Heat treated	60 200	414 1379	98 250	676 1724	28 8	180 480	0.28	7.75	7.75
Stainless steel Type 430	Fe Bal. Cr 16 C 0.12	Annealed C.R. 45%	45 111	310 765	70 115	483 793	30 2.5	150 217	0.28	7.75	7.75
Stainless steel Type 431	Fe Bal. Cr 16 Ni 1.75 C 0.20	Annealed Heat treated	85 150	586 1034	120 195	827 1345	25 20	250 400	0.280	7.75	7.75
Stainless steel Type 440A	Cr 17 C 0.7	Hardened and tempered	240	1655	260	1793	5.0	$51R_c$	0.28	7.75	7.75
Stainless steel Type 446	Fe Bal. Cr 25 C 0.35	Annealed	50	345	80	552	30	165	0.27	7.47	7.45
Invar	Fe Bal. Ni 36	Annealed Hot rolled	42 50	290 345	70 75	483 517	41 37	130 140	0.289	8.00	8.09
Stainless steel Type 312	Fe Bal. Cr 29 Ni 9	Hot rolled	45	310	90	621	30	170			
Cast 28 Cr 10 Ni alloy ACI Type HE	Fe Bal. Cr 29 Ni 9	Cast[2]	45 (.5%)	310	85	586	10	200	0.277	7.67	7.67
Stainless steel Type 330	Fe Bal. Ni 36 Cr 16	Hot rolled Cold drawn[1] Cold drawn[3]	55	379	100 80 150	690 552 1034	35	200	0.284	7.86	7.86
A 286	Ni 26 Cr 16 Mn 1.4 Mo 1.3 Ti 2 Al 0.2 V 0.3 Fe Bal.	Sol. treated, quenched, and aged	105	724	153	1055	20.0	$24R_c$	0.286	7.92	7.9

Melting range		Specific heat (32-212 °F) (0-100 °C)		Coeff. of thermal exp. (32-212 °F) 10^{-6} (0-100 °C) 10^{-6}		Thermal conductivity (Room temp.)		Electrical resistivity		Tensile modulus of elasticity	
°F	°C	Btu/lb•°F	J/kg•K	in./in.•°F	mm/mm•°C	Btu•in./h•ft²•°F	W/m•K	ohms•cir mil/ft	ohms•mm²/m	10^3 ksi	10^3 MPa
		0.110	461	61.	3.4			492	0.818	28.0	193
2700-2790	1482-1532	0.11	461	6.13.4	117	17	493	0.819	29.0	200	
				6.0	3.3	150	22	360	0.598	29.0	200
2700-2790	1482-1532	0.11	461	5.1	2.8	173	25	340	0.565	29	200
2600-2700	1427-1482	0.11	461	5.5	3.1	173	25	420	0.698	29	200
2650-2750	1454-1510	0.11	461	5.5	3.1	173	25	330	0.548	29	200
2600-2750	1454-1510	0.11	461	5.1	2.8	165	24	360	0.598	29	200
2600-2700	1427-1482	0.11	461	6.5	3.6	140	20	420	0.715	29	200
2500-2750	1371-1510	0.11	461	5.6	3.1	140	20	360	0.598	29.0	200
2550-2700	1399-1482	0.12	502	5.7	3.2	145	21	405	0.673	29	200
2600	1427	0.123	515	0.6	0.3	73	11	480	0.798	21	145
		0.14	586	10.5	5.8					22	152
2515	1379	0.11	461	6.3	3.5	90	13	600	0.997		
2500-2600	1371-1427	0.110	461	9.4	5.2	105	15	548	0.911	29.1	201

Metal	Nominal composition (%)		Condition and temper	Yield strength ksi	Yield strength MPa	Tensile strength ksi	Tensile strength MPa	Elongation in 2 in. (50 mm) (%)	Hardness	Density lb/in³	Density 10^3 kg/m³	Specific gravity
Cast 35 Ni 15	Fe	Bal.	Cast[2]	38 (.5%)	262	69	476	10	192	0.290	8.03	8.04
Cr alloy	Ni	36										
ACI Type HT	Cr	16										
Iron-silicon	Fe	Bal.	Cast[2]	17	117	17	117	0	500	0.253	7.00	7.00
alloy	Si	14.5										
	C	0.8										
	Mn	0.35										
"Durichlor"	Fe	Bal.	Cast[2]	17	117	17	117	0	500	0.254	7.03	7.04
	Si	14.5										
	Mo	3										
	C	0.8										
"Durimet T"	Fe	Bal.	Cast[4]	35	241	70	483	40	140	0.283	7.83	7.85
	Ni	22	Hot rolled[4]	45	310	85	586	50	140			
	Cr	19										
	Mo	2.5										
	Si	1										
	Cu	1										
Beryllium			Hot pressed	32	221	45	310	2.0	90	0.067	1.85	1.8
			Extruded	40	276	82	565	16.0				
Columbium			Annealed sheet	30	207	50	345	35.0	75	0.309	8.55	8.6
Titanium			High purity	20	138	35	241	50.0	90	0.163	4.51	4.2
			Commercial pure	75	517	90	621	22.0	225			
Zirconium			Extruded, air			65	448	24.0	80R_b	0.235	6.50	6.5

Melting range		Specific heat (32-212 °F) (0-100 °C)		Coeff. of thermal exp. (32-212 °F) 10⁻⁶ (0-100 °C) 10⁻⁶		Thermal conductivity (Room temp.)		Electrical resistivity		Tensile modulus of elasticity	
°F	°C	Btu/lb•°F	J/kg•K	in./in.•°F	mm/mm•°C	Btu•in./h•ft²•°F	W/m•K	ohms•cir mil/ft	ohms•mm²/m	10³ ksi	10³ MPa
		0.135	565	9.7	5.4					23	159
2300	1260	0.12	502	3.6	2.0	360	52	380	0.632		
2350	1288	0.12	502	3.6	2.0	360	52	380	0.632		
2650	1454	0.12	502	7.8	4.3	145	21	580	0.964	22.7 27.8	157 192
2345	1285	0.475	1989	6.9	3.8	1090	157	35	0.058	40.0	276
4380	2416	0.065	272	4.0	2.2	360	52	79	0.131	15.0	103
3075	1691	0.139	582	4.7	2.6	1190	171	288	0.479	15.5	107
3325	1829	0.067	281	2.9	1.6	116	17	270	0.449	11.3	78

INDEX